Lecture Notes on Data Engineering and Communications Technologies

Volume 30

Series Editor

Fatos Xhafa, Technical University of Catalonia, Barcelona, Spain

The aim of the book series is to present cutting edge engineering approaches to data technologies and communications. It will publish latest advances on the engineering task of building and deploying distributed, scalable and reliable data infrastructures and communication systems.

The series will have a prominent applied focus on data technologies and communications with aim to promote the bridging from fundamental research on data science and networking to data engineering and communications that lead to industry products, business knowledge and standardisation.

** Indexing: The books of this series are submitted to ISI Proceedings, MetaPress, Springerlink and DBLP **

More information about this series at http://www.springer.com/series/15362

Natalia Kryvinska · Michal Greguš
Editors

Data-Centric Business and Applications

Evolvements in Business Information
Processing and Management (Volume 2)

 Springer

Editors
Natalia Kryvinska
Department of e-Business, Faculty
of Business, Economics and Statistics
University of Vienna
Vienna, Austria

Michal Greguš
Department of Information Systems,
Faculty of Management
Comenius University
Bratislava, Slovakia

ISSN 2367-4512 ISSN 2367-4520 (electronic)
Lecture Notes on Data Engineering and Communications Technologies
ISBN 978-3-030-19068-2 ISBN 978-3-030-19069-9 (eBook)
https://doi.org/10.1007/978-3-030-19069-9

This Springer imprint is published by the registered company Springer Nature Switzerland AG
The registered company address is: Gewerbestrasse 11, 6330 Cham, Switzerland

Preface

With the second volume, we continue to analyze challenges and opportunities for doing business with information. Evolvements in business information processing and management include divergent viewpoints—technological as well as business and social.

Explicitly, starting with the chapter "Towards Project Documentation in Agile Software Development Methods", the author's survey agile methodologies of software development that reduce the amount of required documentation. As the authors discover—the teams working with such projects cannot afford to completely abandon it, because resignation could lead to chaos and undermine the whole enterprise. So, before and during the work on the project, some documentation has to be created. The problem presented in the paper concerns the proposal of project documentation templates according to the agile approach of software development.

In the next chapter titled "Advanced Information Technologies and Techniques for Healthcare Digital Transformation and Adoption in Ophthalmology", acquiring and processing a large number of patient data, enhanced communication with patients as roles of advanced information technologies, and techniques that can be adopted in healthcare digital transformation are discussed. The aim was to contribute to a specific interdisciplinary research narrowed to a branch of medicine. The research consists of several stages such as literature investigation, thematic analysis, classification, questionnaire, and interviews.

The chapter authored by Michał Pawlak and Aneta Poniszewska-Marańda "Software Testing Management Process for Agile Approach Projects" presents the issue of testing methodology based on advice and best practices advocated by experts in the field of testing. The method is intended to provide a step-by-step instruction of managing testing activities in a project environment. Given the popularity of agile development methodologies, the presented approach is mainly intended for use in agile environments. Further, the authors claim that there are many software development methodologies, each following own philosophy, guidelines, objectives, and good practices. All of them include a dedicated phase of testing and fixing of defects. Unfortunately, the software test management does not have an opportunity of having so many different methodologies available to the test managers. Ability to plan, design, and create efficient tests is the most critical ability for any good tester. Test design techniques, also testing techniques, are used to identify and select the best test cases with the most efficient test conditions.

In the work called "Analyzing User Profiles with the Use of Social API" the authors concern the scope of software engineering and data analysis using artificial intelligence methods. The authors analyze (define and categorize) information that they can get for the help of the presented techniques and available technologies, including the development of their models of access to user data, as well as the methods of interpretation and use of the data obtained. Besides the smart pro-applications proposed by the application can help to get the most out of the service, reaching a larger group, which obviously creates new business opportunities.

The work entitled "Self-similar Teletraffic in a Smart World" emphases building of the future smart world by using recent advances in networking science and practice, which range from traffic engineering and control to innovative wireless scenarios and include key challenges arising from the upcoming interconnection of massive numbers of devices, sensors, and humans. The authors bridge the gap between performance modeling and real-life operational aspects, as well as they leverage measurement data to provide a better understanding of the wired and wireless networks' operation under realistic conditions. Specifically, the authors examine self-similar teletraffic in communication networks, capable of carrying diverse traffic, including self-similar teletraffic, and supporting diverse levels of Quality of Service (QoS). Advanced self-similar models of sequential and fixed-length sequence generators and efficient algorithms simulating self-similar behavior of teletraffic processes in communication networks are developed and applied.

The next chapter "Assessment of the Formation of Administration Systems in the Enterprise Management" advances a method for assessing the level of administration systems formation in the enterprise management. The method is based on the multi-criteria approach and the use of the trapezoidal membership function of confidentiality, availability, and integrity parameters of the administration systems in certain fuzzy term sets, which enables the adoption of managerial decisions to ensure a higher level of compliance of such systems. It describes peculiarities of testing actual state of administration systems in the enterprise management as the preconditions for assessing their level of formation. The chapter substantiates expediency of isolating confidentiality, availability, and integrity as grounds for identifying gaps between the "as is" and "to be" of administration systems in the enterprise management. It suggests using fuzzy logic tools and mathematical apparatus when assessing the level of formation of administration systems. Method for assessing the level of such formation has been applied into practice.

In the chapter authored by Bianka Chorvátová and Peter Štarchoň "Franchising and Customer Experience Management in Telecommunication Industry Phenomenon—Literature Review (2011–2018)", the latest literature available in a specific database that discusses the connection between franchising, telecommunications, and customer experience management is provided. A classification of 48 up-to-date articles and publications is made with a focus on its research area and researched problems. Franchising is a phenomenon largely discussed for about 50 years now. Its types and techniques are similar all over the world, but cultural and legislative differences are specific for every country. Nowadays, the phenomenon of franchising is expanding also in telecommunications companies. They change from state-controlled monopolies and start to penetrate also the private market. As the competition is growing, customer experience management is more important to focus on than ever before.

In the chapter "Conceptualization of Predictive Analytics by Literature Review", a detailed literature analysis on Predictive Analytics, mainly in articles published in relevant journals during the selected time period, from 2010 till now, is arranged. Various databases were used in order to find the most relevant articles for this topic. Articles were systematically analyzed regarding the author (authors), year of publication, the area of research, output, and journal, where the article was published. The main contribution of this article is evidence of the most relevant articles related to Predictive analytics, which can be used for every reader and also an overview, where, or in which fields Predictive analytics is applied and how was used during the past years in various researches.

The chapter called "Information Processing from Unemployment Rates: Evidence from Spain, Switzerland and the European Union" describes the behavior of the unemployment rates and actions taken to control it in the country with the most turbulent unemployment rates as it is Spain, in the country with the lowest and steadiest unemployment rates as it is Switzerland, and in the European Union which represents the countries with average unemployment rates. Actions on the labor market that could have impacted unemployment are also described. The hypothesis of hysteresis was validated for the univariate series of gender inequality and the unemployment rates using the LM test, and on panel data using ILT test. Employing Pesaran CD test, the cross-dependence of the series of distinct characteristics, in both, unemployment and the unemployment gender inequality series was confirmed.

In the next work named "Innovative Activities Development of Industrial Enterprises in Ukraine", the main problems and perspectives of the innovation sector of the Ukrainian economy development are explored. Usage of econometric tools is proven. The main obstacles to the introduction of innovations in the activities of industrial enterprises of Ukraine are institutional factors. Thus, the described analytical toolkit is the basis of the method for the strategic monitoring of entrepreneurship innovation development.

The chapter "Determinants of Employment in Information and Communication Technologies and Its Structure" identifies potential determinants affecting the employment in ICT in short run and long run. The authors examine its structure according to subsectors, gender, age, and education in EU countries. They find significant differences between countries in the structure of ICT employment and also identify several trends. Furthermore, the authors apply correlation, panel Granger causality tests, and panel co-integration regression in order to find potential determinants of employment ICT. The results show a positive effect of ICT skills and education on employment in ICT.

In the research performed at the "Simulation Model of Planning Financial and Economic Indicators of an Enterprise on the Basis of Business Model Formalization" methodical and applied aspects of the simulation model implementation to the system of planned financial and economic indicators of an enterprise are considered. It is performed with a usage of a formal description of its main business processes. By means of both classical economic-statistical methods and the newest methods of analysis, in particular neural network technologies, the dynamics of key factors of the simulation model are studied, which are crucial for the formation of metrics of such planning indicators. Possibility of implementing invariant scenarios for the development of financial and economic situation with a strict identification of the system of financial and economic indicators and the dynamics of their metrics on a certain horizon of planning is demonstrated.

The paper entitled "How Global Is Your Business? A Business Globalization Index to Quantify a Business' Globalization Degree" defines the scope of globalization with regards to its meaning for businesses and within this scope, a measure is proposed to determine a firms' degree of globalization. Starting with a review of scientific work, a multilayered, modular approach is proposed built on three core aspects of globalization. By using a vector-based form, this index is able to provide both: a general measure for a company's degree of globalization and a detailed evaluation of a company's different states with regards to economical, spatial, and regulatory aspects of globalization. The index developed is validated using a sample of four multinational companies (MNC). Results show, that firms, which are deemed global, only reach half of the overall possible degree of globalization.

In the work headed "Stress Testing Corporate Earnings of US Companies", the authors define a stress testing framework for corporations. Stress testing is well known in banking sector, however not that well established in the enterprise risk management. The authors believe stress testing, if adapted for corporations, could play a key role in their risk management as you cannot manage what you cannot measure. The authors test the proposed framework on the US publicly traded companies using two events of interest, the dot-com bubble of 2001 and financial crisis of 2008. They use OLS, elastic net and partial least squares regressions (PLSR) to predict the change of corporate earnings during these events. Results suggest by authors proposed factors in combination with elastic net or PLSR are able to predict movements of earnings.

A further chapter "From Information Transaction Towards Interaction: Social Media for Efficient Services in CRM" authored by Vivien Melinda Wachtler aims to illustrate the efficient services provided by CRM and social media common work. First, it provides an overview of the most important definitions that are used throughout this chapter. Second, it gives a picture about the recent changes in CRM, focusing on customers' behavior changes. Third, it provides readers with information about social CRM and summarizes the new and improved services as a result of CRM and social media common work. Besides, it introduces the top 10 CRM software and their interaction to social networks.

Then, the work called "Assessment of eCall's Effects on the Economy and Automotive Industry" explores a Pan-European automatic emergency call system for cars—eCall. All new types of vehicles used within the European Union must be equipped with hardware enabling the operation of eCall. Despite being targeted by many researchers and official authorities, there remain many questions regarding the effects of this technology. Building on the expected monetary and nonmonetary benefits of eCall, as well as other projections about the system's capabilities, in this study, the authors used the Delphi method to address unanswered issues and to examine the logic behind earlier projections of eCall's effects on both the automotive industry and the economy. In addition to presenting the findings from three bilingual rounds of the Delphi method with 16 experts from the automotive industry, the authors also assess the public perception of this technology and its future development. Besides, the authors present the opinions of experts regarding eCall's potential to ensure annual monetary benefits of €20 billion, based on the prices of cars augmented by the implementation of eCall and its future link with autonomous cars.

And, the final chapter "Data-Centric Business Planning—A Holistic Approach" provides a well-planned and structured business plan by using different approaches to fine grain essential aspects of starting a business. It not only states a detailed description of the business idea, but also considers all marketing as well as financial aspects. Approaches such as a SWOT analysis, marketing mix, competitor-, financial-, and scenario analysis, give a detailed overview of the various types of data use, interpretation, and presentation. Based on the increasing popularity of starting own business, the demand of having a well-prepared plan arises. Not only in order to attract investors, but also to generate a solid basis for a successful start of a business. Collection, choice, analysis, and presentation of numerical data are indispensable activities to make cost-effectiveness possible from the very beginning.

Natalia Kryvinska
Department of e-Business, Faculty of Business
Economics and Statistics
University of Vienna
Vienna, Austria
natalia.kryvinska@univie.ac.at

Department of Information Systems, Faculty of Management
Comenius University
Bratislava, Slovakia

Michal Greguš
Department of Information Systems, Faculty of Management
Comenius University
Bratislava, Slovakia
michal.gregus@fm.uniba.sk

Contents

Towards Project Documentation in Agile Software Development Methods

Aneta Poniszewska-Marańda, Arkadiusz Zieliski and Witold Marańda

Abstract Agile methodologies of software development reduce the amount of required documentation but the teams working according with them can not afford to completely abandon it, because the resignation could lead a chaos to the project that would undermine the whole enterprise. So, before and during the work on the project, some documentation has to be created. The problem presented in the paper concerns the proposition of templates of project documentation, according to agile approach of software development.

1 Introduction

Agile software development methods have been around the world since 2001, when Agile Manifesto was published. Since then, despite the existence of well-described traditional waterfall methodologies for many years, they are very good and they are often chosen by developers teams. Their success is based on their basic characteristics:

- People and interactions between them over processes and project tools.
- Working software over detail documentation.
- Cooperation with client over contract negotiations.
- Responding to changes over strict adherence to the plan.

These characteristics make that much less documentation (if any) is created in projects based on the agile methodology than in the classic approach. Another positive effect of using this approach to project management is that the project is really open to new ideas all the time, making it much more flexible.

A. Poniszewska-Marańda (✉) · A. Zieliski
Institute of Information Technology, Lodz University of Technology, Lodz, Poland
e-mail: aneta.poniszewska-maranda@p.lodz.pl

W. Marańda
Department of Microelectronics and Computer Science, Lodz University of Technology,
Lodz, Poland
e-mail: witold.maranda@p.lodz.pl

© Springer Nature Switzerland AG 2020   1
N. Kryvinska and M. Greguš (eds.), *Data-Centric Business and Applications*,
Lecture Notes on Data Engineering and Communications Technologies 30,
https://doi.org/10.1007/978-3-030-19069-9_1

Methodologies of agile family are in fact a certain pattern of behavior, they do not define a rigid pathway and they allow their adaptation to own developers needs. Therefore, they do not specify the complete set of documents that are needed to create during the project—it is left to the teams managers [1].

However, even the smallest design documentation, is often needed—many factors are influenced, including: human memory—no one is able to memorize the assumptions of the whole project, there is a need of clear definition of responsibilities for certain project fragments in a team and there is a need to set a schedule of the works.

Agile methodologies of software development reduce the amount of required documentation but the teams working according with them can not afford to completely abandon it, because the resignation could lead a chaos to the project that would undermine the whole enterprise. So, before and during the work on the project, some documentation has to be created. The main purpose of its materials is to define: what?, for what? and how? the produced system is supposed to do—the developer team without this knowledge is not able to meet the client's expectations.

The problem presented in the paper concerns the proposition of templates of project documentation, according to agile approach of software development. The proposed documents are based on documents already existing in other methodologies, but they have undergone some modifications to facilitate and accelerate their writing and use.

The presented paper is structured as follows: Sect. 2 gives the outline of types of documents used in agile software development methodologies. Section 3 deals with proposed set of documents for project development in agile methodology. Section 4 presents the tests documenting in agile development while Sect. 5 describes the set of tests documents.

2 Types of Documents in Agile Methodologies

Agile methodologies of project management generally do not define the list of documents that should be created during the enterprise realization. This situation is quite opposite of classical methodologies, where often the description of the methodology or its good practices contain the list of documents that have to be filled, for example PRINCE2 method [2, 3] or PMBoK method [4, 5].

However, it can not be said that project teams are left alone here, because each methodology can use documents derived from other methodologies, and only in addition creates some own documents, which the best suit the specific project case. In addition, some of agile methodologies point some documents specific to them, such as Scrum backlog or features list to develop/non-functional requirements/risk analysis for further work in DSDM [6, 7].

The project documentation can be divided into two basic parts [8, 9]:

- organizational documentation,
- technical documentation.

2.1 Organizational Documentation

The task of organizational documentation is to define the main goals, deadlines, groups of people affected the project, the opportunities, threats and the responsibilities in the project, the documents describing the financial part, and all others that help to organize the team's work. The agile approach includes in this group among others the backlog and non-functional requirements. As these methods do not impose the specific documents, they can be borrowed from the classical project management approach. The vision document from OpenUP [9] or Project Initiating Document from PRINCE2 [2] can be of interest.

Vision document contains closely defined main assumptions of described product, their version common to all stakeholders. It should be short and specifically describing what we expect from the project result. There are also boundaries in which the product has to include—they are often set by the budget. It should also be written in a simple language so that it can be understood by the client and every member of the team [10]. This document may be subject to change during the project development [11, 12].

Project Initiating Document—its purpose is to determine the direction and scope of the project [8, 13]. It is intended to answer the basic questions posed when accepting the product:

- What does the project intend to accomplish?
- Why is it important?
- Where will the product be manufactured?
- Who will manage the project?
- How and when does all this happen?

This document is much more extensive than Vision document, however, some of its sections, such as: Business Case, Risk Register, Quality or Communication Management Strategy, may be useful when accepting the project by the decision-making bodies on the client's side.

Backlog—document from Scrum method. It contains a list of short and brief descriptions of functions provided by the system, so-called *user stories*. They are most often created based on the template [14]:

As $\langle user - type \rangle$, I want $\langle action - description \rangle$ to $\langle description - of - result \rangle$

Thanks to this convention, this document becomes very transparent and it also contains the most needed information [14]:

- defines who the user is,
- describes what he wants to do,
- specifies why does he want to do it?

These data are sufficient to determine what must be done within a given project—it defines the functional requirements as well as product boundaries [10, 15]. The backlog concept also includes the table on which the stories are deployed and their

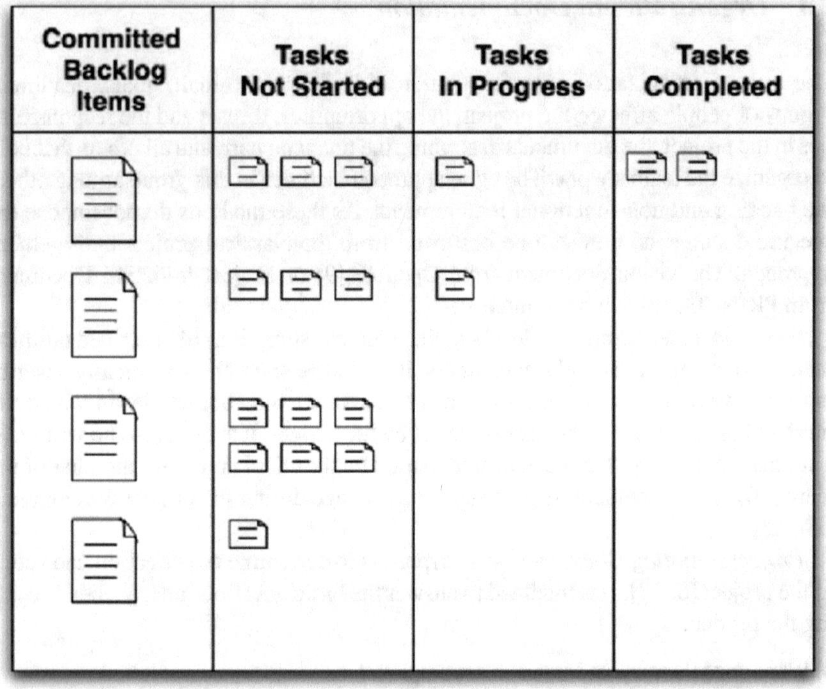

Fig. 1 An example diagram of backlog table [9, 14]

flow from left to right illustrates the progress of their work. The table can be organized in many ways—an example is given in Fig. 1.

2.2 Technical Documentation

Technical documentation is intended to describe the technical aspects that will make it easy to get to know the project by new team members, for example at stage of support for the entire system, after completing the production stage—when development team is being replaced by maintenance team, not always related to the system authors. It also aims to make it easier to work by its target users. Its collection can include:

- executive documentation,
- UML diagrams, describing the model and project of created system,
- sketches and graphic designs of user interfaces,
- instruction manual,
- documentation of the code,
- scenarios of tests,
- other materials useful to the team.

These documents are produced at every stage of product lifecycle and they can be continuously changed so that they correspond to the actual and current direction of developed software. This part of documentation is no regulated by agile methods, so it is entirely arbitrary on the part of design teams to select the documents and their structure.

3 Set of Documents for Project Realization in Agile Software Development Methodology

It is difficult do fix what should be a complete design documentation in agile software development methodology. Answer this question does not exist, because in different situations, different sets of documents will be needed and the required level of detail and therefore their complexity will be needed [9, 16].

This section describes the proposed set of documents, which are the guidelines to facilitate and accelerate the work with project documentation in agile software development methodology. This set contains both organizational and technical documents and it is also not a closed group of documents. This collection can be extended by other needed documents. The set contains the following elements:

1. Vision document.
2. Backlog.
3. Description of use cases.
4. Functional specification.
5. Technical specification.
6. Tests documentation.

They are presented in the following subsections.

3.1 Vision Document

Vision document presents a vision that brings the creating system and motivates the need for its creation. It is supposed to be short, but it must contain as many details as possible:

- product destiny—answer to the question "what system has to do",
- information on what the proposed product will differs itself from the competition,
- information for whom it is intended,
- main functional and non-functional requirements—description of what and how the system is supposed to do more detailed than in first point of this document.

Such construction has to allow on early stage of the project to quickly understand the issue by the person having to discuss the proposal for creation and implementation of the system, and at subsequent stages for those who need to understand the meaning of the project.

This document may be modified during the project lifecycle. According to the principles of agile methodology, the project changes can take place at any point in the manufacturing process. Consequently, the vision originally given for implementation and approved does not necessarily reflect the final product. It is intended only to help get started by identifying the initial requirements for client system and therefore it needs to be updated with project changes.

3.2 Backlog

Backlog document contains a list of users' stories that contribute to created system. However, it is intended to present a bit more information and present them in an uniform way. An element that differs it from its classical form, apart from additional information defined in it, is the fact that stories are not have a sequence form, but they consist of normally complemented elements of the sentence, so "who", "what" and "why".

The whole document contains the following fields (Table 1):

1. *ID Story*—it is supposed to allow the easy identification of a story, which can be referenced in other documents, comments in program code and elsewhere in the documentation. The following identifier's template is suggested: *Sxxx* where *S* is an abbreviation of term "Story" and *xxx* is a sequential number, starting with 001. Unfortunately, the numbers themselves do not carry too much information, but collecting the story information in transparent manner allows to easily and quickly search the documents using these short tags.
2. *As ...*—the column in which the first part of user's story, specifying the person for whom the action will be executed on the system, is located. The names of persons who carry out the actions should also be systematized within the whole project, as this parameter really determines the actor of the action. By simply writing it in a separate column, it is easy to group the user stories by system actors.
3. *I would like to ...*—middle part of classic user stories, describes how large action has to be performed in the given case. This element, in a certain simplification, can be defined as a specific use case of designed system.
4. *in order to ...*—the last part of user story, determining the result of performed action. This element can later provide the base for creating the acceptance tests that verify if the created system works as the client expected.
5. *Value*—a field to determine the value/priority of a given user story. This parameter can be used in generating process of burndown charts to determine the remaining work to be done over a time, in the whole project or in a single iteration.
6. *Person*—person or persons responsible for creating the functionality described by the user story are filled in this field.

Table 1 Structure of a backlog

ID Story	As ...	I would like to ...	in order to ...	Value	Person	Module	Action name	Issue
Sxxxx	*One of the system actors*	*Description of action to be performed*	*Purpose of the action*					
ID story according to Sxxxx schema	Determine by which user (actor) of the system the proposed action should be performed	Short description of action to be performed	Brief description of action purpose—what is the purpose of result, whether it is the target product	Point value of story compared to the others—which are the more important	Person responsible for implementing the mechanism described in the story	Module name under which the proposed action will be performed	Action name connected with the feature described in story	ID under which the story is available in task management system

7. *Module*—an optional field to determine to which system's module a given story should be assigned. Of course, it is useful if created project consists of many smaller subsystems.
8. *Action name*—more precisely the name of main method from the program code that executes this action. The field is also optional and it can be used to facilitate the searching of code for a given system functionality. Marking in this document to which module the action is assigned and by which method is executed makes it significantly easier, and at the same time, speeds up finding a particular piece of code during the process of introducing the changes or improvements in a given feature.
9. *Issue*—the optional field to allow to enter the ID task used to identify the story in task management system, such as Jira, Trac or Bugzilla. It is intended to allow the easy finding the information about the process of creation or testing a given feature in such a system if it is used in a project.

Based on this document, before each iteration the backlogs for iteration are created by copying the selected stories into a new file or by distinguishing the same entries from one iteration. The second method is recommended because it reduces the number of changes that need to be made when modifying a single entry and it is also legible. Another way is to add one more column with iteration number in which the system part is being developed.

3.3 Description of Use Cases

The purpose of *description of use cases* is collecting the scenario for use cases listed in the backlog, so the "big actions". It is based on a spreadsheet, with the difference that it can be a single file, one file containing several sheets or several files. The last two variants are designed to sort the entries and separate them from the iteration—there is no problem with later modification of entries, because everyone is located only in one place.

This document is created at the beginning of each iteration and describes the scenarios for all user stories scheduled for a given phase. Based on it, the programmers create the code of the functions and testers prepare the test cases and their scenarios in the next stages. The structure of document sheet is as follows (Table 2):

1. *Story*—identifier of described story (use case) is placed in this field, it is given only once in the line beginning the scenario.
2. *Flow*—column defining whether it is a main or alternative scenario. The basic scenario is defined by word "main" and the alternative is called from the step number of main scenario, where the paths are split followed by defining the next alternative from the step, for example for the first alternative scenario of step "3" it will be the mark "3a".

Table 2 Structure of description of uses cases

Story	Flow	Step	When ...	Then ...
Sxxxx	*main*	*next number*		
Story ID from backlog to which the use case applies	Term identifying a variant of action course in given use case	Step number under variant. If there is a branch, step numbering resets and starts again from 1	Description of start state of step, e.g. situation that has happen in order perform the action	Description of sequence of operation execution at a given step

3. *Step*—next step number in the given use case variant is entered here. One row in a sheet reflects one item. Numbering starts at 1 and counts from the beginning under each scenario variant—the first step in alternative route 3a will be 1, not 4.
4. *When ...*—this column provides the initial condition for the step. It could be for example another action or state of input data / modified data in the previous stage. This state, or more accurately the correctness or inability of the input arguments, is most often the point of scenario fork. The exemplary values of this field may be: "Unregistered user wants to register" or "If entered data is correct".
5. *Then ...*—this column contains the desired behaviour of the system, which is a brief description of what happens after the state from the field "when ...". However, it must be precise enough for the person writing the functionality to know exactly how to program this step of the scenario.

The appropriate diagrams can be added to this document if it will be useful for development team—in this way the next column can be added to determine the file name with the diagram describing this use case.

3.4 Functional Specification

Functional specification document serves to describe all the functionalities available in the designed system. It is also one of two the most detailed descriptions provided in this proposed set of documents. The format for writing a functional specification is pretty facilitative except for the division into parts for the single functions. Each functionality should be a single point or chapter, depending on the preferences of the team and the size of creating document.

An important feature of this document is detail the information what a particular part of the system has to do and how to operate—it is very important, because the programmer must know what is expected of him based on this set of documents. This is also the place where specific names of elements used in the application code, such as constants, names of classes, methods, names of tables and fields in database or error codes returned by the application, must appear. Of course, this detail can not change into coding in the documentation, because such document will not really make much sense. Including the detailed descriptions of used classes, constants, fields or tables from database in this document is intended to eliminate the naming problems and allow the safe creation of several related system elements by different persons or teams.

3.5 Technical Specification

The *technical specification* document describes all the technical aspects of the project—these about the manufacturing process, designed to help the developers to set up the development environments or to write uniform code, as well as those needed when deploying a finished product in a test or production environment. It is the second of the most detailed documents in the proposed documentation set, next to the functional specification.

The form of this document is optional, it is only important to contain the required data. However, it is recommended to divide it into sections that group the similar aspects, for example the "conventions" section, which contains diagrams and instructions for naming the files, methods and similar elements.

The technical specification should include at least the following elements:

- *List of technologies*—set of all technologies that comprise the entire system, such as programming languages (Java, C++), libraries (Spring, Hibernate), servers (Apache Tomcat), database systems (Oracle, MySQL). If the certain component is available in multiple versions, the used version number should be specified, the addresses from which it can be downloaded, or ready-made parts of libraries code to import it.
- *Description of environments*—full descriptions of environments in which the application will be running at every stage, i.e. creation, testing, working. The information on what operating systems are required (Debian, SELinux) and configuration information, network settings (DNS settings) should be located here.
- *Tools*—list of all tools along with numbers required during system development, such as IDE (Eclipse, IntelliJ, Microsoft Visual Studio), code revision control system (Git, SVN), code build automation systems (Maven, Gradle, Jenkins) and other tools supporting the software development and maintenance (Oracle SQL Developer, SoapUI).
- *Architecture*—description of architecture on which the creating system is to run. It should consist of at least two parts: logical part describing the connections between the system layers and physical part to present the hardware platform on which the product will operate. For example, in the case of a web application, the logical architecture can describe the division into layers (according to MVC pattern) and physical architecture represents the layout of servers, connections between them, along with the information where that part of application will operate.

- *Conventions*—set of guidelines determining how to write the system code (names of variables, methods, classes), character encoding format (UTF-8), commenting style, directory structure.

This list does not contain all the issues relevant to write in the technical specification. It is only a list of elements that appear in most IT projects, and each of them is different, so the above list can be modified as needed. The good practice also is to include UML diagrams in sections where it is easily describe some things with a picture rather than in words, e.g. system architecture.

The presented above proposal of documentation for projects carried out in accordance with agile methodologies, meets the basic assumptions of these concepts, i.e. their small number and thus a small amount of work required to prepare them. Often, there are the documents created with the development of the whole project, thanks to which they are not created "on stock" at the very beginning of the project according to the original assumptions that may change during the process of creating the product.

The proposed documents, namely vision document, backlog, use cases, functional specification, technical specification, and their form is open to change, and thus can be adapted to the needs of a specific project, for example by adding the new sections not listed in this proposition. Nevertheless, some of them are closely related to each other by expanding and refining the issues discussed in the documents, which is why they together form a coherent description of the entire project.

4 Documenting the Tests in Agile Software Development

Agile methodologies do not provide the suggestions on how to document the tests— neither the scenario descriptions nor the reports of their execution. It is therefore normal to create own document structures here and to borrow their templates from other methods. Test documentation can consist of the following elements [8, 13]:

- *Test plan*—document describing the objectives and methods of testing the project, the test schedule, the items to be tested.
- *Test cases and test scenarios*—these documents accurately describe which cases have to be covered by the tests, as well as complete diagrams showing the steps of particular test case, so that the person doing the test knows exactly what to do.
- *Test scripts*—this group can include any codes of unit tests created by developers, but also all scripts created by testers for various types of mechanisms to simplify and automate the testing process, for example the codes to use the Selenium framework.
- *Sets of test data*—test data can also be included in the test documentation, because if the tests are to be valid (especially in the case of re-testing after modification or repair of detected errors), they should be done on close or the same data sets and parameters of tested system. Another useful aspect of early preparation of sets

of test data is that the testing process itself can be accelerated because the tester is not responsible for preparing the data, which is not always a simple task, for example in the case of testing the specialized systems, not every tester must know the exact operating principles of the test process, which risks the preparation of erroneous test data, and thus the false results of performed tests. On the other hand, the preparation of such data by specialists in the field guarantees a correct answer to the question whether the system correctly processes the data.

- *Reports covering the functionality by scenarios and/or test scripts*—these reports show in which scope the given product can be tested with currently prepared test database. The bigger is coverage, the greater number of errors will be detected. However even 100% of coverage of a project by tests is not able to ensure that the prepared product will be free of errors.
- *Test reports*—each test has to be completed with a report describing the test result. It not only confirms the performance of a particular test, but also its details, e.g. in case of automated tests—the number of performed tests, the number of positive and negative tests. In addition, this report, in the case of a negative test, is intended to allow the person fixing the error to reproduce it in order to facilitate the repair process.

4.1 Set of Testing Documents in Agile Methodology

Proposed set of *test documentation* consists of a small number of templates—there are just three types of documents. One of them is a test schedule to determine what tests and on which elements are to be performed at a given moment. The second one is a template of test case description. It is a descriptive document that contains the detail characteristics of the test, aimed at presenting the objectives and the course of a test. The last one is the test report to summarize the performed test—it clearly shows what the error was and under what conditions it was created in the case of a negative test, and in the opposite case, only to report the positive test result.

The *tests schedule* is a document intended to be a backlog for testers, therefore its form is close to it. The recommended form of its writing is a spreadsheet with the following columns:

- *Test ID*—analogous to a similar field proposed in the backlog of design documentation, the identifier is *Txxx*, where *T* is the abbreviation for the word "Test", and *xxx* is the sequential test number starting with 001.
- *Module*—name of the module being tested.
- *Type*—specification of test type, indicating whether it is for example a regression test, an integration test, an acceptance test or another one.
- *Date*—date when certain test is to be performed. It does not have to be the exact moment in time (e.g. May 10, 2017), but only the location during the project lifecycle, e.g. "3 days after the sprint no 2" or "after the work on module A".
- *Person*—person responsible for carrying out the test.

- *Name*—brief description of the test, accurate enough so that the person reviewing the schedule did not have to look for a description in the details of the test cases.
- *Status*—this column displays the current test status: *pending, in progress, passed, failed*. It has to clear indicate of what has already been done and with what effect. A detailed report of test result can be found in the document of test case description.

Test case description is a document showing a single test case. It is not only a scenario of test case, but also a complete set of information about the entire planned test—starting with the selection of time at which the test can run, across the scenario, and finishing the specification of location of data needed to complete the test. This document is composed of the following parts:

- *Number and name*—in the form of document header, these data come from the test schedule and allow to combine the entries with the descriptions of test cases.
- *Initial conditions*—determination when a test can be performed, e.g. after full implementation of the activity from server X.
- *Scenario test case*—this section contains a complete and accurate description of test execution—all required prerequisites: required configuration, next steps of the test, expected result.
- *Location of resources*—this section can be omitted, it is only needed for tests that require additional materials. These materials include: source codes of automatic test, sets of input data to perform the test.

This document is written in form of text document—it can be one large file, containing all the descriptions, updated by all testers or one separate file per each test.

Test report—a document intended to give a clear summary of test performance. Due to the specification of the issue, it may have two variants: test completed positively or test completed negatively (the whole test or only one test component completed negatively).

The introduction of document in both variants is the same and covers the following parts:

- *Test number and name*—these data come from the test schedule and it is used to combine all the documents associated with a single test.
- *Order number of the test*—number identifying the performance of this test case. If performance of number "1" ended in error, the test after error correction will have number "2" and so on. This number is intended to order the next performances the same test cases.
- Date of test performance—the exact date of the test, if this process lasts longer than a day, should include a date range.
- Person performing the test—data about the person performing the test case.

The second part of the document describes the course and the completion of test. This part significantly differs in terms of coverage and provided information, depending on the variants mentioned earlier.

In the case of positive test, it is enough to write a summary sentence after the introduction, indicating that the test has been completed successfully. If multiple test sets are available in the test case, the summary should contain a list of them (collections should be uniquely named), with a signature indicating the positive completion for all sets.

In case of negative test, there are two options:

- There are multiple sets of test data involved with a test—the whole list of information sets with the error/errors occurred is presented. This information must be supplemented by a detailed description of the error that occurred with additional configuration information, e.g. demonstrating all user privileges "on which" the tests were performed.
- The test case does not contain data sets, but only passes through specific steps of the scenario—it should give information in which stage the error occurred and the set of parameters that were set on the elements, on which the tested functionality depended on.

5 Project Documentation According to Proposed Set of Documents

This section presents the proposition of project documenting scheme in accordance with agile methodology. The proposed set of documents consists of the following elements: vision document, backlog, description of use cases, functional specification, technical specification, test schedule, description of a test case, test report.

Vision Document

The vision document has no imposed rigid form, but due to the data it should contain, it is worth dividing it into paragraphs of at least four sections.

The first one should specify the name of the designed product (it can be a code or temporary name, but it is important that it appears—it is easier to refer to the project itself later or start building its brand outside the company). In this part we should also answer the question "*what should the system do?*".

The second section shows what the product should be different from its competition (if it exists). We can also provide here the key functions that are not necessarily differentiating the elements from existing solutions.

The next section is to present the target group of users of the designed system. If we would like to use any additional division of functions relative to end users, it is worth mentioning here.

The last part of the document is devoted to the main functional and non-functional requirements posed before the designed system. In addition to describing these requirements (this description, apart from their writing, may also contain a short motivation), it is worth to expand the first section a bit more. This part should not be too detailed, because a separate document is provided for these elements.

The vision document should be a short document (1–2 A4 pages), so that it can be read quickly, but at the same time content enough for the reader to have some

Table 3 Template for description of use cases with commentary

Story	Flow	Step	When ...	Then ...
Sxxxx	*main*	*next number*		
Story ID from backlog to which the use case applies	Term identifying a variant of action course in given use case	Step number under variant. If there is a branch, step numbering resets and starts again from 1	Description of start state of step, e.g. situation that has happen in order to perform the action—user will send a form	Description of sequence of operation execution at a given step, e.g. what is to happen with the form sent by the user
Field filled only once—in the line beginning the description of a story	Flow is filled in analogy to the story's identifier—if it does not change, it is not given			
	1a Side variants are determined from the number of the step in which they arise and are supplemented with the letter "a". Other variants starting in the same step get the next letter of the alphabet			

idea about the presented product. This form of document can be useful, for example, when trying to find an investor or co-worker idea.

Backlog

A template of Backlog document is presented in Table 1. All user's stories are saved in one document one below the other. This document is to be a base for planning subsequent iterations of the project, therefore it should always be up-to-date.

Description of Use Cases

template of this document is presented in Table 3. As with the backlog, all descriptions of the stories (use cases) are saved in one file. To preserve the readability of the document between the lines describing the different cases, it is recommended to leave one or two empty lines.

Functional Specification

The functional specification, similar to the vision document, has no specific form. This document should in sufficient way describe all the functions of the designed system. It can be the basis on which the producer (as well as the developers) will be billed from the created product.

It is recommended that it be divided into chapters, covering groups of functions related to each other, for example: "1. User management, 2. Editing documents, 3. Import/export of documents". Their subsections would be descriptions of individual mechanisms, e.g.: "1.1. User registration, 1.2. Login users". The description should be sufficient, so it means the subsections should contain all details of the issue.

Technical Specification

The form of technical documentation is optional, however, in order to clearly separate the content recommended for this document, it is worth to use the division into chapters, containing:

1. *List of technologies* with which the system is to be built: language(s) of programming, libraries, servers, databases.
2. *Description of the environments* for which the planned system should be written, tested, run and planned, and their configuration.
3. *Tools* planned for the use in the process of manufacturing and maintaining the created software, e.g. IDE, version control system, automation of code building.
4. *Architecture*—description of the architecture according to which the designed system is to be built. This description should include both logical description (characteristics of system layers and connections between them) and physical (description of the hardware platform) parts.
5. *Conventions*—a list of guidelines on how to write the code, as well as naming of all elements related to the project (source files, resource files, etc.).

Description of a Test Case

The description of a test case can be constructed according to one of two models:

- One document containing descriptions for all test cases.
- Each test case has its own document containing its description.

  Regardless of the chosen form, the description of one test case is as follows:

- Txxxx: The name of the test case.
- Initial conditions: this section contains all the conditions that must be met before the testing process can be undertaken.
- Scenario: here should be a full scenario of the described test. In case the tested functionality has many paths of its transition, each such path should be described in a separate variant.
- Location of resources: this section should give the locations where all the materials needed to perform the test have been placed. If the test does not require such data, this section may be omitted or supplemented with the text: "not applicable".

Test Report

The test report should contain the following elements:

- *number and name of the test*—it is to enable an easy and quick connection of the document with the test schedule and description of the test case. Test number: test order number The number of the next test approach.
- *Test date*—the date the test was performed.
- *Person performing the test*—name and surname of the tester.

The second part of the document depends on the success or failure of the test. If the test ended positively, it is closed in a short summary of this fact. In the case of an error, the procedure is similar, but a detailed description of the error must be given as well as at what moment and at which configuration the error occurred, as well as, if possible, the stack from the application logs. A good addition to reports describing errors are also attachments with views, illustrating the error of course, if it is possible to do so.

6 Conclusions

The proposed set of documents was prepared based on various types of documents available both for agile software development methodologies and for classic approach methodologies. The main goal of developing this set was to be within the framework defined by agile approach and to be simple to implement within project groups experienced in agile software development methods. So, they are just the most needed documents that define the product itself and work on it. The whole set consists of eight types of documents:

- Vision document—presents the vision of designed project.
- Backlog—list of all actions that can be performed on the system.
- Description of use cases—detailed descriptions of control flow within the actions specified in backlog.
- Functional specification—full description of system behaviour in all actions foreseen in the project.
- Technical specification—description of technical issues related the environment in which the finished product will have to work.
- Tests schedule—plan of tests of creating system to maintain the product quality.
- Test case description—scenarios of planned tests.
- Test report—full descriptions of results of performed tests together with the information to help trace and fix the detected errors.

The number of proposed documents is small and their content does not have to be very extensive to be sufficient. Its fragments can only be written before work on a particular element. In addition, they cover the entire project from the process of determining what the product is supposed to do, until it is fully tested. The proposed set of documents as well as the agile methodologies themselves are just a kind of signpost and developers do not need to stick to such a structure in order to use them.

The presented set of documents can be adapted to the requirements of a specific project and the habits of a development team.

Despite the fact that agile methodologies of software development aim at minimizing the documentation being created, in the case of tests it is necessary. It aims to set a strict definition of when and how the creating software to be tested. It is also intended as a basis for proving the functioning correctness of created system, e.g. though the test reports.

The presented set of document templates can also be a great help to teams which want to move from the classical methodology of project management to agile methods and they can not found themselves in the reduced number of prepared documents. It may also be useful in teams that, due to the simplicity of agile methodologies, work in harmony with them, however, due to the lack of determined ways of documenting the project, they have done it in a chaotic way or skipped this element of project lifecycle.

References

1. Cadle J, Yeates D (2007) Project management for information systems. Prentice-Hall
2. Office of Government Commerce (OGC) (2009) An introduction to PRINCE2: managing and directing successful projects
3. AXELOS (2009) Managing successful projects with PRINCE2, 2009 edn
4. Project Management Institute (2009) A guide to the project management body of knowledge (PMBOK guide), 4th edn
5. Duncan WR (1996) A guide to the project management body of knowledge. Project Management Institute, USA
6. DSDM Consortium (2007) DSDM Atern handbook, UK. DSDM Consortium
7. Kryvinska N (2012) Building consistent formal specification for the service enterprise agility foundation. Soc Serv Sci J Serv Sci Res Springer 4(2):235–269
8. Strode DE (2005) The agile methods: an analytical comparison of five agile methods and an investigation of their target environment. Massey University, Palmerston North, New Zealand
9. Lane D, Coffin R (2006) A practical guide to seven agile methodologies, Part 2, Devx.com
10. Wysocki RK (2009) Effective project management: traditional. Agile, extreme. Wiley Publishing Inc
11. Kaczor S, Kryvinska N (2013) It is all about services—fundamentals, drivers, and business models. Soc Serv Sci J Serv Sci Res Springer 5(2):125–154
12. Gregus M, Kryvinska N (2015) Service orientation of enterprises—aspects, dimensions, technologies. Comenius University in Bratislava. ISBN: 9788022339780, 2015
13. Kerzner H (2003) Project management: a systems approach to planning, scheduling and controlling. Wiley
14. Schwaber K, Sutherland J (2011) The scrum guide. Game Rules, Scrum.org
15. Kryvinska N, Gregus M (2014) SOA and its business value in requirements, features, practices and methodologies. Comenius University in Bratislava. ISBN: 9788022337649
16. Molnr E, Molnr R, Kryvinska N, Gregus M (2014) Web intelligence in practice. Soc Serv Sci J Serv Sci Res Springer 6(1):149–172

Advanced Information Technologies and Techniques for Healthcare Digital Transformation and Adoption in Ophthalmology

Robert Furda and Michal Greguš

Abstract Acquiring and processing a large number of, including different formats, patient data, and enhanced communication with patients are roles of advanced information technologies, and techniques that can be adopted in healthcare digital transformation. The aim of this chapter is to contribute in specific interdisciplinary research that was narrowed to one branch of medicine, ophthalmology. Our research consisted of several stages such as literature investigation, thematic analysis, classification, questionnaire, and interview. In central part of the research we dealt with selected use cases and schematic proposals of multi-perspective architecture in digital transformation for ophthalmology. Because some stages of our research were already published that in this chapter we communicate in details only the results of literature investigation and questionnaire. At the stage of literature investigation we examined five preselected searchable database sources and via questionnaire we scanned the awareness and opinions of the members from Slovak Ophthalmological Society.

1 Introduction and Background

The domain of healthcare has always been at the forefront of research interest in the deployment of advanced information technologies and techniques (AITaT) in optimizing solutions for customer needs, forming and resolving business problems, taking business decisions, and implementing business. The amount of data that is processed on daily basis in healthcare is exponentially growing. The natural consequence was that the healthcare data gained a dimension of Big Data. Current scientists permanently research the adoption of various statistical and artificial intelligence techniques to address the processing of such massive sets of data. Another dimen-

R. Furda (✉) · M. Greguš (✉)
Faculty of Management, Department of Information Systems, Comenius University, Odbojarov 10, 820 05 Bratislava, Slovak Republic
e-mail: Robert.Furda@fm.uniba.sk

M. Greguš
e-mail: Michal.Gregus@fm.uniba.sk

© Springer Nature Switzerland AG 2020
N. Kryvinska and M. Greguš (eds.), *Data-Centric Business and Applications*,
Lecture Notes on Data Engineering and Communications Technologies 30,
https://doi.org/10.1007/978-3-030-19069-9_2

sion is the data collection capability in obtaining the patient's data or in gathering important information from the medical history of patients [83].

More studies regarding healthcare and AITaT described potential benefits and achieved business value that presented various findings, among others, missing consensus about the operational definition of Big Data. For natural language processing and clinical decision support the Big Data analytics technique is most widely used, but often the promised business value was elusive and the potential component of the evaluation to maximize the commercial value of the projects was missing [47, 72, 80].

Current AITaT such as deep data analysis, Deep Learning, and Big Data Analytics therefore can be used to inventory assets, to develop new drugs, to process patient records, to evaluate clinical trials, and to manage healthcare costs. The methodology of information processing used in healthcare is the same as in other sectors but their end use and materiality are tailored to the specifics, and needs, of healthcare facilities, organizations, research laboratories, pharmaceutical companies, and other collaborative workplaces. It is generally known that data collection has greatly improved in recent years, but the introduction of the latest technologies for their processing still lags behind the expectations of the practice [40, 59].

Among the AITaTs, emerging technologies include the recent development of three-dimensional (3D) printing that can create models of things, shapes, objects and structures often also applied in the healthcare sector. Creating 3D objects in almost any shape, defined in a CAD (Computer-Aided Design) file, is entered to 3D printer that is controlled by CAD data and is building an object vertically layer by layer. In medical practice for preparation of CAD data the practitioners often use two-dimensional radiographic images, magnetic resonance imaging or computer tomography [28, 37, 38, 110].

New technical challenges in data processing appeared by deployment of the Internet of Things (IoT) applications that produce high volume of data. The limited network bandwidth needed to decentralize processing because real-time data analytics processing must happen at the right place and at the right time close to the location where the data is generated. In addressing the challenge the type of architecture called "fog computing" was optimized for the distribution of computational, networking, and storage capabilities [16].

The change is critical point during digital transformation that is also a very extensive and enterprise-wide exercise, fundamentally transforming business with a focus on processes, capabilities, acceleration, innovation, and revenue generation, or even a complete change in the business model. Healthcare management should also develop digital strategies that minimize risk, use standard systems, promote innovation and transformation change [11, 50].

The literature describes several definitions of Big Data and we argue that each of them is correct. Big Data includes the ability to manage a huge amount of data of different format and character, with right speed, and in the right time frame that the outcomes of analyzes and computation could be achieved in real time [51].

Nowadays, Big Data in healthcare is actively used to change the way decisions are made and can handle a massive explosion of data that is found in many medical organizations. Changing the entire healthcare ecosystem brings cost-effective measures, better resources, and measurable value across the world. The amount of collected data is growing and the lack of proper treatment about this data brings a lot of problems. The sustained data collection brings a lot of challenges especially by handling of personal health information that cannot be shared. According to KPMG (2012), 70% of healthcare data is generated in Europe and North America. Already in 2015 the world produced more than 20 Exabyte healthcare data [92, 93].

Big Data analytics offers also great insights in healthcare domain. Traditional database architectures are not enough to face the Big Data challenges that are now entering healthcare organizations. Big Data analytics play an important role in achieving prediction capabilities in healthcare, in solving problems, and a healthcare issue arises [93].

The Big Data, the huge amounts of data containing different data types, needs to use also specific techniques to provide their classification, clustering, association, and regression. These techniques are covered by Data Mining that is one of the most motivating areas in research and becomes increasingly popular in healthcare organizations. Data Mining has the goal to extract usable patterns and knowledge from Big Data, plays an important role for uncovering new trends in healthcare organizations, and is useful for all the parties associated with this field [2, 29].

Data Mining can be used to help in finding medicines for existing illnesses, detect patterns of genetic diseases, and the origin of just discovered diseases around the world. By introducing data mining techniques also healthcare can finally gain control over the lack of available records. Data mining can be used in care of patients and can help in healthcare visions, strategies, plans, and administration. Using these methods hospitals and health insurance providers can save millions of dollars, of administrative pains, and, most importantly, a myriad of lives [114].

The linkage between healthcare services and cloud computing technologies have recently attracted attention not only on migration of IT systems and the management of distributed healthcare data but also the misuse of the information that is hidden in the data. It is therefore important to review the data-based services on healthcare migrated to cloud computing, which, for example, can predict patients' future length of stay at the hospital [119].

Enterprise Architecture's role is to describe an approach that manages rational efforts such as development or implementation of a plan to provide the highest quality solution to customer satisfaction. It is powered by the IT application framework and describes how to increase efficiency and effectiveness of business. The Enterprise Architecture's foundation is to analyze the current business situation and define relevant goals that should be achieved in the future. Enterprise Architecture Management addresses both large and small changes, whether IT, business processes or data. It helps to understand what exists in the business now and how it will look in the future, where changes are being made and when, and creates artifacts to model these changes so that it is clear to everyone [108].

To achieve a proven business value from Enterprise Architecture is to get a strong mandate from the organization. Development projects must be in line with Enterprise Architecture that should provide a focus on services for business functions and strategic management of the organization. Enterprise Architecture can support digital transformation on multiple levels and at the same time. Critical success factors should be taken into account: the transformation program must be aligned with Enterprise Architecture, Enterprise Architecture team must be oriented to practical and flexible services, and the organization must be changed from business-unit oriented information technology planning to process-oriented planning [44, 56, 62, 102].

In addressing specific impediments of digital transformation process in healthcare we focused on research of Enterprise Architecture framework for investigating the behavioral and active structural aspects that apply to business and application layers. We excluded the technology layer. We included an analysis of, including inherent relationship between, selected scopes such as strategy, business and education to standard elements such as process, service, and function of an application. We presented classification and assignment of individual impediments that indicate to healthcare managers where they may potentially struggle during different stages of digital transformation. Among others, the impediments classifications facilitate strategic planning and managerial decisions during implementation of the emerging information technologies and techniques. In addition, the idea was to contribute to the successful implementation of healthcare digital transformation, thereby delivering business value within healthcare sector [33].

The development and delivery of information gathering services from different sources require the architecture of automation capable of communicating with customers and using the necessary information. Related artifacts cover requirements such as collaboration, coordination, measurement of customer satisfaction, information sharing, and text analysis [74].

On one hand there are visions of the AITaT researchers but on other hand there are visions within healthcare. In order to review and evaluate the current state, awareness and opinions, we have decided to explore the overlapping AITaT application within the business transformation space and healthcare that has been narrowed to ophthalmology branch. Ophthalmology is the branch of medicine that deals with the prevention, diagnosis, and treatment, of eye diseases.

The business transformation space cannot miss the digital dimension that includes key components of digital transformation. An integral part of this space is Enterprise Architecture with comprehensive approach in design of centralized customer support systems and possible use cases. That can include Big Data analytics tools, intelligent web applications, and a variety of text, speech and image analysis techniques. Among other things, research results in this field can facilitate digital transformation in areas that include the specified use cases in ophthalmology branch. They include, for example: taking anamnesis of patients, analyzing image data on diabetic retinopathy, diagnosing age-related macular degeneration, analyzing the incidence of intraocular tumors, using mobile ophthalmological clinics, and others [31, 35, 36, 61].

2 Aims and Methods

The aim of our research was to contribute within the specific interdisciplinary research that was narrowed to one branch of medicine, ophthalmology. Our research consisted of several stages, such as literature investigation, thematic analysis, classification, questionnaire, and interview. In the central part of the research, we dealt with selected use cases and schematic proposals of multi-perspective architecture of digital transformation for ophthalmology.

In this chapter we summed up and presented predominantly our results of two research stages such as literature investigation and questionnaire because the other research stages we already published. However, to present the overall picture about our research we also included in brief all research stages that reflected the original research question:

What benefit can we expect from digital transformation in ophthalmology?

In order to respond to our research question we determined that the research goal in the healthcare environment was focused on key program pillars such as knowledge, decision, change, and value, with patient in the middle (Fig. 1).

The first three stages helped us to gain the knowledge and become familiarized with the subject area that helped us to identify problems. The identified problems awareness facilitated us to define objectives of a solution. We looked for solutions based on Enterprise Architecture methodology in selecting suitable use cases that cover some of specific processes, and schematic proposals of multi-perspective architecture of digital transformation for ophthalmology. The achieved results we demonstrated by publishing partial outcomes. We evaluated the business value and usability in practice via method of questionnaire and interview as research tools of a qualitative nature (Fig. 2). Finally we discussed the applicability of use cases, specifications of suggested AITaT deployment, and potential added value by applying some principles of value chain framework model [5, 64, 67, 75, 88].

Fig. 1 Key program pillars of research

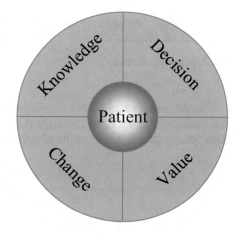

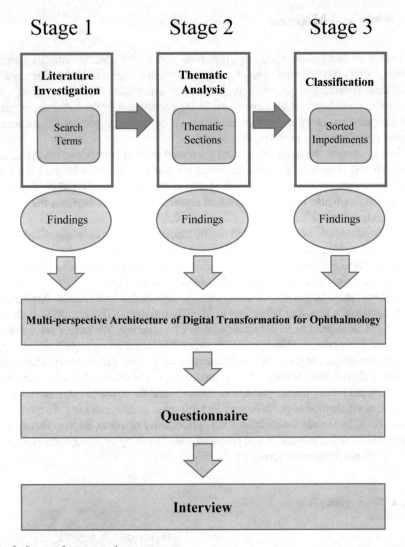

Fig. 2 Stages of our research

For the presentation of our results in the stages of literature investigation and the questionnaire we used summary tables, different types of charts, and textual descriptions. The research results of other stages we presented in brief summaries.

2.1 Method of Literature Investigation

At the beginning of research the literature investigation allowed gain the proper information. The proper methodology should be selected to summarize and synthesize the current and future research to acquire a good understanding on available information. This can also help in assistance of learning process for a topic. For the first stage of our research we found, and selected, for literature investigation two appropriate information sources with the approach developed by Bauer and Strauss in 2016 and Kryvinska et al. in 2013 [8, 63]. This methodology allowed us to be quickly familiarized with a subject area and to pass through a plenty of publications available over the Internet.

To familiarize with the interdisciplinary issues (in the meaning of the AITaT and ophthalmology branch) through the latest available publications, as recommended, we defined appropriate keywords, and created final groupings. From the all groupings we selected five search terms (ST1–ST5) that we used in searching of wanted publications:

1. "Big Data Analytics",
2. "Big Data Analytics" + Healthcare,
3. "Big Data Analytics" + Ophthalmology,
4. "Big Data" + Ophthalmology,
5. "Data Mining" + Healthcare.

To find the proper publications we selected the following five searchable database sources (SDS1–SDS5):

1. IEEE Xplore
2. SpringerLink
3. Wiley online library
4. Emerald Insight
5. IGI Global.

We searched each selected search term in each selected searchable database source. The collected list, that contains not only the number of search results per searchable database source, we analyzed. The analysis examined in each offered list of results included if the publication title has been shown clearly its relevance to the current search term. If yes then we recorded the particular publication. If not then we analyzed the abstract (or available full text). Then, if relevance to the current search term has been shown in abstract then we recorded the particular publication. If not then the next result was selected for the analysis until two publications per search term we recorded. One of the conditions to record a publication was if the particular publication is one of three base media types such as chapter of book, journal article, and conference paper (Fig. 3).

In the case the searching did offer no relevant results, we did not record anything. The information about the selected publications we recorded in supplemental table that contained the following items: searchable database source, search terms,

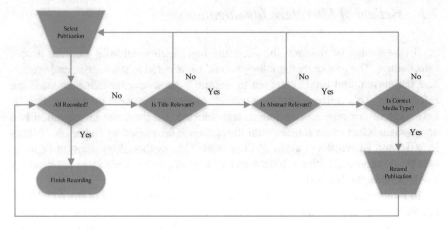

Fig. 3 Process of recording particular publications

media type, authors, title, the name of conference, journal, or publisher, the year of publishing, and the number of authors.

From a statistical point of view, we presented in results the analysis of two data sets: the number of search results and the characteristics of the selected publication from the supplemental table.

2.2 Method of Thematic Analysis

During second stage of our research in order to broaden the understanding of interdisciplinary issues and visually organize gained information we continued the consequent research in using the mixed mode of quantitative and qualitative analysis. In this stage we took advantage of our knowledge and skills in already finished literature investigation. We created one thematic analysis and reused the advices and recommendations published by Braun and Clarke, as well as advices based on the mixed research methods offered by Peek et al. [15, 85]. We decided to gain the information from another source, comparing to the first stage, we selected searchable database source Google Scholar [115]. The search terms we combined from the keywords such as healthcare, digital strategy, digital transformation, and digitalization. We prepared the data corpus and applied fourteen steps (Fig. 4) of thematic analysis—already published in mean time [32]. We defined six initial codes, such as process, information and communication technology/information technology (ICT/IT), Big Data, service, algorithm, and architecture. The selected textual components we merged into sets of logical textual sequences that we reported in thematic discussion.

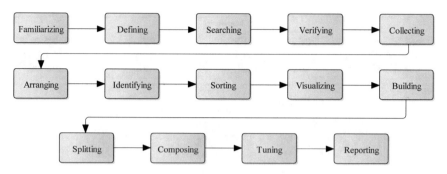

Fig. 4 Sequence of steps in the thematic analysis (adapted according to [32])

2.3 Method of Classification

In this stage of our research we used the principles of the classification method. The method focuses only on some important data and increases the difference between the samples in the classified data corpus [69]. Among others, the classification strategy can help in production of new knowledge in designing innovative artifacts, in improving and understanding the information systems behavior, that was facilitated by Design Science Research (DSR) methodology [86].

Through a new literature investigation during the second stage of our research, the thematic analysis, we gained the data corpus that we analyzed by looking for published impediments (including challenges, gaps, and others). The classified impediments that occurred already in the process of healthcare digital transformation can be helpful to management to make right decisions during strategy formulation and defining goals [33, 91].

We reused selected components of the information technology standard of well-known Enterprise Architecture framework TOGAF—The Open Group Architecture Framework [104]. Once the list of impediments has been compiled we assigned individual impediments according selected elements of Enterprise Architecture such as process, service, and function of an application. The final classification of assigned impediments we grouped according to the defined scopes such as strategy, business and education [33].

The extract of DSR flow documented the build process: from requirement through elements up to scopes (Fig. 5).

2.4 Method of Design Science Research

Method of DSR focuses on the development and performance of architectural artifacts with the explicit intent of improving the functional performance of the artifact. This method is typically applied to categories of artifacts, including algorithms,

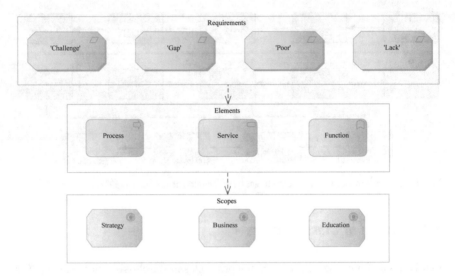

Fig. 5 The extract of DSR flow (adapted according to [33])

human-computer interfaces, design methodologies (including process models), and languages. Its application is most prominent in engineering science and informatics, but is not limited to them, and it can be found in many other disciplines and areas. The framework covering information systems research and the roles and techniques can be specified in the form of guidelines to achieve a better understanding, implementation and evaluation of information systems research and its outcomes [48, 82, 109]. Research in the field of natural sciences follows a conventional design. Its homogeneity facilitates the recognition and evaluation of results. A similar format was missing in original DSR. This problem was solved by Peffers et al. [86].

The mapping between Enterprise Architectures before and after change should be done layer by layer. This means to understand and define how to provide the transition from existing artifacts such as functional or architectural components or services of a base information ecosystem to the wanted target artifact [20].

The methodology and recommendations based on Enterprise Architecture and framework TOGAF we used in selecting use cases, and in preparation of schematic proposals projected architecture as well. Some of them we already published [30, 31, 34]. The achieved results and potential support from practice we simulated via method of questionnaire—to follow appropriate data collection in answers and its analysis. The final recommendations we gained via method of structured interview that helped us to determine whether the deployment would be beneficial.

2.5 Method of Questionnaire

The questionnaire is a relatively convenient way to collect information from a large number of people within a defined time frame. The questionnaire proposal is extremely important to ensure the collection of accurate data and to make the results interpretable and generalizable [52]. In origin, a questionnaire is a quantitative research tool consisting of a series of questions (or other types of calls) for the purpose of gathering data from respondents. Although questionnaires are often intended for statistical analysis of answers, it is not always true. Questionnaires have advantages compared to other types of surveys, for example, they are inexpensive, they do not require much effort, and often have standardized responses/answers to select that greatly make simpler the data compilation. In our case the target was to reach a lot of respondents in a short time [41, 52]. Research on health services is increasingly using standard data-driven questionnaires that can be compared across studies. Clinical trials, for example, usually include measures to identify patients about illness, satisfaction with services, or a quality of life related to health [13].

We created one questionnaire with ten questions that were tuned to gain the responses from ophthalmologists in Slovak Republic. The questionnaire implementation was done using the Google Forms tool (at no extra cost). We set to allow anonymous responses. Thanks to cooperation with Slovak Ophthalmological Society we have been granted permission to send our questionnaire to e-mail addresses of Slovak ophthalmologists.

From statistical point of view we present in results the analysis of each question's answers of respondents, and in addition one brief comparison of two groups of respondents that differ in number of years they have been active in ophthalmology branch.

2.6 Method of Interview

In the last stage of our research we investigated more deeply the business value and usability of proposed solutions in practice. To do that we used research tool of a qualitative nature: the method of interview. This research tool is an individual interview between the researcher and the respondent who communicate with one another (e.g. face to face, audiovisual system, etc.). The purpose is to obtain additional information to supplement, clarify, and verify the amount of quantitative information that has been obtained. The researcher usually has predefined circles of questions that responders respond during the interview either in a defined order or in a different order. The interview is usually flexibly adapted to the content of the interview. Such a conversation is either a sound recording or, in the case of time options, the contents of the responses are written to the record sheet [24].

In our research, the interview succeeded in gaining an insight into our proposals documented in the use case descriptions. We had two opportunities to interview a number of foreign experts from the ophthalmology branch to verify which of our proposals make sense, how and where to apply, and to confirm to what extent they can bring business value into practice.

3 Results and Findings

3.1 Brief Results Introduction

We document a brief introduction of results in each research stage:

– In the first stage of our research we provided the literature investigation at the end of 2016. The selected searchable database sources offered in total 190,701 search results. We collected the information about search results for each defined search terms and for each searchable database source. The table with search results we extended by two additional rows about the web page link, where the searchable database source was accessible, and the remark about the searching conditions, if not default. In parallel we collected data to the supplemental table that included information about searchable database source, search terms, media type, authors, title, the name of conference, journal, or publisher, the year of publishing, and the number of authors. Due to the voluminous of the supplemental table as a whole we published in this chapter only selected data.
– In the second stage of our research we applied focused literature investigation to select appropriate publication for data corpus creation. We processed fourteen steps activity on the data corpus. We depicted the associative relationship of sub-codes to the initial codes. The relevance and representativeness of initial codes to healthcare digitalization we confirmed in composition of forty logical textual sequences [32].
– In the third stage of the research we analyzed the same data corpus that we gained during the second stage. In the results we depicted also the relationships of the selected Enterprise Architecture elements such as process, service, and function of an application, and the scopes such as strategy, business, and education.
– Within the DSR phase we focused on use case descriptions such as the automated data gain for composing a medical history, automatic ocular disease diagnosis, and others. For example, already known automated detection processes for ocular disease was not yet widely available [120]. We did not find a relevant analysis about business value from digital transformation point of view. Among others, we discussed about usability of medical services that can be offered, for example, through cloud computing, and they intercepted aspects of behavior and the active structure elements of the Enterprise Architecture and application layer. We created several basic schematic designs of target architectures.
– After developing of architecture proposals we needed to evaluate the business value and usability of selected use cases in practice. We used research tool questionnaire that we applied during the third quarter of 2017. We received 128 responses out of

265 emailed ophthalmologists (48.3%). The responses we collected over a two-and-a-half-month period. The time stamp of the responses documented the most of the responses (96%) we received during the first thirty days after sending the emails to respondents.

– From the proposals the questionnaire confirmed the selection of two selected use cases that we verified in brief interviews with five foreign experts from ophthalmology branch. We gained the valuable information about where the proposals can be applied and what is still needed to specify that added value is tangible in practice.

3.2 Statistical Results of Literature Investigation

The total number of search results of the literature investigation was 190,701 (Table 1). Incidentally, by applying one search term, "Big Data Analytics" + Ophthalmology, the "Emerald Insight" database returned no one relevant result.

We used two graphs to visualize our search results: the number of search results per search term and per searchable database source (Figs. 6 and 7). The most search results we found for the search term "Data Mining" + Healthcare (72%), and only 1% for "Big Data Analytics" + Ophthalmology. The searchable database source "IEEE Xplore" offered nearly two-times more search results than other searchable database sources (63% vs. 37%). On the opposite side the searchable database source "Emerald Insight" offered 67-times less in search results (1%) than the best one "IEEE Xplore".

The first specific view on the number of search results was the visualization for each searchable database source per search term via five pie charts (Fig. 8). The searchable database source "IEEE Xplore" offered significantly the most search

Table 1 Summary of search results per search term and searchable database source

Search term	Searchable database source					
	IEEE Xplore	SpringerLink	Wiley online library	Emerald Insight	IGI Global	Total
"Big Data Analytics"	1,750	11,596	4,947	884	2,252	21,429
"Big Data Analytics" + Healthcare	2,200	1,578	842	104	609	5,333
"Big Data Analytics" + Ophthalmology	1,820	99	73	0	8	2,000

(continued)

Table 1 (continued)

Search term	Searchable database source					
	IEEE Xplore	SpringerLink	Wiley online library	Emerald Insight	IGI Global	Total
"Big Data" + Ophthal-mology	19,675	2,837	2,629	57	84	25,282
"Data Mining" + Healthcare	94,003	16,966	22,436	726	2,526	136,657
Total	119,448	33,076	30,927	1,771	5,479	190,701
Remark	Basic search		All content	Articles and Chapters	All Products	
Web Page	http:// ieeexplore. ieee.org/ Xplore/ home.jsp	http://link. springer. com/	http:// onlinelibrary. wiley.com/	http://www. emeraldinsight. com/	http:// www.igi-global. com/	

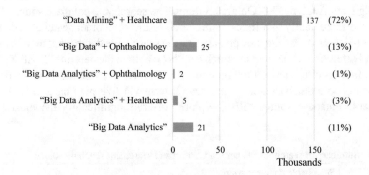

Fig. 6 Visualization of search results per search term

results (from 41 to 91%) in the case of four of five search terms, however for the search term "Big Data Analytics" offered only 11% of the search results. The searchable database source "SpringerLink" offered significantly the most search results for the search term "Big Data Analytics" (54%). Significantly the lowest number of search results from the percentage point of view (0%) we obtained with two searchable database sources: "Emerald Insight" and "IGI Global", specifically both for two search terms: "Big Data Analytics" + Ophthalmology and "Big Data" + Ophthalmology.

The second specific view on the number of search results is the visualization for each search term per searchable database source via five pie charts (Fig. 9). The visualization of the search results confirmed that two of five search terms significantly

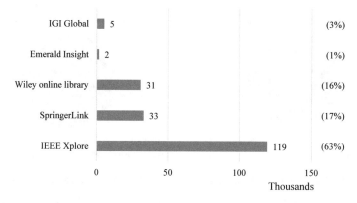

Fig. 7 Visualization of search results per searchable database source

dominated: "Big Data Analytics" and "Data Mining" + Healthcare. All searchable database sources for these two search terms offered jointly from 80 to 91% search results. However one non-proportional search result was obtained in the searchable database source "IEEE Xplore" where for the search term "Big Data Analytics" it offered only 1%. For the rest three search terms the searchable database sources offered significantly lower numbers. And especially one of the five search terms, "Big Data Analytics" + Ophthalmology, the searchable database sources offered only 0% in four cases and 2% in one case.

The second part of the statistical evaluation of the literature investigation was based on the selected search results. The information about particular publications we collected in supplemental table, but as mentioned, due to its voluminous we published only selected data.

In the list of particular publications of the selected search results we expected to have fifty records in the supplemental table. Because of one search term, "Big Data Analytics" + Ophthalmology applied for searching in the Emerald Insight database source, returned no one result (see Table 1) the number of items in the supplemental table was only forty-eight (96% of original expectation). To document the base contents of the supplemental table we created the simplified list (Table 2) that consisted of the search term code (ST1–ST5), the searchable database source code (SDS1–SDS5), the first author surname, the first words of the publication title, and the link to the reference list.

The numbers of media types per searchable database source (Table 3) showed that the media type of journal article contributed almost to half of all (48%), of book chapter at exactly one third (33%), and at least was of conference papers (19%) per searchable database source.

The numbers of media types per search term showed that just one search term, "Big Data Analytics" + Ophthalmology, was not offered in media type of journal article (Table 4).

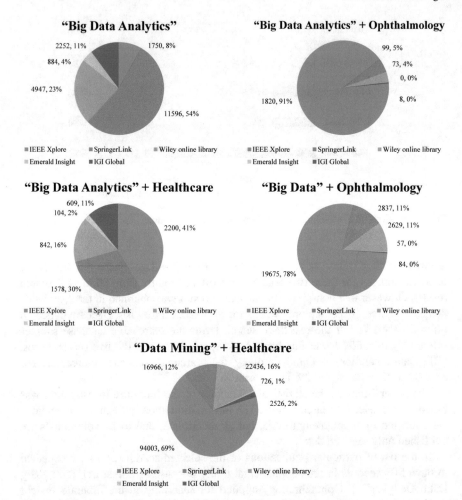

Fig. 8 Set of pie charts with number of search results per search term

The numbers of year of publishing per searchable database source showed that the selected search results, they were published in 2016, contributed almost to half (48%) of all (Table 5). One selected search result, relatively old (published 1988), was offered by the searchable database source "Emerald Insight". The newest selected search result was from year 2017, offered by the searchable database source "IGI Global". It indicated pre-published registration. From the years of publishing three of them (2014, 2015, and 2016) contained thirty six selected search results what was exactly three quarters of all (Fig. 10). The searchable database source "IEEE Xplore" contained all selected search results exclusively from year 2016.

The number of media type per number of authors showed that the selected search results have the mostly two authors (29%), one and three authors have ten publication

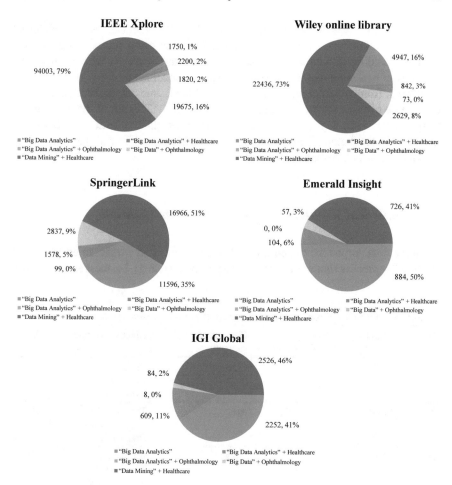

Fig. 9 Set of pie charts with number of search results per searchable database source

titles (both 21%), and the most, twenty authors, appeared in just one of the selected search result what was the media type conference paper (Table 6).

We assessed that the above statistics regarding the not published supplemental table brought well imagine about its contents, all recorded information regarding the selected search results.

3.3 Statistical Results of Thematic Analysis

In the outcome of the literature investigation we documented 6,012 search results, from which we selected 60 (1%) for the data corpus—twenty (33%) conference articles and forty (67%) journal articles. We formed six thematic sections for each initial code. From all thematic sections we collected 165 sub-codes that we sorted to two groups according to relation either to one or two initial codes. We visualized the relations in two mind mappings. In the report we documented forty logical textual sequences, from which eight of them described the theme "Digitalization and Process", nine of them the theme "Digitalization and ICT/IT", four of them the theme "Digitalization and Big Data", ten of them the theme "Digitalization and Service", five of them the theme "Digitalization and Algorithm", and four of them the theme "Digitalization and Architecture" [32].

3.4 Statistical Results of Impediments Classification

From the data corpus we pulled out 114 impediments that we classified into eight groups. The process element assigned three groups of strategy, business and education

Table 2 Simplified list of particular publications of the selected search results

No	Searching database source	Search term	First author surname	Abbreviated title	Link
1	SDS1	ST1	Chawda	Big Data and advanced analytics tools	[17]
2	SDS1	ST1	Ali	Tutorials: Tutorial I: HPC and Big Data analytics…	[4]
3	SDS1	ST2	Jiang	An intelligent information forwarder for…	[53]
4	SDS1	ST2	Reddy	Predictive Big Data analytics in healthcare	[93]
5	SDS1	ST3	Yin	Automatic ocular disease screening and…	[116]
6	SDS1	ST3	S	DSSM with text hashing technique for text…	[97]
7	SDS1	ST4	Tang	SQL-SA for Big Data discovery polymorphic…	[103]
8	SDS1	ST4	Makonin	Mixed-initiative for Big Data: the intersection…	[70]

(continued)

Table 2 (continued)

No	Searching database source	Search term	First author surname	Abbreviated title	Link
9	SDS1	ST5	Zhang	Building cloud-based healthcare data mining…	[119]
10	SDS1	ST5	Abuwardih	Privacy preserving data mining on published…	[1]
11	SDS2	ST1	Zadrozny	Big Data analytics using Splunk	[117]
12	SDS2	ST1	Guller	Big Data analytics with Spark	[45]
13	SDS2	ST2	Raghupathi	Big Data analytics in healthcare: promise and…	[90]
14	SDS2	ST2	Chowdary	A scalable model for Big Data analytics in…	[19]
15	SDS2	ST3	Bennett	Creating personalized digital human models…	[10]
16	SDS2	ST3	Stefanowski	Final remarks on Big Data analysis and its…	[101]
17	SDS2	ST4	Tynkkynen	An analysis of ophthalmology services in…	[107]
18	SDS2	ST4	Torre-Díez	Decision support systems and applications in…	[106]
19	SDS2	ST5	Mans	Process mining in healthcare: evaluating and…	[71]
20	SDS2	ST5	Kaur	Role of data mining in establishing strategic…	[57]
21	SDS3	ST1	Daniel	Big Data and analytics in higher education…	[21]
22	SDS3	ST1	Schoenherr	Data science, predictive analytics, and Big Data…	[99]

(continued)

Table 2 (continued)

No	Searching database source	Search term	First author surname	Abbreviated title	Link
23	SDS3	ST2	Hiller	Healthy predictions? Questions for data…	[49]
24	SDS3	ST2	Kune	The anatomy of Big Data computing	[66]
25	SDS3	ST3	Robson	Guardian angels: knowing our molecules, drug…	[95]
26	SDS3	ST3	Robson	Architecting IT all	[94]
27	SDS3	ST4	Knutsson	Modified big-bubble technique compared to…	[60]
28	SDS3	ST4	Keane	Deep anterior lamellar keratoplasty versus…	[58]
29	SDS3	ST5	Gaber	Data stream mining in ubiquitous…	[39]
30	SDS3	ST5	Bellazzi	Predictive data mining in clinical medicine…	[9]
31	SDS4	ST1	Qin	Big Data analytics with swarm intelligence	[89]
32	SDS4	ST1	Mueck	How leading organizations use Big Data and…	[76]
33	SDS4	ST2	Lemieux	Meeting Big Data challenges with visual…	[68]
34	SDS4	ST2	Agrawal	Analytics based decision making	[3]
35	SDS4	ST3	–	–	
36	SDS4	ST3	–	–	

(continued)

Table 2 (continued)

No	Searching database source	Search term	First author surname	Abbreviated title	Link
37	SDS4	ST4	Schnieden	Audit and performance indicators: a case study...	[98]
38	SDS4	ST4	Vinekar	IT-enabled innovation to prevent infant...	[111]
39	SDS4	ST5	Chen	Using data mining to segment healthcare...	[18]
40	SDS4	ST5	Gebremeskel	Combined data mining techniques based...	[42]
41	SDS5	ST1	Brajesh	Big Data analytics in retail supply chain	[14]
42	SDS5	ST1	Deka	Big Data predictive and prescriptive analytics	[23]
43	SDS5	ST2	Ranjan	Big Data applications in healthcare	[92]
44	SDS5	ST2	Darrel	The benefits of Big Data analytics in the...	[22]
45	SDS5	ST3	Dobre	NoSQL technologies for real time (patient)...	[25]
46	SDS5	ST3	Waegemann	mHealth: history, analysis, and implementation	[113]
47	SDS5	ST4	Singh	The application of rough set theory and near...	[100]
48	SDS5	ST4	Obaidullah	Bangla and Oriya script lines identification...	[81]
49	SDS5	ST5	Barazandeh	Knowledge discovery and data mining...	[6]
50	SDS5	ST5	Wang	Benefits and barriers in mining the healthcare...	[114]

Table 3 Numbers of media types per searchable database source

Media type	Searchable database source					
	Emerald Insight	IEEE Xplore	IGI Global	SpringerLink	Wiley online library	Total
Journal article	8	1	2	4	8	23
Book chapter	0	0	8	6	2	16
Conference paper	0	9	0	0	0	9
Total	8	10	10	10	10	48

Table 4 Numbers of media types per search term

Media type	Search term					
	"Big Data Analytics"	"Big Data Analytics" + Healthcare	"Big Data Analytics" + Ophthalmology	"Big Data" + Ophthalmology	"Data Mining" + Healthcare	Total
Journal article	4	6	0	7	6	23
Book chapter	4	3	6	1	2	16
Conference paper	2	1	2	2	2	9
Total	10	10	8	10	10	48

Table 5 Number of selected search results in years of publishing per searchable database source

Year of publishing	Searchable database source					
	Emerald Insight	IEEE Xplore	IGI Global	SpringerLink	Wiley online library	Total
1988	1	0	0	0	0	1
2008	0	0	0	0	2	2
2009	1	0	0	1	0	2
2011	1	0	0	1	1	3
2012	0	0	1	1	0	2
2013	0	0	0	1	0	1
2014	2	0	1	1	2	6
2015	1	0	0	3	3	7
2016	2	10	7	2	2	23
2017	0	0	1	0	0	1
Total	8	10	10	10	10	48

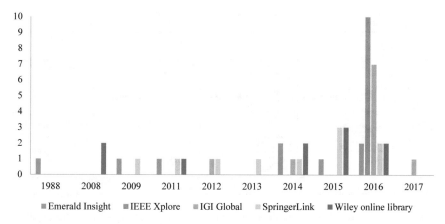

Fig. 10 Number of selected search results in years of publishing per searchable database source

Table 6 Number of authors per media type	Media type	# of Authors						
		1	2	3	4	5	6	20
	Journal article	4	5	7	3	3	1	
	Book chapter	5	7	2	2			
	Conference paper	1	2	1	1	2	1	1
	Total	10	14	10	6	5	2	1

scope with 64 (56%) impediments. The service element assigned three groups of strategy, business, and education scope with 33 (29%) impediments. The function of an application element had assigned only two groups of strategy and business scope with 17 (15%) impediments. We documented that the gained information built on the selected components of the information technology standard in well-known Enterprise Architecture framework TOGAF can be reused in the process of managerial decision making [33].

3.5 Results of Design Science Research

The business goal in healthcare digital transformation was the deployment of AITaT in the cases they could bring the business value. Our first direction was to gather knowledge from Big Data and to design basic processing flow with the view on application functionality (Fig. 11).

The process of managerial decision making can be more effective if the knowledge about possible impediments of digital transformation in healthcare is available. We collected and classified known impediments they influenced AITaT deployment. The classification was done according strategy, business, and education scope [33].

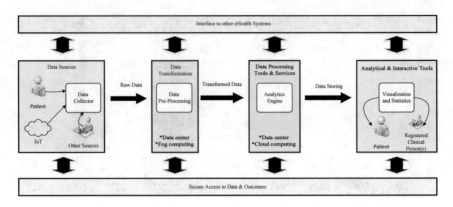

Fig. 11 Conceptual architecture of Big Data analytics in healthcare (adapted according to [30, 34, 90])

Table 7 List of use cases, according individual segments or the combinations, with potential change in deployment of AITaT

Segment	Current state	Future state
Textual data	Manual input to medical history	Automated input via special application
Textual data	Manual collection of medical history from different data sources	Automated collection of medical history from different data sources
Image data	Manual recognition of anatomical difference in ocular adnexa and eye globe	Automated analysis of anatomical difference in ocular adnexa and eye globe
Textual and image data	Manual ocular disease diagnosis	Automated ocular disease diagnosis
Textual and image data	Manual therapy proposal	Automated therapy proposal
3D printing	Decision about stereotactic planning without 3D eye model	Decision about stereotactic planning with printed 3D eye model

We identified the problem in insufficient AITaT deployment in healthcare, specifically in ophthalmology branch where input or processing of medical data was not using any of the AITaT in deployment. Our objective was to define solution that can follow the vision to deploy AITaT but only in the cases they can bring a business value. We identified a set of use cases they can demonstrate a business value in either ophthalmology processes or in healthcare generic processes used also in ophthalmology branch. We took into account three different segments and their combinations such as text data, image data, and 3D printing option. The current and future state of selected use cases we classified to different segments (Table 7).

In healthcare, the natural prerequisite was always the orientation to patient that the actor-centric approach can identify within healthcare system the participated types

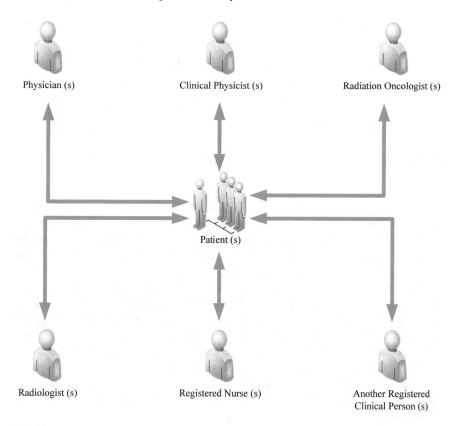

Physician (s)   Clinical Physicist (s)   Radiation Oncologist (s)

Patient (s)

Radiologist (s)   Registered Nurse (s)   Another Registered
                                                                           Clinical Person (s)

Fig. 12 Actor-centric approach for the use case of stereotactic radiosurgery process

of actors. The development and implantation of primary healthcare services must facilitate the exchange of information and communication among all involved actors [7, 112]. Among others, we analyzed the state of the actor-centric approach in the use case of stereotactic radiosurgery process, in which the participation of more actors is mandatory in successful provisioning of the operation or its preparation (Fig. 12).

Enterprise Architecture management is based on a collaborative decision-making approach that is built on methods and techniques of adaptive case management, and is improving architectural decision-making. The process of identification and decision making in a complex environment is knowledge-based and is strongly dependent on involvement of stakeholders. The next decision-making step is based on case data that is based on Enterprise Architecture model and other findings appeared in the previous steps. So the knowledge gained during each step contributes to the case decision. For improving the effectiveness of managerial decision making is used a decision support system that can also offer benefit to Enterprise Architecture management in improving of architectural decision-making. For a multi-perspective architecture environment can be created an extension to a decision support system, called Enter-

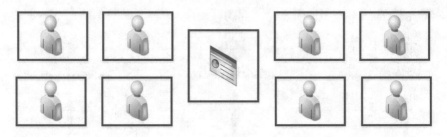

Fig. 13 Example: enterprise architecture cockpit (adapted according to [54, 55])

prise Architecture cockpit, that is a kind of facility or device through which multiple considered viewpoints on system can be consulted, applied simultaneously, and linked each other. This way the multi-perspective architectural dependencies and the impacts of changes are displayed in parallel. There the changes that appeared in one view are appearing in other views as well [54, 55, 121].

The idea of multi-perspectives of a collaborative Enterprise Architecture cockpit we accommodated for the healthcare. The managerial decision in healthcare making of multiple participants, they accessed the cockpit regarding considered viewpoints can facilitate the decisions, consultancy, and can be applied simultaneously. We estimated the healthcare was proper source for evaluation in the points of usability and fit with the organization (Fig. 13). The multi-perspective architecture accommodated specifically the ophthalmology branch offered to all registered clinical persons multiple viewpoints on the system with possibility for simultaneous consultation and consideration.

3.6 Statistical Results of Questionnaire

For the questionnaire we created ten multiple choice questions. The first nine questions were single answer questions (required to choose just one answer). The last question was multi answer question (required to choose at least one answer) with the option to add additional comment. The details about each question, results, and short evaluation we describe in the following.

Question No 1: *How many years have you been active in ophthalmology branch?*
We defined to choose just one of five options to answer:
Less than 5/5–10/10–20/20–30/More than 30

In the questionnaire we did not try to get information about age because it is sensible information, though the questionnaire was anonymous. Instead to determine the age of ophthalmologists we decided to ask for number of years of practice in the field, in our opinion, it better describes the ophthalmologist's experience. Whatever the length of practice is often age-related. In general, older people are more conservative and less likely to tolerate change. So it is assumed that the promotion of

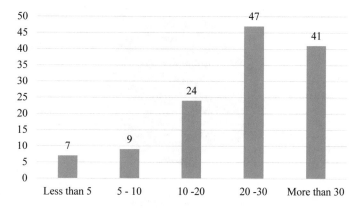

Fig. 14 Answers to question no. 1: *How many years have you been active in ophthalmology branch?*

new trends and technologies is not their primary concern. So such information on years of practice has helped us compare in analysis the trends and evaluations of the following questions [41, 52].

The results of the answers to question no. 1 show that most participating ophthalmologists (68.75%) have been active in the field for more than 20 years (Fig. 14). The individual intervals contain the following results: 7 (5%) have less than 5 years of activity, 9 (7%) have 5-10 years of activity, 24 (19%) have 10–20 years of activity, the most in individual interval have 47 (37%) 20–30 years of activity, and 41 (32%) have over 30 years of activity.

Question No 2: *Approximately how many patients do you communicate, investigate, operate, and consult, etc., on weekly basis?*

We defined to choose just one of three options:

Less than10/10–50/More than 50

Further information on the length of practice of the ophthalmologist was, in our opinion, to obtain information on the number of patients who have ophthalmologists serviced within a certain time interval. We chose the one-week interval as the most appropriate.

The results of the answers to question no. 2 show that neither ophthalmologist works on weekly basis with less than ten patients (Fig. 15). The vast majority of ophthalmologists communicate (investigate, operate, consult, etc.) with more than 50 patients on weekly basis 100 (78%). The rest 28 (22%) communicate with less than 50 but more than 10 patients on weekly basis—the ophthalmologists are specialized to time consuming or complicated cases.

Question No 3: *Approximately how many % of your patients need an accompanying person, e.g. the blind, visually impaired, non-governmental, etc.?*

We defined to choose just one of four options:

Less than 5%/5–20%/20–50%/More than 50%

The blind, visually impaired and non-governmental patients are accompanied by another person when they are examined. The role of the accompanying person is

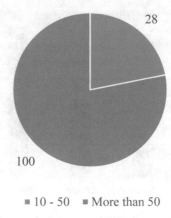

■ 10 - 50   ■ More than 50

Fig. 15 Answers to question no. 2: *Approximately how many patients do you communicate, investigate, operate, and consult, etc., on weekly basis?*

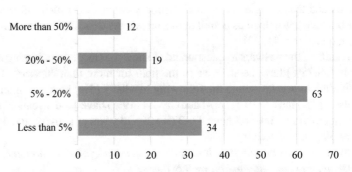

Fig. 16 Answers to question no. 3: *Approximately how many % of your patients need an accompanying person, e.g. the blind, visually impaired, non-governmental, etc.?*

to ensure that the patient has a seamless involvement in the investigation, and that person often acts as a mediator for the ophthalmologist. In such a case, the use of the AITaT could play an important role in the transfer of information between the patient and the doctor in terms of time and quality [65].

The results of the answers to question no. 3 show that less than 5% of the patients require an accompanying person, as reported by 34 (27%) ophthalmologists, from 5 to 20% of patients require an accompanying person, as reported by the majority 63 (49%) ophthalmologists, from 20% to 50% of patients require an accompanying person, as reported by 19 (15%) of ophthalmologists, and more than 50% of patients require an accompanying person, as reported by 12% (9%) of ophthalmologists (Fig. 16).

Question No 4: *Approximately how much time (in %) do you spend with the patient if you need to prepare his or her medical history, personal details, the reason for visiting, other procedures, etc.?*

We defined to choose just one of four options:

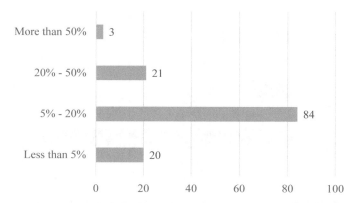

Fig. 17 Answers to question no. 4: *Approximately how much time (in %) do you spend with the patient if you need to prepare his or her medical history, personal details, the reason for visiting, other procedures, etc.?*

Less than 5%/5–20%/20–50%/More than 50%

The initial activity of examining each patient is to build a medical history. This activity plays a very important role in the next investigation process to correctly process and properly evaluate the information obtained from the patient, accompanying person, accompanying documentation, and other available resources. It requires accuracy and consistency [65].

The results of the answers to question no. 4 show that 20 (16%) of ophthalmologists need less than 5% of the time during overall examination of patient, 83 (65%) ophthalmologists need from 5% to 20% of the time during overall examination of patient, 21 (16%) ophthalmologists need 20% to 50% of the time during overall examination of patient, and 3 (2%) ophthalmologists need more than 50% of the time during overall examination of patient (Fig. 17).

Question No 5: *Do you also use electronic data to compose a medical history, e.g. on your computer, central database, etc.?*

We defined to choose just one of three options:

Yes/No/Sometimes

Electronic data collection offers direct input of patient data at the initial point of contact, for example, if assembling medical history by using electronic patient record (EPR) systems. Currently a large spread of this form was implemented, the practice proved many advantages, and it has a clear potential for replacing the patient's data collection in traditional paper form [65, 105].

The results of the answers to question no. 5 show that 87 (68%) ophthalmologists use electronic data to compose a medical history, 26 (20%) ophthalmologists use sometimes electronic data to compose a medical history, and 15 (12%) ophthalmologists use no electronic data to compose a medical history (Fig. 18).

Question No 6: *Would you welcome to have the possibility if a medical history will be prepared by patient, or his/her accompanying person, based on a special application, done prior arrival?*

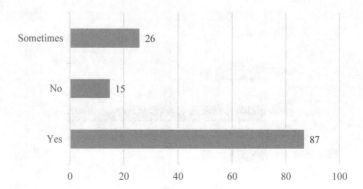

Fig. 18 Answers to question no. 5: *Do you also use electronic data to compose a medical history, e.g. on your computer, central database, etc.?*

Fig. 19 Answers to question no. 6: *Would you welcome to have the possibility if a medical history will be prepared by patient, or his/her accompanying person, based on a special application, done prior arrival?*

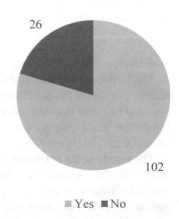

We defined to choose just one of two options:
Yes/No

AITaT offer opportunities for developing applications that are able to use artificial intelligence elements. Although the development of such applications is costly, the preparation and evaluation of the underlying data for the current medical history of the patient will soon be optimized, prior to the medical examination. The need for such applications is proven by health records audits that show that patient records sometimes misinterpret the underlying, often documented, facts such as the patient's history or information on the patient's medication. The lack of commonly accepted terminology and ontology makes it difficult to exchange and to interconnect even well-recorded information [84].

The results of the answers to question no. 6 show that 102 (80%) ophthalmologists would welcome such an application. However, 26 (20%) ophthalmologists have not shown interest in such an application (Fig. 19).

Question No 7: *How much time would you save if the patient, or his or her accompanying person, prepares medical history in electronic form?*

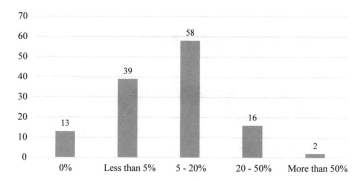

Fig. 20 Answers to question no. 7: *How much time would you save if the patient, or his or her accompanying person, prepares medical history in electronic form?*

We defined to choose one of five options:

Less than 5%/5–20%/20–50%/More than 50%/0%, the medical history I prepare by myself, or my nurse, or my assistant, etc.

The preparation of medical history is currently based on the experience of each ophthalmologist and, to a large extent, is dependent on the prescribed forms of the individual workplaces. These forms are intended to standardize. In the case of enforcement of such standards it is important to take advantage of the possibilities offered by AITaT because of the availability of electronic health records continues to expand. If the medical history of the patient is recorded and regularly updated in electronic form then it includes comprehensive data on past medical meetings such as hospitalization, surgery, pregnancy, birth, and others. It also contains information about past health problems and, if appropriate, about family history or events (for example, alcoholism or parental divorce). Such facts, already recorded, are important not only in general for good patient care but they can also be important in adjusting the severity [77, 78].

The results of the answers to question no. 7 show that 13 (10%) of ophthalmologists estimated that it does not save time if the patient or his/her accompanying person prepares a medical history in electronic form. 39 (30%) of ophthalmologists estimated that they save less than 5% of the time, 58 (45%) of them estimated to save 5–20% of the time, 16 (13%) of them estimated to save 20–50% of the time and only two (2%) of them estimated to save more than 50% of the time (Fig. 20).

Question No 8: *Are you using your "e-card", e.g. eHealth card, or eID chip card, for electronic communication?*

We defined to choose just one of three options:

Yes/No/I do not have such an "e-card"

In order to respond to the special needs of the outpatient sector and to facilitate the exchange of information on medicinal products, for example, the Austrian e-Medics system has been designed as a national electronic health record system based on the Austrian "e-card" network as a platform for unique patient identification and safe exchange of health data [46].

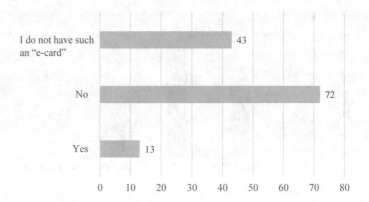

Fig. 21 Answers to question no. 8: *Are you using your "e-card", e.g. eHealth card, or eID chip card, for electronic communication?*

However, in Slovak Republic from January 2018 the Electronic Healthcare system was deployed that is under the supervision of National Health Information Centre (state-funded organization founded by the Ministry of Health of the Slovak Republic) [79]. The Electronic Healthcare system is mandatory for every medical practitioner. Also the ophthalmologists, to whom we sent our questionnaire, are obliged to use the "e-card" of the system these days. The results of our questionnaire we obtained before deployment of this system so the answers of the questions no. 8, and no. 9, we evaluated by the original date.

The results of the answers to question no. 8 show that an "e-card" was used for electronic communication by 13 (10%) ophthalmologists, 72 (56%) ophthalmologists did not use their own "e-card" for electronic communication, even though they own it, and 43 (34%) ophthalmologists have no "e-card" (Fig. 21).

Question No 9: *Would you welcome to have the possibility if you can use your "e-card" in access to patient data, e.g. from a healthcare insurance company, other health care facilities in Slovak Republic or the European Union, etc.?*

We defined to choose just one of three options:

Yes/No/I already have such an "e-card"

The results of the answers to question no. 9 show that 2 (2%) ophthalmologists already have an "e-card" in access to patient data, 23 (18%) ophthalmologists do not have an interest about an "e-card" in access to patient data, but 103 (80%) ophthalmologists would welcome to use an "e-card" in access to patient data (Fig. 22).

Question No 10: *Would you welcome to have the possibility, in the context of digitalization,*

We defined to choose minimum one of four options:

A: Of automatic diagnosis of the textual patient data, e.g. medical history;

B: Of automatic diagnosis of the patient image data, e.g. magnetic resonance imaging, computed tomography, scanning, etc.;

C: Of automated therapy proposal;

D: Other.

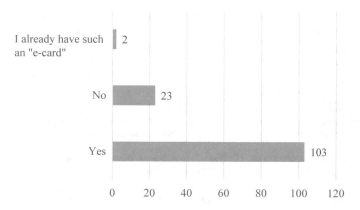

Fig. 22 Answers to question no. 9: *Would you welcome to have the possibility if you can use your "e-card" in access to patient data, e.g. from a healthcare insurance company, other health care facilities in Slovak Republic or the European Union, etc.?*

An automated diagnosis, including medical images, has been in active area of research among academics for the last thirty-forty years, at the time of medical imaging was shifting from film to electronic images. In field of ophthalmology about twenty years ago there it was used a sophisticated image management system (structured analysis of retina) that automatically provided a kind of images diagnosis, compared images, measured selected key features in images, annotated contents of image, and searched for images that reported similarities in content. Ophthalmologists and physicians in other areas expected to be able to rely on images in reducing repetitive work, to provide help for doctors in difficult diagnoses or with unknown illnesses, and to manage images in large image databases [43].

Even though digitalization is moving forward in healthcare, the reality showed that such systems are still missing in local ophthalmology branch. The results of the answers to question no. 10 show that 115 (90%) ophthalmologists supported the digitalization trend in deploying the AITaT for automated diagnosis or automated therapy proposal. Thirteen (10%) ophthalmologists responded only the options "D"—seven of them responded only "No". Four of 115 ophthalmologists, they selected option A, B, C, or their combination, also inserted a comment to the option "D" (Fig. 23).

The options of the question no. 10 we evaluated also from cumulative point of view with aim to document the number of results per option. The cumulative results showed that the most preferable is option "B" although the most complicated to deploy in practice (Fig. 24).

To complement the base questionnaire statistics we provided one short comparison of the selected results between two groups, according the years an ophthalmologist has been active in ophthalmology branch. There we selected the groups they lay on opposite sides. The compared results are related to the number of patients communicated on weekly basis and interest for a special application that can prepare medical history of patient.

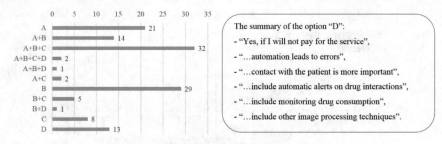

Fig. 23 Answers to question no. 10: *Would you welcome to have the possibility, in the context of digitalization,...? (Answers A, B, C and D in combinations.)*

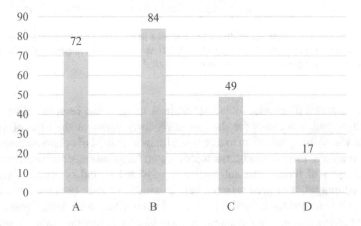

Fig. 24 Answers to question no. 10: *Would you welcome to have the possibility, in the context of digitalization,...? (Answers A, B, C and D in cumulative.)*

In the first group, active in branch more than 30 years, was 41 ophthalmologists from which

- 33 (81%) communicated with more than 50 patients on weekly basis. These thirty three ophthalmologists communicated with more than 50 patients on weekly basis and 29 (88%) of them did have an interest about an "e-card" in access to patient data.
- 35 (85%) need more than 5% of the time during overall examination of patient.
- 26 (63%) use electronic data to compose a medical history.
- 35 (85%) would welcome a special application to prepare medical history of patient.
- 19 (46%) estimated to save more than 5% of the time.

In the second group, active in branch less than 5 years, was 7 ophthalmologists from which

- 2 (29%) communicated with more than 50 patients on weekly basis. Although these two ophthalmologists communicated with more than 50 patients on weekly basis they did have no interest about an "e-card" in access to patient data.
- 7 (100%) need more than 5% of the time during overall examination of patient.
- 6 (86%) use electronic data to compose a medical history.
- 6 (86%) would welcome a special application to prepare medical history of patient.
- 6 (86%) estimated to save more than 5% of the time.

The result of the questionnaire confirmed that the members of ophthalmology society in Slovak Republic were prepared and supported the potential changes that can bring the healthcare digital transformation in adoption of AITaT into practice.

3.7 Results of Interview

We carried out a total of five structured interviews with experts from ophthalmology branch. Interviews were conducted in late May and early June 2018, during one international ophthalmology congress and one ophthalmology meeting. Respondents were from Great Britain, Italy, Greece, Switzerland and Denmark. Organizing the interview throughout the participants' agenda was not easy, and that was the reason we did not complete one interview. The environment was loud, overwhelmed, and inappropriate for voice recording, so we wrote the contents of the responses to the record sheet. The interviews ranged from 5 to 25 min and were structured in two parts. In the first part we briefly explained the aim of interview and gathered general information about the interviewee that was similar to the questionnaire. In the second part we discussed about interviewee view on an AITaT deployment in ophthalmology branch in focus to the use cases we selected in DSR stage.

Brief summary of responses of each interviewee:

1. About 22 years active in branch, communicating with more than fifty patients per week, more than 50% of patients need accompanying person, needed from 5 to 20% of the time during overall examination to build a medical history, not using electronic data, but like the idea of special application to prepare a medical history, supporting e-card deployment and applications for automated therapy proposal, the deployment of AITaT in practice recommended for the automated data gain for composing a medical history, a sophisticated image management system based on Big Data recommended for learning purposes or at the starting of a career exclusively, because of doctor takes the responsibility.
2. About 40 years active in branch, communicating with more than fifty patients per week, preferable communicating with patient, no supporting electronic data or e-card, the AITaT deployment was not supported due to the missing standards.
3. About 35 years active in branch communicating with more than fifty patients per week, preferable communicating with patient, the most of the time providing surgery, no time for electronic data (the work of his assistant), arguing that

artificial intelligence is not yet at an appropriate level and can create barriers to decision-making.
4. About 25 years active in branch communicating with more than fifty patients per week, focusing purely on surgery, supporting the trends of AITaT—unfortunately the interview was interrupted.
5. About 10 years active in branch communicating with more than fifty patients per week, working with electronic data of patients, using e-card to access data, fully supporting all the trends of AITaT we discussed, especially supporting the automated data gain for composing a medical history, because of a distinct and constant drop in staff, such as nurses and assistants, has a long-term tendency.

The proposals of AITaT adoption in practice that we described during interview were supported by all interviewees with additional recommendations and comments. The interviewees supported our direction and confirmed the understanding about changes during healthcare digital transformation. In addition, the interviewees suggested some recommendations about: where a real business value can be expected, and estimations about how long the change could be applied within practice.

4   Discussion and Reflection

The amount of scientific publications is increasing rapidly every year. We practically verified that staggering amount of information contained in publications such as experimental results, and statistics was exponentially growing, and therefore it was quite difficult to obtain relevant information. This problem has recently hampered progress in science, making it more difficult to achieve results, and thus to produce valuable products. However, access to a structured query-able databases enabled scientists and engineers to quickly discover, search, and compare available information [118].

At present, each research cannot miss in first stages efficient and effective information search, collection, and analysis. We analyzed and statistically evaluated the search results they helped us to define the best searchable database in number of search results. In our first literature investigation the searchable database source IEEE Xplore offered in total the most search results for all five defined search terms (63%) and in list of particular publications of the selected search results offered the most recent information.

Although our results have favored one searchable database source there is no doubt that other searchable database sources are also usable. The particular example is our next stage of the research where we selected according recommendations the searchable database source Google Scholar, to gain the knowledge and data corpus for analysis [31, 115].

In the first two stages of our research we searched six different searchable database sources. Five of them in the first stage offered in total 190,701 search results, from which we selected 48 relevant publications to familiarize with the interdisciplinary issues. In the second stage we documented 6,012 search results, from which we selected 60 (1%) for the data corpus [32]. If we compare with other literature reviews, one about empirical research on Electronic Health Record implementation documented 364 initially identified articles, from which was 21 (5.8%) articles analyzed [12]. Another source about the state of Electronic Health Records documented 2,356 abstracts, plus papers and books, identified through cited references [27]. Comparable results assured us that our method and the results of literature investigation were the right starting point for our research.

In our questionnaire we focused on the specific branch of medicine, ophthalmology, in using the web based accessible survey form. Using this survey method we received 128 responses out of 265 emailed ophthalmologists (48.3%) that we collected over a two-and-a-half-month period. Similar case of a questionnaire for medical students towards e-health that focused on the knowledge, attitudes and expectations was distributed to 136 respondents. They finally collected 100 surveys (74%). This result was better if compare our result but the group of respondents was more homogenous—students of one faculty [26].

In each business area the goal is to maintain long-term and sustainable success in a rapidly changing and uncertain environment. Significant improvement in business is reflected in business transformation when it is also necessary to change the organizational culture of the company. In the case that the business change has digital context then the digital essence of AITaT deployment is often referred to digital transformation. Such a transformation has been often slowed down by strict regulations, reluctance on the part of stakeholders in health care to promote change, ignoring cultural change, and human factors in the increasingly technological world. New technologies contribute to an equitable partnership between patients and professionals, and they also help bring innovation into medicine practice [73].

With statistical results of the questionnaire we gained a certain overview that allowed us to map the current state in the ophthalmology branch in Slovak Republic and the readiness for digital transformation. In brief we summarized our findings of the questionnaire in the following:

- The number of active years in ophthalmology branch had no influence on the readiness for the AITaT deployment.
- The most of the respondents handle more than fifty patients per week so we briefly calculated if the most of ophthalmologists would save more than 5% for preparation of medical history. Then after deployment of an automated preparation the ophthalmologists could handle two-three patients more per week or could communicate minimum 5% of the time more with a patient. Note: The possibility of communicating longer with the patient during investigation was currently strong demand of doctors so the described business change within the digital transformation can satisfy this requirement.
- To deploy any AITaT services in automatic diagnosis based on the patient textual, or image, data or in automated therapy proposal was natural demand of ophthalmologists so our research direction was supported in ophthalmology branch.

However, efforts to improve screening and eye disease diagnosis using modern image and data analysis techniques need to be more widespread, although intelligent systems for analyzing eye diseases through cloud computing are already in place. Image information input for automatic analysis and assessment should use advanced model classification algorithms. Access to data analysis can be available through mobile apps or web portals that enable effective screening, eye disease monitoring, and early diagnostics [116]. In our finding of the interview we documented that the ophthalmologists support the goal but in practice applied for education purposes, for young ophthalmologists, and for general ambulance, including a kind of mobile ophthalmology clinic [61].

The Electronic Health Record data was continuing to expand with an enormous number of patients that was leading to a large amount of stored data (achieving the dimension of Big Data) and can be transformed into knowledge. New key components of AITaT for analyzing these data such as data mining techniques, natural language processing, or automated summarization and visualization methods, are at the forefront [87, 96].

Among others, in our research we focused on benefit that can be expected during digital transformation process in ophthalmology branch. The digital transformation is expecting also a business value from AITaT deployment. We investigated the digital transformation process that was described and evaluated from perspective of Enterprise Architecture and TOGAF framework principles.

In using different research methods our goal was to identify problem in ophthalmology branch where the possible AITaT deployment is still missing. We defined objectives to find a solution that can demonstrate a business value. Whether the proposed solution could be effective and efficient we verified in gaining the opinions of ophthalmologists. They confirmed and helped to specify where and how could be the selected use cases beneficial.

The use case of automated data gain for composing a medical history used a specific application was strongly recommended because of on one side a distinct and constant drop in staff, such as nurses and assistants, has a long-term tendency and on other side the gain the patient data from other connected sources can be faster and reliable.

The use case of automatic ocular disease diagnosis used a specific application was rated very positively by the interviewees. However, everyone stressed that, in the end, the doctor will decide because has the main responsibility. But for education purposes, for young ophthalmologists, and for general or mobile ophthalmology ambulance was a sophisticated image management system recommended as excellent solution.

The functional AITaT deployment could be facilitated by the multi-perspective architecture that ensured multiple viewpoints on the system to allow simultaneous consultation and consideration for final decision.

The selected use cases were the part of primary activity of ophthalmology processes in healthcare organization so the business value could achieve the real margin after AITaT deployment.

5 Conclusion

Challenges in the modern period triggered deployment of AITaT even in segments that have been defeated. Health care is specific and difficult to deploy because it is intended for the patient. The patient service depends on many factors, including the cultural change in the segment, so the deployment of AITaT is complicated. Our first research steps, starting with literature research, have shown that extracting valuable information has not been an easy task. However, the methods used helped us to correctly analyze the information in order to achieve the set goals. We have demonstrated and reported on our findings that could help in the process of healthcare digital transformation and in the deployment of AITaT. All such efforts can facilitate a better quality of life.

References

1. Abuwardih LA et al (2016) Privacy preserving data mining on published data in healthcare: a survey. In: 2016 7th international conference on computer science and information technology (CSIT), pp 1–6
2. Aggarwal CC, Zhai C (2012) Mining text data. Springer Science & Business Media
3. Agrawal D (2014) Analytics based decision making. J Indian Bus Res 6(4):332–340
4. Ali HH (2016) Tutorials: Tutorial I: HPC and Big Data analytics in biomedical informatics. In: 2016 international conference on high performance computing simulation (HPCS), pp 1–17
5. Aziz HA, Guled A (2016) Cloud computing and healthcare services. J Biosens Bioelectron 7(3):1–4
6. Barazandeh I, Gholamian MR (2016) Knowledge discovery and data mining applications in the healthcare industry: a comprehensive study. In: E-health and telemedicine: concepts, methodologies, tools, and applications, pp 1097–1118
7. Batchelor S et al (2015) Understanding health information seeking from an actor-centric perspective. Int J Environ Res Public Health 12(7):8103–8124
8. Bauer C, Strauss C (2016) Location-based advertising on mobile devices. Manag Rev Q 66(3):159–194
9. Bellazzi R et al (2011) Predictive data mining in clinical medicine: a focus on selected methods and applications. Wiley Interdiscip Rev Data Min Knowl Discov 1(5):416–430
10. Bennett M, Quigley A (2011) Creating personalized digital human models of perception for visual analytics. In: User modeling, adaption and personalization. Springer, Berlin, Heidelberg, pp 25–37
11. Bharadwaj A et al (2013) Visions and voices on emerging challenges in digital business strategy. Manag Inf Syst Q 37(2):633–661
12. Boonstra A et al (2014) Implementing electronic health records in hospitals: a systematic literature review. BMC Health Serv Res 14
13. Boynton PM, Greenhalgh T (2004) Selecting, designing, and developing your questionnaire. BMJ 328(7451):1312–1315
14. Brajesh S (2016) Big Data analytics in retail supply chain. In: Handbook of research on strategic supply chain management in the retail industry, pp 269–289
15. Braun V, Clarke V (2006) Using thematic analysis in psychology. Qual Res Psychol 3(2):77–101
16. Byers CC, Wetterwald P (2015) Fog computing distributing data and intelligence for resiliency and scale necessary for IoT: The Internet of Things (ubiquity symposium). Ubiquity 4

17. Chawda RK, Thakur G (2016) Big Data and advanced analytics tools. In: 2016 symposium on colossal data analysis and networking (CDAN), pp 1–8
18. Chen J, Liu SS (2009) Using data mining to segment healthcare markets from patients' preference perspectives. Int J Health Care QA. 22(2):117–134
19. Chowdary SH et al (2016) A scalable model for Big Data analytics in healthcare based on temporal and spatial parameters. In: Proceedings of first international conference on information and communication technology for intelligent systems, vol 1. Springer, Cham, pp 117–122
20. Committee on Future Information Architectures, National Research Council (2011) A two-phase approach to modernization and transformation of business and information ecosystems. National Academies Press (US) (2011)
21. Daniel B (2015) Big Data and analytics in higher education: opportunities and challenges. Br J Edu Technol 46(5):904–920
22. Darrel A et al (2016) The benefits of Big Data analytics in the healthcare sector: what are they and who benefits? In: E-health and telemedicine: concepts, methodologies, tools, and applications, pp 842–875
23. Deka GC (2014) Big Data predictive and prescriptive analytics. In: Handbook of research on cloud infrastructures for Big Data analytics, pp 370–391
24. Dicicco-Bloom B, Crabtree BF (2006) The qualitative research interview. Med Educ 40(4):314–321
25. Dobre C, Xhafa F (2017) NoSQL technologies for real time (patient) monitoring. In: Medical imaging: concepts, methodologies, tools, and applications, pp 932–961
26. Edirippulige S et al (2006) Knowledge and perceptions of e-Health: results of a survey of medical students in Sri Lanka. In: 2006 international conference on information and automation, pp 437–439
27. Evans RS (2016) Electronic health records: then, now, and in the future. Yearb Med Inform Suppl 1:S48–S61
28. Farahani N et al (2017) Three-dimensional imaging and scanning: current and future applications for pathology. J Pathol Inform 8
29. Finlay S (2014) Predictive analytics, data mining and Big Data. Palgrave Macmillan UK, London
30. Furda R et al (2017) [Digitalization in healthcare—perspectives and objectives in use of advanced information technologies]. Zdravotníctvo a sociálna práca (Health Soc Work) 12(4 Suppl):128–130
31. Furda R, Gregus M (2017) Big Data, analytics and other technologies in healthcare digitalization. Zdravotníctvo a sociálna práca (Health Soc Work) 12(3):4–14
32. Furda R, Gregus M (2017) Conceptual view on healthcare digitalization—an extended thematic analysis. IJBDAH 2(1):35–54
33. Furda R, Gregus M (2019) Impediments in healthcare digital transformation. IJARPHM 4(1):21–34
34. Furda R, Gregus M (2017) [Selected aspects of advanced information technologies in healthcare digitalization]. Presented at the Progressive approaches and methods to increase the efficiency and performance of organizations, Vratna dolina
35. Furdova A et al (2012) One-day session LINAC-based stereotactic radiosurgery of posterior uveal melanoma. Eur J Ophthalmol 22(2):226–235
36. Furdova A et al (2015) Cause of blindness in diabetic patients in the world and in Slovakia. In: New trends in current health nursing, health, economics and health. Samosato, pp 156–162
37. Furdova A et al (2017) Early experiences of planning stereotactic radiosurgery using 3D printed models of eyes with uveal melanomas. Clin Ophthalmol 11:267–271
38. Furdova A et al (2018) 3D Printing planning stereotactic radiosurgery in uveal melanoma patients. In: 3D printing. IntechOpen, Rijeka, pp 155–171
39. Gaber MM et al (2014) Data stream mining in ubiquitous environments: state-of-the-art and current directions. Wiley Interdiscip Rev Data Min Knowl Discov 4(2):116–138
40. Gastaldi L, Corso M (2012) Smart healthcare digitalization: using ICT to effectively balance exploration and exploitation within hospitals. Int J Eng Bus Manag 4

41. Gault RH (1907) A history of the questionnaire method of research in psychology. Pedagog Semin 14(3):366–383
42. Gebremeskel GB et al (2016) Combined data mining techniques based patient data outlier detection for healthcare safety. Int J Mach Learn Cybern 9(1):42–68
43. Goldbaum M et al (1996) Automated diagnosis and image understanding with object extraction, object classification, and inferencing in retinal images. In: Proceedings of 3rd IEEE international conference on image processing, vol 3, pp 695–698
44. Gregus M, Kryvinska N (2015) Service orientation of enterprises—aspects, dimensions, technologies. Comenius University in Bratislava, Bratislava
45. Guller M (2015) Big Data analytics with Spark: a practitioner's guide to using Spark for large scale data analysis. Apress, New York, New York
46. Hackl WO et al (2014) Crucial factors for the acceptance of a computerized national medication list. Appl Clin Inform 5(2):527–537
47. Haddad P et al (2015) Evaluating business value of IT in healthcare: three clinical practices from Australia and the US. Stud Health Technol Inform 216:183–187
48. Hevner AR et al (2004) Design science in information systems research. MIS Q 28(1):75–105
49. Hiller JS (2016) Healthy predictions? Questions for data analytics in health care. Am Bus Law J 53(2):251–314
50. Hornford D et al (2017) The seven levers of digital transformation. https://publications.opengroup.org/w17d
51. Hurwitz J et al (2013) Big Data for dummies. Wiley, Hoboken, NJ
52. Jenn NC (2006) Designing a questionnaire. Malays Fam Physician 1(1):32–35
53. Jiang P et al (2016) An intelligent information forwarder for healthcare Big Data systems with distributed wearable sensors. IEEE Syst J 10(3):1147–1159
54. Jugel D et al (2015) Modeling decisions for collaborative enterprise architecture engineering. In: Persson A, Stirna J (eds) Advanced information systems engineering workshops. Springer International Publishing, pp 351–362
55. Jugel D, Schweda CM (2014) Interactive functions of a cockpit for enterprise architecture planning. In: 2014 IEEE 18th international enterprise distributed object computing conference workshops and demonstrations, pp 33–40
56. Kaczor S, Kryvinska N (2013) It is all about services-fundamentals, drivers, and business models. J Serv Sci Res 5(2):125–154
57. Kaur H et al (2012) Role of data mining in establishing strategic policies for the efficient management of healthcare system—a case study from Washington DC area using retrospective discharge data. BMC Health Serv Res 12(S1):12
58. Keane M et al (2014) Deep anterior lamellar keratoplasty versus penetrating keratoplasty for treating keratoconus. Cochrane Database Syst Rev 7, CD009700
59. Khuntia J et al (2014) The University of Colorado digital health consortium initiative: a collaborative model of education, research and service. J Commer Biotechnol 20(3):31–37
60. Knutsson KA et al (2015) Modified big-bubble technique compared to manual dissection deep anterior lamellar keratoplasty in the treatment of keratoconus. Acta Ophthalmol 93(5):431–438
61. Krasnik V et al (2017) Prevalence of age-related macular degeneration in slovakia and associated risk factors: a mobile clinic-based cross-sectional epidemiological survey. In: Seminars in ophthalmology, pp 1–6
62. Kryvinska N (2012) Building consistent formal specification for the service enterprise agility foundation. J Serv Sci Res 4(2):235–269
63. Kryvinska N et al (2013) The S-D logic phenomenon-conceptualization and systematization by reviewing the literature of a decade(2004–2013). J Serv Sci Res 5(1):35–94
64. Kryvinska N, Gregus M (2014) SOA and its business value in requirements, features, practices and methodologies. Comenius University in Bratislava, Bratislava
65. Kumarapeli P, de Lusignan S (2013) Using the computer in the clinical consultation; setting the stage, reviewing, recording, and taking actions: multi-channel video study. J Am Med Inform Assoc 20(e1):e67–e75

66. Kune R et al (2016) The anatomy of Big Data computing. Softw Pract Exp 46(1):79–105
67. Leimeister JM et al (2014) Digital services for consumers. Electron Mark 24(4):255
68. Lemieux VL et al (2014) Meeting Big Data challenges with visual analytics: the role of records management. Rec Manag J 24(2):122–141
69. Li Z et al (2017) Optimization to the Phellinus experimental environment based on classification forecasting method. PLoS ONE 12:9
70. Makonin S et al (2016) Mixed-initiative for Big Data: the intersection of human + visual analytics + prediction. In: 2016 49th Hawaii international conference on system sciences (HICSS), pp 1427–1436
71. Mans RS et al (2015) Process mining in healthcare: evaluating and exploiting operational healthcare processes. Springer
72. Mehta N, Pandit A (2018) Concurrence of Big Data analytics and healthcare: a systematic review. Int J Med Informatics 114:57–65
73. Meskó B et al (2017) Digital health is a cultural transformation of traditional healthcare. Mhealth 3
74. Molnar E et al (2017) Business schools and RIS3—enterprise architecture perspective. In: 2017 15th international conference on emerging elearning technologies and applications (ICETA), pp 1–7
75. Molnar E et al (2014) Web intelligence in practice. J Serv Sci Res 6(1):149–172
76. Mueck S et al (2015) How leading organizations use Big Data and analytics to innovate. Strat Leadership 43(5):32–39
77. Nass SJ et al (2009) Overview of conclusions and recommendations. National Academies Press (US)
78. Networks I of M (US) C on RHD et al: Health databases and health database organizations: uses, benefits, and concerns. National Academies Press (US)
79. NHIC: eHealth. http://www.nczisk.sk/en/eHealth/Pages/default.aspx
80. Nichol MB (2006) The role of outcomes research in defining and measuring value in benefit decisions. J Manag Care Pharm 12(6 Suppl B):S19–23
81. Obaidullah SM et al (2016) Bangla and Oriya script lines identification from handwritten document images in tri-script scenario. IJSSMET 7(1):43–60
82. Offermann P et al (2009) Outline of a design science research process. In: Proceedings of the 4th international conference on design science research in information systems and technology. ACM, p 7
83. Ohm F et al (2013) Details acquired from medical history and patients' experience of empathy—two sides of the same coin. BMC Med Educ 13:67
84. Patel VL et al (2009) The coming of age of artificial intelligence in medicine. Artif Intell Med 46(1):5
85. Peek STM et al (2014) Factors influencing acceptance of technology for aging in place: a systematic review. Int J Med Inform. 83(4):235–248
86. Peffers K et al (2007) A design science research methodology for information systems research. J Manag Inf Syst 24(3):45–77
87. Pivovarov R, Elhadad N (2015) Automated methods for the summarization of electronic health records. J Am Med Inform Assoc 22(5):938–947
88. Porter ME (1985) Competitive advantage: creating and sustaining superior performance. Free Press
89. Qin Q et al (2016) Big Data analytics with swarm intelligence. Ind Manag Data Syst 116(4):646–666
90. Raghupathi W, Raghupathi V (2014) Big Data analytics in healthcare: promise and potential. Health Inf Sci Syst 2(1):3
91. Rajarathinam V et al (2015) Conceptual framework for the mapping of management process with information technology in a business process. Sci World J
92. Ranjan J (2014) Big Data applications in healthcare. In: Impact of emerging digital technologies on leadership in global business, pp 202–214

93. Reddy AR, Kumar PS (2016) Predictive Big Data analytics in healthcare. In: 2016 second international conference on computational intelligence communication technology (CICT), pp 623–626
94. Robson B, Baek OK (2008) Architecting IT all. In: The engines of Hippocrates. Wiley-Blackwell, pp 323–387
95. Robson B, Baek OK (2008) Guardian angels: knowing our molecules, drug and vaccine design, medical decision support, medical vigilance and defense. In: The engines of Hippocrates. Wiley-Blackwell, pp 389–467
96. Ross MK et al (2014) "Big Data" and the electronic health record. Yearb Med Inform 9(1):97–104
97. Chiranjeevi HS et al (2016) DSSM with text hashing technique for text document retrieval in next-generation search engine for Big Data and data analytics. In: 2016 IEEE international conference on engineering and technology (ICETECH), pp 395–399
98. Schnieden H, Grimes M (1988) Audit and performance indicators: a case study in ophthalmology. J Manag Med 3(4):301–314
99. Schoenherr T, Speier-Pero C (2015) Data science, predictive analytics, and Big Data in supply chain management: current state and future potential. J Bus Logist 36(1):120–132
100. Singh KR et al (2016) The application of rough set theory and near set theory to face recognition problem. In: Handbook of research on advanced hybrid intelligent techniques and applications, pp 378–413
101. Stefanowski J, Japkowicz N (2016) Final remarks on Big Data analysis and its impact on society and science. In: Big Data analysis: new algorithms for a new society. Springer, Cham, pp 305–329
102. Tamm T et al (2015) How an Australian retailer enabled business transformation through enterprise architecture. MIS Q Exec 14:4
103. Tang X et al (2016) SQL-SA for Big Data discovery polymorphic and parallelizable SQL user-defined scalar and aggregate infrastructure in Teradata Aster 6.20. In: 2016 IEEE 32nd international conference on data engineering (ICDE), pp 1182–1193
104. The Open Group: TOGAF Version 9.1. https://books.google.sk/books/about/TOGAF_Version_9_1.html?hl=sk&id=m11eAgAAQBAJ
105. Thriemer K et al (2012) Replacing paper data collection forms with electronic data entry in the field: findings from a study of community-acquired bloodstream infections in Pemba, Zanzibar. BMC Res Notes 5:113
106. de la Torre-Díez I et al (2015) Decision support systems and applications in ophthalmology: literature and commercial review focused on mobile apps. J Med Syst 39(1):174
107. Tynkkynen L-K, Lehto J (2009) An analysis of ophthalmology services in Finland—has the time come for a public-private partnership? Health Res Policy Syst 7(1):24
108. Uhl A, Gollenia LA (2012) Business transformation management methodology. Gower Publishing, Ltd
109. Vaishnavi V, Kuechler W (2004) Design research in information systems
110. Ventola CL (2014) Medical applications for 3D printing: current and projected uses. Pharm Ther 39(10):704–711
111. Vinekar A (2011) IT-enabled innovation to prevent infant blindness in rural India: the KIDROP experience. J Indian Bus Res 3(2):98–102
112. Wadmann S et al (2009) Coordination between primary and secondary healthcare in Denmark and Sweden. Int J Integr Care 9 (2009)
113. Waegemann CP (2016) mHealth: history, analysis, and implementation. In: M-health innovations for patient-centered care, pp 1–19
114. Wang J et al (2012) Benefits and barriers in mining the healthcare industry data. IJSDS 3(4):51–67
115. Wright K et al (2014) Citation searching: a systematic review case study of multiple risk behaviour interventions. BMC Med Res Methodol 14:73
116. Yin F et al (2016) Automatic ocular disease screening and monitoring using a hybrid cloud system. In: 2016 IEEE international conference on Internet of Things (iThings) and IEEE

green computing and communications (GreenCom) and IEEE cyber, physical and social computing (CPSCom) and IEEE smart data (SmartData), pp 263–268

117. Zadrozny P, Kodali R (2013) Big Data analytics using Splunk: deriving operational intelligence from social media, machine data, existing data warehouses, and other real-time streaming sources. Apress, New York, N.Y.

118. Zaki N, Tennakoon C (2017) BioCarian: search engine for exploratory searches in heterogeneous biological databases. BMC Bioinform 18

119. Zhang P et al (2016) Building cloud-based healthcare data mining services. In: 2016 IEEE international conference on services computing (SCC), pp 459–466

120. Zhang Z et al (2014) A survey on computer aided diagnosis for ocular diseases. BMC Med Inform Decis Mak 14:80

121. Zimmermann A et al (2017) Multi-perspective digitization architecture for the Internet of Things. In: Abramowicz W et al (eds) Business information systems workshops. Springer International Publishing, pp 289–298

Software Testing Management Process for Agile Approach Projects

Michał Pawlak and Aneta Poniszewska-Marańda

Abstract There are many software development methodologies, each following its own philosophy, guidelines, objectives and good practices. All of them include a dedicated phase of testing and fixing of defects. Unfortunately, the software test management does not have a luxury of having so many different methodologies available to the test managers. Ability to plan, design and create efficient tests is the most critical ability for any good tester. Test design techniques, also testing techniques, are used to identify and select the best test cases with the most efficient test conditions. The paper presents the issue of testing methodology based on advice and best practices advocated by experts in the field of testing. The method is intended to provide a step by step instruction of managing testing activities in a project environment. Given the popularity of agile development methodologies, the presented approach is mainly intended for a use in the agile environments.

1 Introduction

Software testing is a process which goal is to control and report defects in a software. It does not matter whether it is an embedded system or an application, each kind of software undergoes testing. It helps to reduce the possibility of defects which affect the effectiveness and efficiency of software.

However, in order to be conducted correctly, the process of software testing must be organized and managed in an appropriate way. It is not an easy task due to the complexity of involved processes. It is important to note that testing is not a separate activity, but is closely connected with development. Testing must be managed in such a way to complement and add a value to the development. Furthermore, testing becomes a highly automated process. Various types of tests can cover many

M. Pawlak · A. Poniszewska-Marańda (✉)
Institute of Information Technology, Lodz University of Technology, Lodz, Poland
e-mail: aneta.poniszewska-maranda@p.lodz.pl

M. Pawlak
e-mail: michal.pawlak@edu.p.lodz.pl

© Springer Nature Switzerland AG 2020
N. Kryvinska and M. Greguš (eds.), *Data-Centric Business and Applications*,
Lecture Notes on Data Engineering and Communications Technologies 30,
https://doi.org/10.1007/978-3-030-19069-9_3

functionalities, from single lines of a source code to communication between different components of a system. However, each project faces the problems of time and resource limit, which affect the development process and overall quality. Test managers must be able to overcome all these difficulties and organize the best possible testing process.

It is widely believed that a testing process includes only performing tests, which involve executing software and evaluating results. In reality, the process of testing is more complex and consists of different activities. According to *International Software Quality Board* (*ISTQB*) [1–3] the process includes: planning, control, defining test conditions, designing, choosing test cases, executing test cases, evaluating results, evaluating exit criteria, reporting the process and the system under test, closing the testing phase. In addition, relieving documents (including source code) and static analysis is included in the testing process. ISTQB [1, 4] also defines the following objectives of testing:

- finding defects,
- gaining confidence about the level of quality,
- providing information for decision-making,
- preventing defects.

The first point is the most basic and obvious objective of testing. Finding defects and fixing them is the easiest way of increasing software quality. The second point refers to evaluation of whether the software works as expected and meets the requirements. Third point considers data that can be gathered during the testing process. Number of found defects, percentage of requirements covered by tests or current quality level can all be taken into account during the decision making process related to the software. The last very important point advocates that testing can prevent the defects. Proper planning and designing of tests, ensuring testability of the produced code, thorough review of requirements and inclusion of quality attributes into them from the start, can prevent the defects and increase overall quality of the produced software.

The paper analysis the chosen aspects of testing methodology based on best practices advocated by experts in the field of testing providing the step-by-step approach of managing testing activities in agile project environment.

The paper is composed as follows: Sect. 2 presents the outline of testing process in software development project, Sect. 3 deals with the main aspects of software testing management while Sect. 4 presents the proposed approach for testing management process in agile software projects, named *Kungfu Testing*.

2   Testing Process for Software Projects

Software testing is a very broad term that includes wide variety of topics. They range from technical like testing techniques and measurements, to more organizational like planning and management of testing. Figure 1 presents the scope of topics,

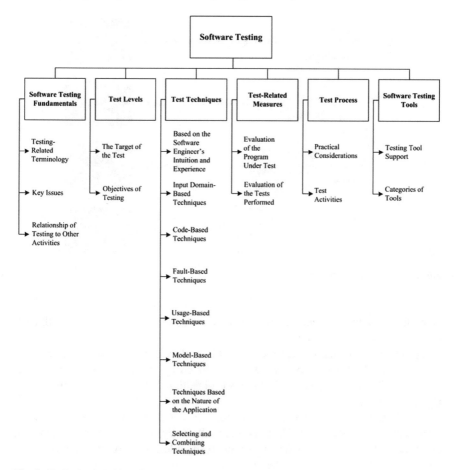

Fig. 1 Topics included in software testing by SWEBOK [4]

from which software testing is composed, proposed by Software Engineering Body of Knowledge (SWEBOK) [4–6]. Software testing can be divided into six topics, each describes a different knowledge are related to testing. The first topic, *Software Testing Fundamentals*, covers concepts, terminology, rules of testing. It also describes relationships between testing and other processes in software design. *Test Levels* refer to testing activities conducted on different levels of abstraction and their different objects and purposes. The topic *Test Techniques* describes a classification of basic test design techniques used for planning and generating tests. *Test-Related Measures* cover methods and measurements for monitoring and controlling the progress of testing and the product quality. *Test Process* refers to the testing process understood as a combination of concepts, techniques and measurements in a single controlled process. The topic covers test planning, evaluation and organization. Finally, *Software Testing Tools* describes tools that can be used for automation of some of the testing

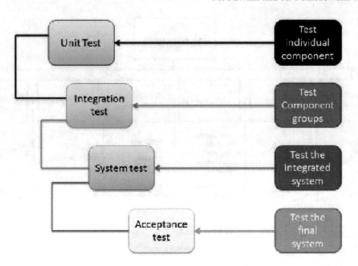

Fig. 2 Levels of software testing [1, 4]

activities. Detailed descriptions of the testing areas that can be supported with tools is also included in this topic.

Software testing can be classified by different levels, mostly related to development phases [1, 2, 4]. They can be differentiated by the object of testing and its purpose. There can be many different levels of testing, however the most typical are presented in Fig. 2.

Unit tests verify the smallest independent and testable parts of the source code, which are referred to as units. The tests are usually made in isolation from one another, but often tested components depend on other modules. Solution to this problem is application of mock objects. The unit tests are usually written and conducted by programmers, sometimes with the help of software testers.

Integration testing is performed to verify the interactions between software components and systems. It is important to note, that it includes testing interfaces, hardware-software integration and, what is often forgotten, system configurations. Integration testing is based on system architecture, but use cases can also be used in that role. This type of testing is usually performed by programmers and testers. Most projects are composed of many modules that have to interact together. It creates a problem of in what order the various components should be created and tested.

System testing validates the whole integrated system against the initial project goals. This type of testing is based on functional and non-functional specification. Testers assess system behaviour, configuration and overall quality. This is done by execution of all test cases from the user point of view (end-to-end testing) and testing non-functional system characteristics like security, reliability, performance etc. In addition any documentation, like manuals and user documentation should be tested as well.

Acceptance testing is designed to validate the final product. It is typically done by the customer to assess whether the software satisfies the acceptance criteria and not to search for defects. This level of testing should build trust in the software and its non-functional attributes. Although the tests are performed by the customers, professional testers should be present in order to gather the valuable feedback.

There are many others classifications of tests and types of tests but the purpose of this paper is not to present all of them.

3 Management Aspects of Software Testing Process

Software testing is a very complex and difficult process. It is important to remember that software testing is not an independent process. It is limited and conditioned by the development process and its methodologies. A good example of it would be a testing in agile methodologies. Large part of testing is performed by developers so there is a risk of testers repeating tests, if communication between both groups is lacking. Frequently changing specification can prevent automation of testing. These are just a few examples of how development can influence testing. However, the relationship between these two processes is two-sided, testing can force changes in a project when some critical defect was identified, or can delay a release if quality standards were not met [3, 6–10].

For these reasons, a test manager is not only responsible for implementing a specific test approach, but must also identify all stakeholders of the testing process to manage and satisfy their requirements. The stakeholders may include but are not limited to: product owner, client, testers, developers, business analysts.

3.1 Management and Organization of Testing in Software Development Projects

From the point of view of management, software testing can be divided into two main phases, each split into stages [11, 12].

During the *planning phase* test manager must obtain the information what to test, how long the testing will take, what strategy of testing will be utilized and finally who will conduct the testing and how. The required information is acquired during risk analysis, test estimation, test planning and test organization stages respectively. It is important to note, that in this case the risk analysis means any method of analysis which helps to determine the important areas to test. It may consist of requirement-based testing, testing based on operational profile or any method presented further in this section.

The *execution phase* on the other hand, is concerned with obtaining the information about performance of the testing process itself, whether the testing proceeds according to the schedule, dealing with expected and unexpected issues and evaluation and improving the whole testing process. These activities are conducted in test monitoring and control, issue management and test report and evaluation stages respectively.

Each stage has its own set of methodologies which help facilitate the activities required to complete its goals.

3.2   Challenges of Software Test Management

Managing software testing is not an easy task. Managers must constantly deal with many different types of challenges. Some may be caused by people and some by project itself. In general, the challenges the managers face can be connected with: project limitations and conditions, stake-holders, software production cycle, testing and human factor:

- project limitations and conditions,
- stakeholders,
- software production cycle,
- testing,
- human factor.

Challenges of project limitations and conditions are inherent to any business project and testing is no exception. Because the process of testing can be considered as a separate project, it faces usual issues and limitations of business projects, namely time, budget and scope. Managers who want to complete testing task assigned to them must often make trade-offs. For example, by reducing the number of tests and tested areas (scope) tests can be completed in time and on budget, but at possible cost of final quality.

Second set of challenges comes from the project stakeholders. Every one of them is interested in the results of testing and has their own expectations and requirements with respect to it. Often, needs of various stakeholders are contradictory or even impossible in time, scope and budget of the project. Test managers must identify the stakeholders, their requirements and relationships between them, so the best possible service can be provided by the testing team. Test manager must be able to navigate a net of stakeholder relations and be able to identify:

- with whom a continuous communication must be maintained,
- to whom the reports must be delivered and what information they should contain,
- to whom different kinds of issues must be reported to.

Software testing is not an independent process. It always applies to the product from the development process, which have a large influence on the organization of testing process. Various methodologies give the testing different roles, require different approaches and trade-offs. Demonstrating it, in the waterfall model the testing is conducted when all features are designed and implemented. The testing is squeezed at the very end of the production cycle. Usually, there is not enough time to test and fix the software. It results in traditional fight between testers and project managers, whether the project can be shipped in the current state or not, whether reliability can be sacrificed for time. On the other hand, in agile methodologies where requirements constantly change and a lot of testing is done by developer themselves, a role of testers may be unclear. These are just a sample of challenges, that can arise from software development models [12–14].

3.3 Software Testing Management

An ideal project has perfectly defined scope with suitable budget and time. There is a perfect synergy between development methodology and testing methodology. Software testing detects all software defects and developers are able to fix all of them in time. As a result, the client receives a product which meets all requirements and is defect-free [14].

Unfortunately, such projects do not exist. While planning testing projects, managers must be prepared for unforeseen events, which will force modifications to their plans. Trade-offs must be constantly made in order to deliver a product on budget and schedule, but also providing customers' satisfaction. In order to accomplish that, thorough analysis must be conducted of every product in order to make reasonable estimations and prioritizations. Various formal methods can help in this process, e.g. risk-based analysis for prioritization or the three-point estimation for estimation.

Test managers must assess their approach to the testing organization and management before the start of every project. Thoughtful analysis of a development method employed in a project will provide insight into how it will affect the testing. Test managers must make suitable adjustments to their plans so the testing will support the development and add-value to the overall project. Furthermore, there are many stakeholders in a testing project. Managers must be aware of them and their expectations in order to provide the best possible service to them.

There exist many tools which will support management and organization of testing. However, before implementing such a tool, a careful and thoughtful analysis of requirements and organizations environment is required. Without it, implementation of a tool may end in a failure and unnecessary costs. All things considered, the job of a test manager requires a lot of planning and awareness of many factors that affect the testing process. And it is no wonder then that turnover of test managers is very high [15–17].

4 Kungfu Testing Approach for Management of Software Testing

Kungfu in its original meaning refers to any skill achieved through a hard work and practice. Therefore, Kungfu Testing can be described as a testing skill obtained from training and experience. Both of those factors are basis of this approach to testing.

Kungfu Testing is a methodology of testing based on advices and best practices advocated by experts in the field of testing. The method is intended to provide a step by step instruction of managing testing activities in a project environment. Given the popularity of agile development methodologies, the presented approach is mainly intended for a use in the agile environments.

The following subsections present the description of *Kungfu Testing* process—the model of software test management presented in the previous chapter. However, firstly an exemplary information system is presented used to illustrate the phases of Kungfu Testing methodology.

4.1 Example Application—ErasmusBooking

ErasmusBooking is an exemplary online service intended to provide all mobility students of Lodz University of Technology (TUL) with information about partner institutions and ability to book a place in any of them. Its main functional requirements are the follows:

- Searching for partner institutions in a provided by a user country.
- Displaying results of the search in a form of a list. Each list item should present of each of the institutions, ranking of the institution, availability and short description of the offer.
- Displaying detailed information about any chosen institution. The information include offered course, languages of the institution, available ECTS points, additional recruitment requirements and housing.
- Logging into the service with a student account.
- Booking a place in the chosen institutions.
- Rating and commenting on the visited institutions.

The functionalities implemented in *ErasmusBooking* are made available via dedicated webpage similar to the hotel booking website *Booking.com*. *ErasmusBooking* offers easy to find and use search interface. The list of search results displays the most important information about each found institution, namely: localisation of each of the institutions, ranking of the institutions, availability, short description of the offer. However, the key functionality available from this subpage is booking of a place in the institution. The booking is made with the student account, so the information about bookings is available to TUL's staff.

From the architecture point of view, *ErasmusBooking* consists of four components:

- database of TUL containing students' personal data and data concerning their mobility,
- *ErasmusBooking Service* which communicates and exchanges data with the database,

- *ErasmusBooking Webpage* which is responsible for providing access to offered services to Erasmus students,
- Graphical User Interface (GUI), displayed by the website and used by users to navigate and use *ErasmusBooking Service*.

4.2   Planning Phase of Kungfu Testing Approach

Every project is different in terms of goals, available resources and targeted market. Software development is no different, however it is possible to divide them into two broad categories: *contract-driven* and *market-seeking*.

In the first case the company's main goal is to fulfil the contract. Typically, in such projects there is some form of requirements list at the beginning of the project. It is bound to change sooner or later, which must be kept in mind by test manager. In the second type there is no single client and no single specification. The company is searching for clients by trying to provide them with the best service. In this case testing may be performed against expectation of many different possible customers.

In both cases, the test manager should take active part in the requirement specification. The main goal is to ensure that testability features will be included in the project's schedule and budget from the very beginning. This will increase the testing team efficiency and effectiveness [13, 15, 18].

Next important activity of the test manager at this stage should be to inquire about possible non-functional requirements, like stability, security etc. They are often overlooked, both by clients and project managers, which results in lack of time and resources for their later inclusion. Making decision-makers aware of them at this stage may cause their inclusion into planning and thus prevent some of further problems. This can also provide the test manager and the test team with more details about required quality level. Of course, it is possible that the client will not be able to provide any information, in case of contract-driven project. Or there will be no single one standard for products for the specified market, in case of market-seeking project. In that case test manager and the testing team will have to find this information elsewhere. There are plenty of other sources of information that can help with defining functional and non-functional requirements [12, 18, 19]:

- user manual drafts,
- published standards and regulations,
- interviews with stakeholders,
- product marketing literature,
- third-party product capability test suites,
- prototypes and lab note on the prototypes,
- BugNet and other websites describing common bugs,
- mailing lists and discussion sites,
- compatible and similar products and their documentation,
- content reference materials.

In case of *ErasmusBooking* exemplary system a list of functional requirements is provided along with a description of the system purpose. No detailed non-functional specification is provided. However, ErasmusBooking is emulating Booking.com, so it can serve as a reference point for assessing the non-functional requirements. Standards and regulations can be found on the website of World Wide Web Consortium (W3C). Additionally, because students are the target users of ErasmusBooking, they can be interviewed in order to find what they expect from such a system.

4.2.1 Analysis Phase of *Kungfu Testing* Approach

Assuming some specification of the product has been collected, it is now important to start making test plans. The purpose of this stage is to determine objectives and scope of testing. In order to do that, the test manager has to analyse the product and identify areas that can be tested in the given project limitations. Kungfu Testing advocates using risk analysis as a way of detecting these areas. There are many different methods of risk analysis but in this methodology *Product Risk Management* (*PRisMA*) is a method of choice [13].

PRisMA was created by Erik van Veenendaal as simple and lightweight method for risk identification and analysis in close cooperation with stakeholders [13]. It advocates "good enough testing" which states that it is impossible to expect zero-defect software, so testing should ensure the product has enough value to be moved into production. In other words it should have the following characteristics:

- have sufficient benefits,
- have no critical problems,
- have benefits outweighing its non-critical problems,
- delaying release would cause more loses than profits.

It is argued that it is not testers' responsibility to make decision whether the product is ready or not. This role belongs to project managers and other stake-holders. Test teams are just information providers for the decision-makers.

The analysis stage can be facilitated using *Product Risk Management* method (Fig. 3). The main idea behind this method is a creation of a product-risk matrix. The matrix contains numeric values of risks of each identified item. Depending on these values, a different testing approach should be employed.

Project information and artefact gathering is a first phase of this method and consists of the following steps:

- gathering input documents—determining and collecting data relevant for product risk analysis, performing static analysis of the obtained documents,
- identifying risk items—listed with own identifier and description; no more than 30–35 risk items in order to keep the process functioning correctly,
- determining impact and likehood factors—described in high-level document for the whole organization,

Fig. 3 Risk evaluation of
PRisMA process [13]

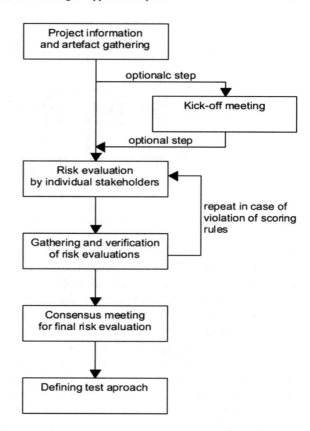

- (optional) defining weights for each factor—the scale from 1 to 5, also called *Likert scale*, is recommended,
- selecting stakeholders—including different roles from business and production, e.g. project manager, developers, system architect and end user,
- defining scoring rules.

Kick-off meeting is to explain the list of risk items, factors influencing them and the whole evaluation process. The test manager should also explain the remainder of the process and should clearly explain what is expected from everyone involved in it. This is also a good moment to discuss two very important criterion for testing, namely suspension and resumption criteria. They refer to the conditions of temporary suspension of testing and resumption of previously suspended testing respectively. It is important to ensure cooperation and commitment of every stakeholder, so their participation will be meaningful.

Risk evaluation by individual stakeholders phase consists of the stakeholders scoring the risk items according to the presented rules. These scores are based on personal stakeholder assumptions which should be also documented.

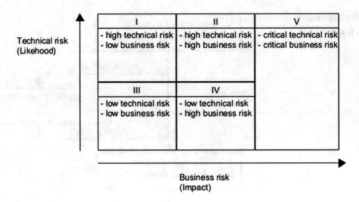

Fig. 4 Example of product-risk matrix for defining test approaches [2]

Gather and evaluating of risk evaluations phase consists of collecting the results of evaluation by stakeholders and checking its correctness. Test manager proceeds to average scores of every risk item and individual scores can be placed in the product-risk matrix (Fig. 4). Problems like unresolved issues and risks that are positioned too close to the centre of the matrix should be discussed in the next phase.

Consensus meeting for final risk evaluation phase can be started after a completion of product-risk matrix. During it a common understanding of perceived product risks should be achieved and may result in modifications of the requirement list, use cases or user stories. The most important thing is that the matrix should be validated with every stakeholder.

Define a differentiated test approach—basing on the risk position in the matrix, the test objects can be identified and prioritized. It defines test scope of the projects, from which necessary test design techniques can be derived.

Taking into consideration the *ErasmusBooking* example, there are a few sources of information that can be used for examination and evaluation of it. Having collected the necessary sources, the risk analysis stage can be started. Firstly, five main risk items are identified for *ErasmusBooking* service:

- lack of communication between components,
- incorrectly displayed data,
- illegible interface,
- business logic defects,
- slow response times.

Next step is to select factors that affect each risk. For *ErasmusBooking* four factors are selected, two for impact:

- visible area (*VisArea*) with weight 1.0,
- business value (*BusVal*) with weight 2.0

and two for likehood respectively:

- system complexity (*SysCom*) with weight 2.0,
- technology incompatibilities (*TechIncom*) with weight 1.0.

With risk items and factors identified, a grading scale must be selected. There is no high-level document describing the evaluation process so it is assumed that any scale can be chosen. Following the guidelines from *Kungfu Testing* the *Likert scale*, from 1 to 5, is selected with no additional constraints for scoring.

At this stage the test manager must conduct stakeholder analysis and choose the ones necessary for risk evaluation. Furthermore, not all of them should be responsible for evaluating of all factors due to their potential lack of expertise. The following stakeholders with corresponding factors are selected for *ErasmusBooking*:

- project manager with all factors,
- business analyst with visible areas and business value,
- system architect with system complexity and technology incompatibilities.

The Kick-off meeting can be conducted with the preparations completed. Each risk item and factor is explained to the stakeholders so they can make informed evaluation. After the meeting risks can be assessed by the individual stakeholders.

4.2.2   Test Estimation Phase of *Kungfu Testing* Approach

The next activity, after the analysis stage, is to estimate time, resources, cost and skill required for completing the test activities (Fig. 5).

Estimation can be done by combining different approaches to the task estimation—the test should be divided into smallest possible tasks, each should be described in detail. Next step is assigning the tasks to specific team members, who then proceed to estimate the best, worst and most likely estimates for their tasks, in order to conduct three point estimation. Although, predictions of all team members should be considered equally important, the fact is the opinion of the person responsible for the given task is the most relevant.

When the group estimation is completed, the test manager should analyse the estimates and consider additional factors that may be involved. Usually a lot of time and resources are spent on meetings, trainings and other non-project related work. Thus, the best way to account for that is to add all estimates together and add additional 25% of their value as a buffer [20].

Identifying opportunities and costs refers to two good practices. advocated in the [15]. The first argues that it is often possible to reduce required time by hiring extra help at specific points in the project. Additional help, even temporary, can reduce others' load and let them focus on more critical activities. The second practice can greatly help in contacts with management. It refers to preparation of a cost model of testing in which a simple simulation of costs can be performed. This will allow the test manager to answer inevitable questions like "what will happen if some tests will be skipped" or "what will be gained through performing this type of testing" [2]. Having model like that puts the test manager in better position when validating,

Fig. 5 Test estimation
process of *Kungfu Testing*

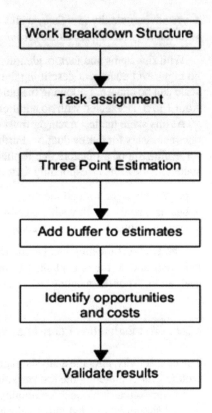

negotiating resources and schedule with management, which is also the last step of
the process.

In *ErasmusBooking* example, the estimation process according to *Kungfu Testing*
starts with dividing large tasks into smallest possible subtasks which can be easily
estimated. For example, at the start of the project the testing team must create a test
specification, create a test plan and build up a testing environment. Each of these
tasks is divided into smaller subtasks and assigned to the specific team members.

4.2.3   Test Planning Phase of *Kungfu Testing* Approach

The objective of this stage is to create a test plan document. It serves as a frame of
reference for testing activities which will be monitored by the test manager. What is
important during its creation is to have a clear purpose of it in mind. A test plan can
be a useful tool for managing tests but it can also be a product itself. Often, detailed
test plans are required for successful software delivery, to satisfy some standards or
norms or may be objects of business transaction on their own.

There is no single proper template of test plans. The best plans are those which
fulfil their role in the project. That is why the most important thing is to understand

each of the fields within it and to properly utilize them. Creating detailed documentation for its own sake is entirely pointless waste of resources and time. However, Kungfu Testing proposes the following fields to be included in plans:

- business background,
- requirements,
- risks,
- test objectives,
- test scope,
- test design approaches,
- task estimations,
- test team,
- test resources,
- suspension and resumption criterion,
- metrics and methods of measurement,
- test deliverables,
- norms, standards and constraints.

Most of the presented above fields are created during the previous stages of the planning process.

Test resources are a summary of all types of resources required for the testing process. They cover software, like testing environment, testing tools, and hardware, e.g. servers, computers and networks. *Test deliverables* is a list of all the documents, tools and other components that have to be created and maintained during the testing process. There may be different deliverables during different stages of testing. Proposed approach advocates for creation of the following documents:

- test plan, before and during the testing,
- test status reports, created after every cycle (or iteration) of testing,
- test results, after completion of the testing process.

Norms, standards and constraints are required if the project uses some of many norms or standards, links to them should be included for easy access. In addition, any constraints and limitations that the project must meet should be included in test plans.

4.2.4 Test Organization Phase of *Kungfu Testing* Approach

In general, a *test team* consists of: *test manager*, who directs and organizes work on the project, *testers*, who create and execute tests, *test administrator*, who manages and maintains a test environment. In addition, developers who create tests as part of TDD can also be considered to be part of testing team. Their contribution is crucial in agile development, where unit and integration tests provide continuous feedback on stability of the system under tests.

There are many techniques and methods of team management and team work. They depend on managers, organization and team members. Specific methods of this topic are outside of the scope of this paper but some details can be found in [2, 3].

First and foremost thing test managers and testers should realize, is that they are service providers. The main goal of testing is to provide information to people responsible for control or ensure quality tasks, so they can make well informed decisions. Test teams should focus on providing the best possible service to their clients, namely their project managers.

Next issue, connected to the previous topic, is that testing groups are not responsible for ensuring the quality of the overall project process. Test teams, in most cases, do not have resources, experience or power required for fixing development processes, nor should it be their responsibility.

A good practice for testing is rotating the testers across features. This practice bring many advantages, firstly it keeps the staff from getting bored form constantly doing the same thing. Secondly, it keeps the testers form becoming overly specialized in single area of the project, which decreases value of such testers. In addition, it prevents knowledge gaps in case such specialists leaves. Thirdly, different testers have different strengths. Although test managers should encourage individual development of their staff—it is sometimes better to assign someone else to the task, if the current employee performs exceptionally slowly and inefficiently. Finally, a good practise is work in pairs. Testers are encouraged to share, explain and react to ideas which bring them to better focus and trigger more ideas. It helps testers concentrate at task at hand and makes the tasks of bug replication and analysis easier.

Testing of *ErasmusBooking* is organised in accordance to the Kungfu Testing recommendations and agile development principles. The testing is conducted in close cooperation with developers who are responsible for low-level tests, by working in the same place and in pairs. Testers provide different point of view combining business perspective and developer perspective on the project.

This ensures that both high- and low-level tests are more valuable and really increase the quality of ErasmusBooking service. Additionally, testers were obliged to change their partner every week in order to familiarize with various aspects of the system and provide help in different areas of the project [5].

4.3 Execution Phase of Kungfu Testing Approach

The test execution phase is concerned with fulfilling the test plan and ensuring that the testing provides the best possible service to stakeholders. This sub-chapter is focusing on providing methods and guidelines for test execution, reporting, evaluation and improvement of testing and development processes.

Test execution phase in *ErasmusBooking* example is conducted in accordance with project development methodology conducted in agile. The tests are conducted during two week sprints. Testing is highly automated and performed in parallel to development so a continuous feedback can be provided. This enables quick fixes and constant improvement of ErasmusBooking service.

4.3.1 Test Monitoring and Control Phase of *Kungfu Testing* Approach

Monitoring and control of testing processes are the most important tasks of test managers at this stage. There are four key parameters that should be monitored during this phase: cost, schedules, resources and quality. They can be monitored by collecting various measurements and metrics. Cost, schedules and resources are relatively easy to follow, test managers must only regular check the state of their budget, verify whether work is delivered on time and resources are in required quantity. Measuring quality of work and performance of staff is a different and difficult problem.

The test managers should consider reading the artefacts created during testing like: bug reports, code, documentation. Having done so, test managers can evaluate individual strengths and weaknesses of their staff. It is possible to create a scale for various skills and asses them for individual team members. It creates multidimensional information, on which training and coaching plans can be developed. This practice is not about micromanagement but about learning what people do and how to make them more efficient.

ErasmusBooking uses several measurements which counterbalance others' side effects to report progress. They measure progress by answering the following questions:

- How much of ErasmusBooking has been tested?
- How much of the planned testing has been done?
- How many problems have been found, and how many are critical?
- How many defects were overlooked by the testing team?
- How much of the testing is being blocked by unfixed defects?

The first question is answered by number of lines of code covered by tests, while the others are simple numeric values. At the end of each time interval a status report, containing all of these indicators is prepared. The report is presented and discussed with the testing team in order to identify the sources of potential problems.

4.3.2 Issue Management of *Kungfu Testing* Approach

The test teams are not responsible for controlling and ensuring quality of the whole development process. However, it does not mean reporting such issues should not be performed. After all, there may be problems with the testing process itself and it is not always in test managers' power to fix them.

There are three steps of issue management. The first step is to *record* every encountered issue—it should have assigned priority, tracking status and owner responsible for it. Next step is to *report* documented issues to a high-level manager. The final step is *control* and it is a responsibility of a project manager to monitor the reported issues and possibly find solutions for them (if they are worth resolving within the scope of the given project).

If the issue is important, the main idea is to gather the support in form of stakeholders and enough evidence that it is critical to fix the problem [20]. The following steps should be considered:

- finding other stakeholders, who might be affected be leaving the issue unresolved,
- performing additional research to find more serious consequences and wider range of processes affected by the problem,
- searching for similar issues in other organizations, it may be possible that they faced something similar and fixed it, which will prove that it is serious issue,
- researching possible solutions and providing recommendations.

4.3.3 Report and Evaluation Stage of *Kungfu Testing* Approach

Reports are the primary product of most testers. They shape the perception of testers in eyes of their receivers. It is test managers' job to ensure that reports generated by their staffs are [8]: actually written (not just reported verbally), numbered (for identification purposes), simple (they description of only one problem at a time), understandable (makes the bug more likely to be fixed), reproducible and legible.

Next important point concerning this stage is status reports. The main goal and strength of any testing team is to provide information about projects. *Kungfu Testing* suggests that the following information be included:

- decisions needed,
- bug fixes needed,
- expected deliverables and their dates of delivery,
- unexpected problems,
- team's progress toward completing the planned tasks,
- bug statistics,
- summary information of deferred bugs.

There is no certain and dependable formula for deciding when the testing is done. The best way to asses that is to use experience, skill and judgement of test teams. However, because testing is an information gathering process, theoretically it can be stopped when enough information has been gathered. Enough in this case means that basing on the collected information it may be reasonably assumed that there are no important or critical defects undiscovered problems in the final product.

Release reports should follow the guidelines presented for status reports but should contain information about the whole testing process. There are many templates that can be used, but no matter the form they should include:

- description of testing results,
- list of unfixed bugs,
- 10 worst things the critics may say about a product.

It is difficult to evaluate overall quality of a product and its fitness for the market. The only thing that test team can do is provide information gathered from bug searching and bug reports. The main idea is to report the testing work and its results, not an opinion of the product. List of unfixed bugs helps in the product maintenance and allows clients and technical support to prepare for them accordingly. The list should not only include bugs but also all design concerns and issues found within the project. Finally, it may be entirely possible that including 10 critical things that may be said about the final product may delay the final release. This approach works because if the problems are significant, the marketing department may decide to delay the release on its own in order to have better marketing position.

ErasmusBooking advocates close cooperation between testers and developers. For that reason many defects are fixed instantly after conversation between the tester and the programmer. Defects that are more sever are reported using Trello platform [19]. A standard report contains a build version, defect severity, defect description, steps required for its reproduction and expected results.

5 *Testing Kungfu* Management Supporting Tool

There exist many software management tools on the market. *Testing Kungfu* acknowledges this fact and does not rely on any specific management tool. In fact, the methodology advocates using programs and solutions most suitable for the given project. However, Testing Kungfu contains the processes that require a dedicated solutions for automation. One such process is PRisMA risk analysis conducted in the analysis stage of the proposed methodology. For that reason it was decided to design and implement such a tool.

Testing Kungfu management supporting tool (*Testing Kungfu MST*) was written in Java 8 as a Maven project designed to facilitate the process of PRisMA risk analysis. It automates the calculations needed for measuring impact, likelihood and priority of all identified product risks and returns the obtained values in easy to read and reuse way.

Testing Kungfu management supporting tool fulfils a set of use cases:

- read stakeholder evaluation,
- calculate impact, likelihood and priority of risks (Fig. 6),
- define methods of testing,
- create and display product-risk matrix,
- save calculated results to a file,
- save a product-risk matrix to a file.

Testing Kungfu MST consists of four components (Fig. 7). The first one is named *Stakeholder files* and is required for the risk processing. It is the only external part of the Testing Kungfu MST. The second component is the application itself. It is composed of objects which model stakeholders, risks and factors influencing risks read from the files provided by the stakeholders. The objects enable processing of the evaluations required for calculation of necessary metrics. *Risk analysers and controllers* are parts of the third component. They are responsible for the calculation of impacts, likelihoods and priorities of each risk. Finally, GUI component named *Graphical*

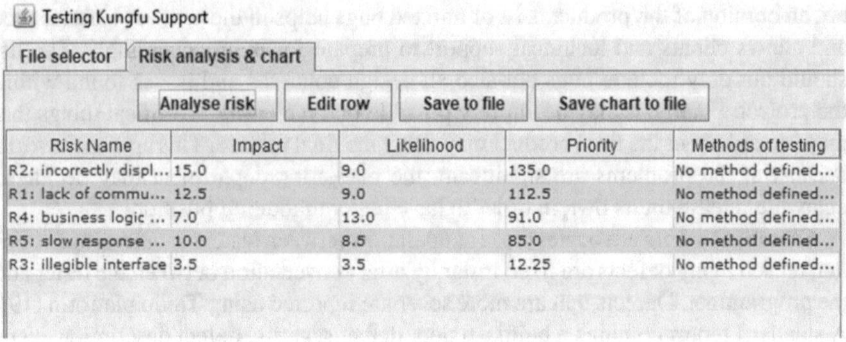

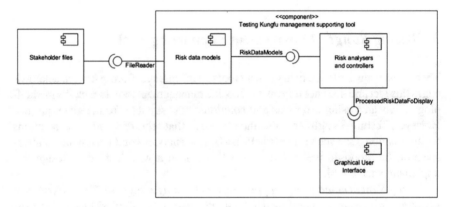

Fig. 6 Results of risk analysis with the use of *Testing Kungfu MST*

Fig. 7 Components of the *Testing Kungfu MST*

User Interface, displays the data to the test manager and allows the generation of the product risk matrix and editing the methods of testing.

6 Conclusions

Software defects and lack of quality may result in losses, not only in money but in human life as well. In order to reduce these risks, software testing must be conducted. Correctly performed, testing does not only mitigate the mentioned dangers but also builds confidence in both software and developers, which in turn bring the profits. Different types of tests accomplish different goals: they ensure delivery of correct functionalities; increase stability, security and reliability of software etc. Each of these goals represents an increase in quality of software which is a result of software testing.

By contrast, the biggest disadvantage of software testing is its costs. People, resources and the testing process itself cost time and money. Usually, there is not enough of either to completely conduct the process of testing. Testing in meaningful way may be difficult, especially in agile development methodologies where everything constantly evolves and changes.

To sum up, software testing has many advantages but it also brings disadvantages and difficulties. However, the profits it generates clearly in a long run outweigh any expenses it caused in a short term.

Kungfu Testing provides test managers with lightweight methods for conducting risk analysis and creating test plans. This creates a frame of reference for testers, who often lack such item especially in agile methodologies, which philosophy of focusing on communication instead of documentation often results in no documentation at all. It is solved by providing methods of creation of such document based on communication and close cooperation between stakeholders. Risk analysis, test plan and test estimates can be updated along with evolution of the overall projects according the current needs and requirements.

Kungfu Testing advice and guidelines for various issues are based on already existing techniques and methods employed by various experts in the field. This assures some level of reliability of them and thus the whole approach. It provides a ready to use set of steps and guidelines for starting and completing testing projects.

Software testing is not an independent process. It always applies to the product from the development process, which have a large influence on the organization of testing process. Various methodologies give the testing different roles, require different approaches and trade-offs. Demonstrating it, in the waterfall model the testing is conducted when all features are designed and implemented. The testing is squeezed at the very end of the production cycle. Usually, there is not enough time to test and fix the software. It results in traditional fight between testers and project managers, whether the project can be shipped in the current state or not, whether reliability can be sacrificed for time. On the other hand, in agile methodologies where requirements constantly change and a lot of testing is done by developer themselves, a role of testers may be unclear. These are just a sample of challenges, that can arise from software development models.

References

1. International Software Testing Qualifications Board (2011) Certified tester—foundation level syllabus. International Software Testing Qualifications Board
2. Roman A (2015) Testing and software quality, models, techniques, tools. PWN
3. Crispin L, Gregory J (2009) Agile testing. A practical guide for testers and agile teams. Pearson Education Inc., Boston
4. IEEE Computer Society (2014) Guide to the software engineering body of knowledge version 3.0. IEEE
5. International Software Test Institute (2017) Software testing levels. http://www.test-institute.org/Software_Testing_Levels.php

6. Zimitrowicz K (2015) Quality of IT projects, software development and testing. Helion, Gliwice
7. Dissanayake P (2014) How to draw a control flow graph and cyclometric complexity for a given procedure. DZone/Performance Zone
8. Kaner C, Falk J, Nguyen HQ (1999) Testing computer software. Wiley
9. Kaczor S, Kryvinska N (2013) It is all about services—fundamentals, drivers, and business models. Soc Serv Sci J Serv Sci Res Springer 5(2):125–154
10. Gregus M, Kryvinska N (2015) Service orientation of enterprises—aspects, dimensions, technologies. Comenius University in Bratislava. ISBN: 9788022339780
11. IEEE Computer Society (2008) IEEE Standard for software reviews and audits. Software & Systems Engineering Standards Committee, New York
12. IEEE Computer Society (2008) IEEE Standard for software and system test documentation. Software & Systems Engineering Standards Committee, New York
13. van Veenedaal E (2011) Practical risk-based testing. Product risk management: the PRISMA method. UTN Publishers, Manchester
14. Kryvinska N (2012) Building consistent formal specification for the service enterprise agility foundation. Soc Serv Sci J Serv Sci Res Springer 4(2):235–269
15. Bryson JM (2004) What to do when stakeholders matter: a guide to stakeholder identification and analysis techniques. University of Minnesota, Mineapolis
16. Kryvinska N, Gregus M (2014) SOA and its business value in requirements, features, practices and methodologies. Comenius University in Bratislava. ISBN: 9788022337649
17. Molnr E, Molnr R, Kryvinska N, Gregus M (2014) Web intelligence in practice. Soc Serv Sci J Serv Sci Res Springer 6(1):149–172
18. Kaner C, Bach J, Pettichord B (2002) Lessons learned in software testing. A context-driven approach. Wiley, New York
19. Black R (2009) Practical tools and techniques for managing software and hardware testing. Wiley Publishing Inc., Indianapolis
20. Kaner C, Bach J, Pettichord B (2002) Lessons learned in software testing. A context-driven approach. Wiley, New York

Analyzing User Profiles with the Use of Social API

Aneta Poniszewska-Marańda, Cezar Pokorski and Vincent Karovič

Abstract This work concerns the scope of software engineering and data analysis using artificial intelligence methods. The aim of the work is to analyze (define and categorize) information that we can get for the help of the presented techniques and available technologies, including the development of their models of access to user data, as well as the methods of interpretation and use of the data obtained. The "smart" pro-applications offered by the application can be built based on the application and help to get the most out of the service, reaching a larger and larger group, which obviously creates new business opportunities.

1 Introduction

Web 2.0 is a term that has risen to the heights of media trends in 2004–2006 as a technological symbol of the Internet-related leadership. However, it did not mean new technology as such, but radical change in the way in which content is published on the Web. The original sources of the Internet and its World Wide Web site consisted of predominant static sites that were once published or regularly updated by their authors. The content of these websites came in most cases almost exclusively from their creators, who also dealt with technical aspects, and in the case of corporations from their employees.

However, in 2001, there was a breakthrough the popularity in this area was very dynamic, because it was noticed how attractive it is for users to leave their trace on the page—comment on the content presented, impressions of our opinion, and express opinions in the same place in which the content itself is presented. It should

A. Poniszewska-Marańda · C. Pokorski
Lodz University of Technology, Lodz, Poland
e-mail: aneta.poniszewska-maranda@p.lodz.pl

V. Karovič (✉)
Faculty of Management, Comenius University, Bratislava, Odbojárov 10, 831 04 Bratislava, Slovakia
e-mail: vincent.karovic2@fm.uniba.sk

© Springer Nature Switzerland AG 2020
N. Kryvinska and M. Greguš (eds.), *Data-Centric Business and Applications*,
Lecture Notes on Data Engineering and Communications Technologies 30,
https://doi.org/10.1007/978-3-030-19069-9_4

be remembered that the mere possibility of contact and sharing of industries on the Internet was nothing new, but it did have the form of controversial statements of groups as a parallel world in relation to WWW resources, available through other protocols and separate from it.

The "Web 2.0" trend meant big changes—now users from passive recipients have become the main source of content in modern websites, which was more attractive for administrators and creators. Users like to share their ideas or knowledge and make them popular. On the other hand, sites creators have been relieved of the obligation to constantly find new and attractive content.

The services that are today popularly called social services have been created in the wake of this trend. Their main interest is often the user himself and everything he wants to post on the web site.

These services are primarily about ideas and functionality oriented in a predetermined order. An example of such services can be Flickr or Picasa, collecting statistics on our music preferences—like Last.fm or Pandora, or analyzing our geographic location (for example, Google's library), sometimes combined with the exchange of information between users about traffic situations (such as Waze).

The most popular services of this type usually take their popularity behind the public API, allowing the community to build and expand their skills, use it, and get it to people who have not heard about the service.

The existence of such services is linked to the emergence of many new opportunities and threats for users. In this work, we analyze the benefits that we can derive from the services and data provided by the most popular social services, as well as a discussion of security mechanisms that aim to minimize the threats of overly easy access to user data.

2 Access to Data in Most Popular Social Networking Services

First, it should be determined what the social services really are, as a key concept for all the work as well as the modern Internet. Quoting some inaccessible Internet sources and an economic article quoted in turn in article [1].

Social services are websites that are an online meeting place for people seeking new knowledge, in which participants exchange all information based on their individual profiles, using a specific interface for this purpose. Thanks to a wide range of shared communication tools, these websites allow us to make contacts in the most convenient forms for users: using discussion forums, message exchanges (both by e-mail, messenger, and through a message system implemented in the site engine), entries to the guest book.

As we can see, the key element distinguished in the very meaning of the concept of a social website is the user's profile (Fig. 1 shows an example) which focuses the whole interaction with other users and this interaction. The most important features

Fig. 1 Preview of user profile on Facebook (view for person not logged in)

of websites of this type are the freedom of individual expression and interaction with other users, which translates into the ease of establishing new contacts with people who will seem interesting to us.

But can computer algorithms help us by choosing proposals and try to indicate this potentially attractive knowledge, having only the data available on these websites? This is what this work is about, discussing what information can be obtained and how to use it in a machine-like way.

2.1 Popularity of Social Networking Sites

Obviously, more users a website has, more likely a user will be satisfied with another user because there are more potential acquaintances. The increase in the number of users is the most commonly determined success marker in this order.

Since 2008, we have seen a continuous and significant increase of interest in this type of services (Fig. 2 shows interest over time). In an era of growing popularity and the development of current and future leaders in this segment of the network, it is particularly important that companies create and promote their brands through these media, service developers benefit from what they offer, and investors are ready to finance promising projects, in social services.

The success of this market segment is also measured by the number of users registered in the website and the growth rate of this number. The absolute record holder in this category service type, which premiered in June 2011, was Google+

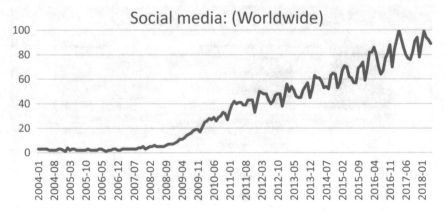

Fig. 2 Growing interest in the term "social media": Numbers are interest in the search with respect to the highest point in the chart for the given area and the time period. The value of 100 is the highest popularity of the term. 50 means it has half popularity. Score 0 means that there was not enough data for that term [2]

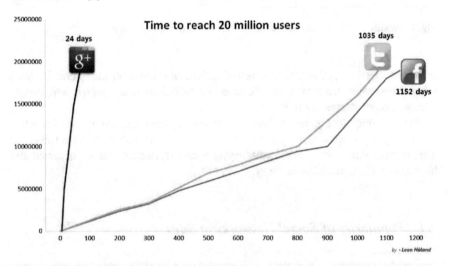

Fig. 3 Google Plus growth rate compared to Twitter and Facebook [4]

(alt Google Plus). Google Plus managed to gain absolute growth that time a staggering pace of a million users a day—less than three weeks after the opening, available only at the invitation of the trial version, there was already an astronomical number of 20 million users (Fig. 3 shows a Google Plus growth rate compared to Twitter and Facebook) [3].

It is also noticeable that the popularity of this type of websites is probably influenced by more complex factors, and their "sociality" and following the Web 2.0 trend do not guarantee success.

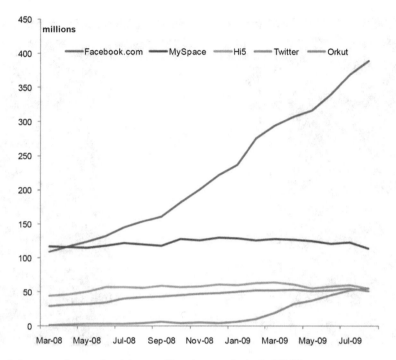

Fig. 4 Increase of users of social networking sites popular in the US [5]

Considering the data (Fig. 4 shows an Increase of users of social networking sites popular in the US) cited in the report [5], it is easy to notice that among some of the portals, which were equally popular in 2008, only Facebook moved to a strict lead and in fact became not only a leader but also a deeply rooted brand in the Internet users' consciousness as a model example of this type of site. Other services that are compared here: Hi5, Myspace, Orkut, have never adopted globally (primarily in Poland), or even turned out to be a fiasco, dying out after a few years or adding negative associations in the environment of Internet users (an example is Myspace, received as a heavy and dappled website only for music bands and their biggest fans).

This list still includes Twitter (Fig. 5 shows an example of user profile on Twitter)—icon of the so-called microblogging[1]—whose popularity is growing, and the service itself is properly identified with this type of service, which has been widely popularized by him. The success of Twitter caused that the network swarmed with neighbors trying to fill its gaps—in most cases these sites do not get the popularity of their original even on the local scale.

---

[1]Microblogging is a description of the phenomenon consisting in publishing very short statuses (usually up to 140 or 160 characters) concerning, among other things, what is being done or thinking, where restricting the length of an entry is enforced by a given platform, and other users subscribe to our statuses (micro-answers) by adding us to the "observed" list.

Fig. 5 Twitter service—profile view of the user (actress Kaley Cuoco)

The element that distinguishes these successful websites like Twitter or Facebook is the open, publicly available API.

API, or Application Programming Interface, which is the second key component of the title of this part of publication work. API is one of the basic concepts in computer science. They are defined as: "a set of standardized queries […] that were defined for the program on which they are called. […] API defines the way a programmer should ask for services from a given platform" [6].

The key to the success of the most recognized social platforms was the intense development of side products created around the main site using the API provided to offer a more convenient or more universal access to its content, expand its functionality or expand the group of recipients.

Co-founder of Twitter, Christopher Isaac "Biz" Stone admits in an interview [7] that their API generates 10 times more traffic on servers than directly referenced from web browsers, and that it is thanks to applications doing "practically everything" in cooperation with Twitter, it has gained so much popularity and increased number of users. Currently, only 30% of messages published on this site are sent via a web browser [9].

2.2   Available User Data

Information about the user, regardless of how they are represented on the website, internally—and therefore also in the API—are stored in separate fields, making up the so-called user profile, i.e. the entirety of how a given user is seen on the site by other users.

Each social network has objects representing individual users—they are the basis for the existence of such a site, as well as certain objects associated with them by means of relations.

Table 1 shows what data on the individual (user) is provided by selected social services through calls to their API methods. It is worth noting that this table contains only the formulas contained directly in the object representing the user in these websites. In order to increase readability, its origin (in the last column) and possible variants of the field name it has been distinguished in each row (when different services have different or very similar forms).

Table 1 omits some fields not about the user, but rather about only the appearance of his profile (like background image or color of links) or signifying settings related to a given account from the point of view of another account that asks for them (for example, informing we download his messages via WWW and SMS).

Several applications can be drawn from Table 1. One of them will be that the most common information (present in many different websites) is the most basic: login, real name, age, gender, basic language and location. The latter is somewhat manually entered by the user, not specific geographic coordinates). We can also collect several Internet addresses, given as a "user's website", under which there may be more information related to it (as well as something entirely random, like pajacyk.pl or greenpeace.pl). However, do not underestimate the magnitude of information that we potentially carry and the remaining fields. For example:

- **Time of registration**—it can tell us to a certain extent how active we are in the Web, how long or how quickly we try to know the news we hear about. We can distinguish so-called early adopters, i.e. people with a passion for early assimilation of any new technologies and services, from an average internet user who registers to the website under the influence of its already large popularity.
- **Location**—understood as a precise geographical location, a coincidence of our stay at the time of the publication of some content or indication of the object nearby, but also as the name of the city, country or time zone—all this information that should not be completely rejected. If we run an advertising campaign and/or we want to reach a certain product or service to a specific group of recipients, these data help determine both the optimal suggestions and hopes where the recipient is reachable (the fifth local time usually has the lowest reach of active users' on-line services).
- **Number of followers (friends)**—this is usually a direct indicator of popularity of a given user on the Web (the number of users who will keep him/her). The more people want to read messages published by a given user, the more he is influential

Table 1 Fields describing the example of user data directly available in the API of Facebook services, Last.fm, Blip, Twitter [8, 10, 11, 14]

Field identifier	Data type	Description	API
id	Integer	Numeric user account ID inside the given platform	Blip Facebook Last.fm Twitter
name—Facebook realname.—Last.fm	Text string	The full name of the user (i.e., name text string Name and surname)	Facebook Last.fm Twitter
first_name	Text string	Name	Facebook
last_name	Text string	Surname	Facebook
gender—Facebook gender—Last.fm sex—Blip	Text string	Sex	Blip Facebook Last.fm
locale—Facebook lang—Twitter	Text string	The language in which the user from the service (Facebook: country code and language code, Twitter: language code)	Facebook Twitter
languages	Array of text strings	The languages in which the user can communicate	Facebook
link url	Text string (URL)	Profile address on this website	Facebook Last.fm
username—Facebook name—Last.fm login—Blip screen_name—Twiter	Text string	Username (login)	Blip Facebook Last.fm Twitter
image	Text string (URL)	The address of the image represented this user (his avatar)	Last.fm
third_party_id	Integer	Numeric identifier, anonymous, but unique for a given user	Facebook
timezone—Facebook utc_offset—Twiter	Number	Shifts in UTC time in hours (Facebook) or minutes (Twitter)	Facebook Twitter
time_zone	Text string, e.g. Warsaw"	The name of the time zone	Twitter

(continued)

Table 1 (continued)

Field identifier	Data type	Description	API
registered created_at	Time stamp	Date of account registration in the service	Last.fm Twitter
updated_time	Text string (ISO-8601)	Time of the last profile update, such as the publication of the last statusor changing data[a]	Facebook
verified	Logical value	Account verification status[b,c]	Facebook Twiter
subscriber	Logical value	Specifies whether the account is an account with extended capabilities, paid	Last.fm
bio	Text string	User's "biography"	Facebook
birthday	Text string in MM/DD/YYYY format (American date format)	Date of birth of the user	Facebook
age	Integer	User's age in years	Last.fm
education	Array of objects containing fields: year, type, school (object with fields: name, id, type and optional: year, degree, specialization tables, class tables, plaques of classmates)	The history of education together with a list of friends from schools and universities	Facebook
email	Text string (RFC 822)	E-mail address	Facebook
hometown	Object containing name and identifier	City of user's origin	Facebook
Interested_in	Array of text strings	Genders that the user is "interested in"	Facebook
location	Facebook: object with name and identifier Twitter: text string	The city of current residence, current geographical location	Facebook
country	Text string (ISO 3166-1 alpha-2)	The code of the country the user comes from	Last.fm
political	Text string	The political views of the user (usually one word)	Facebook

(continued)

Table 1 (continued)

Field identifier	Data type	Description	API
favorite_teams	An array of objects consisting of a name and identifier	List of favorite sports teams	Facebook
quotes	Text string	Favorite quotes	Facebook
relationship_status	Text string	The status of the relationship (with the other person) of the user. He accepts one from values: Single, in a relationship, Engaged, Married, It's complicated, In an open relationship, Widowed, Separated, Divorced, In a civil union, In a domestic partnership	Facebook
religion	Text string	Religious views (short, arbitrary text)	Facebook
significant_other	Object containing the name and identifier	The person with whom the user remains in the connection status (one or the other according to the relationship_status)	Facebook
video_up-load_limits	Object containing restrictions on the length of the video size	The field specifies the maximum duration and maximum dimensions of the video file that can be placed by this user—for sending applications video on his behalf	Facebook
website—Facebook url—Twitter	Text string (URL)	The address of a personal (any) website	Facebook Twiter
work	Array of objects containing fields: employer, location, position, start and end time	Employment history provided by the user	Facebook

(continued)

Table 1 (continued)

Field identifier	Data type	Description	API
playcount	Integer	The number of music plays recorded by Last.fm	Last.fm
playlists	Integer	The number of playlists created by user with songs music	Last.fm
statuses_count	Integer	The number of public statuses (messages) published from account	Twitter
geo_enabled	Logical value	Specifies whether the user uses geolocation services and indicates his status as the current location	Twitter
followers_count	Integer	The number of other users of site who subscribe to the statuses they publish	Twitter
friends_count	Integer	The number of users whose statuses are subscribed to by the user whom field concerns (i.e., he added to "friends")	Twitter
protected	Logical value	Specifies that statuses published by a given user are not visible to everyone else, but only to those who he subscribes to	Twitter

[a]E-mail address. City of user's origin, array of text-strings, which the user is "natively concerned". Does not apply to changes: languages, link, timezone, interested_in, video_upload_limits and "obsolete" fields: favorite_athletes and favorite_teams
[b]Facebook: the user is considered verified when he confirms his phone number
[c]Twitter: the user is verified when the Twitter service will contact him personally and will confirm his identity—this process is used only for well-known personalities, such as actors or politicians

on the Web and probably the smaller the chances that he will be able to "see" the message from the next, new user.

- **The amount of content generated**—Twitter offers us statuses count with the number of messages of a given user, last.fm—playcount with the number of listened (scrobbled[2]) tracks. This information, especially in comparison with the time that has elapsed since the registration of a given user in the service, will allow to assess whether it is a person in each service that is inactive, not very active, very active or extremely harmful, i.e. spamming.
  Quoting from the vbeta.pl website [16]:

The research company Websence Security Labs has published a report on the amount of spam on social networking sites and other Web 2.0 sites. She found that 95% of content created by users in interactive websites is spam. Under this word, Websence understands all unnecessary messages, advertisements, malicious links or applications prepared by hackers. He claims that spam in the community is already more in e-mails (currently 85% of e-mails are classified as unwanted).

- **Name and surname**—inconspicuous fields but representing the data according to which we identify people daily. Therefore, also from the point of view of automatic analysis of user profiles, they can be used as a solid suggestion providing the basis for linking accounts from different websites. Of course, these data do not have to be perfect.

The object representing the user is not the only one in these websites, but rather serves as the starting point for the data of the other types, usually constituting most of the content on the website. In the case of Facebook, it will be, for example, pictures, well-liked websites that represent the worst approximation of the word "everything." On Twitter and Blip, these will be characteristic short messages published by Internet users, in the news about the listened and liked tracks.

In defining these connections between the user and his interests, he collapses, just like it was shown in Table 1, Facebook. In addition, they are often represented in a manner characteristic of a given type of service. Therefore, further links between the object of the user and others are separated into separate tables, whereas in Table 2 only the Facebook service was used, in Table 3 from Twitter.

In order to simplify the recording and increase readability, the structure of these fields is written in the notation defined by the present operation, modeled according to the JSON (JavaScript Object Notation) concept. The content of the object is preceded by its type (or the word Object, if the type is not referred to by the document), and the names of their types are used in the place of specific values. Example:

---

[2]"Scrobbling is a word invented by the creators of the Last.fm site (probably by Richard Jones, founder of Audioscrobbler), meaning creation "creating a history of what we are listening to" [15].

Table 2 Example objects associated with the user object in the Facebook service API and their form [12, 13]

Field identifier	Data type	Form (structure)
| accounts | Accounts, except for the basic one on behalf of which the user can publish accounts information, i.e.: Facebook[a] applications and websites managed by this user | ```
Object {
 data: [
  Object {
   id: string,
   category: string,
   name: string
  }, · · · ]
}
``` |
| activities | "Activities" listed on the user's profile page. These are usually pages (formerly: groups) of Facebook representing any content, e.g. Towel Day or Janusz A. Zajdel Award | ```
Object {
 data: [
  Object {
   id: string,
   category: string,
   create_time: timestamp
  }, · · · ]
}
``` |
| albums | Virtual albums with photos, created by the user | ```
Object {
 data: [
  Object, · · ·
 ] }
``` |
| apprequests | A list of "requests" (pending notifications/requests) of a single application to be used in the context of the application | ```
Object {
 data: [
  Object, · · ·
 ] }
``` |
| books | Books listed as liked/recommended in the user's profile. Field category it does not describe the genre of the book, but always includes the value of "Book"—this is due to the uniform treatment of objects by Facebook and the fact that the category of this object is simply a "book". The list of all liked users is in the likes field | ```
Object {
 data: [
  Object {
   id: string,
   name: string,
   category: string,
   create_time: timestamp
  }, · · · ]
}
``` |

(continued)
```

Table 2 (continued)

Field identifier	Data type	Form (structure)

checkins	Places where your presence has been signaled by the user	```
		Object {
		  data: [
		   Checkin, · · ·
		  ]
		}
		``

events	Events that the user chooses (or considers choosing)	```
		Object {
		  data: [
		   Object {
		    id: string,
		    name: string,
		    start_time: timestamp,
		    end_time: timestamp,
		    location: string,
		    rsvp_status: string
		   }, · · ·]
		}
		``

family	Family connections (if the family is on Facebook)	```
		Object {
		  data: [
		   Object {
		    id: string,
		    name: string
		    relationship: string
		   }, · · ·
		  ]
		}
		``

feed	User's "board". It contains messages published by these people (companies), links to video files, etc., along with information on who of your friends (fans) liked them and commented (and how)	```
		Object {
		  data: [
		   Post, · · ·
		  ]
		}
		``
		(up to 25 posts)

friendlists	Lists of friends (groups) created by you	```
		Object {
		  data: [
		   Object {
		    id: string,
		    name: string
		   }, · · ·
		}
		``

(continued)

Table 2 (continued)

Field identifier	Data type	Form (structure)
friends	People added by the user to friends	```Object { data: [Object { id: string, name: string }, · · · } ```
games	Games that the user has added to the section "Art and Entertainment" of his profile	```Object { data: [Object { id: string, name: string, category: string, create_time: timestamp }, · · ·] } ```
groups	Groups to which the user belongs	```Object { data: [Object { version: integer, id: string, name: string, administrator: boolean, bookmark_order: integer }, · · ·] } ```
home	The stream of notifications presented to the user on the Facebook home page—primarily the activity of his friends	```Object { data: [Post, · · ·] } ``` (up to 25 posts)

(continued)

Table 2 (continued)

Field identifier	Data type	Form (structure)

inbox	The messages in the message inbox	```Object {
		  data: [
		    Thread, · · ·
		  ]
		}```

interests	Interests visible in the user's profile. Under no circumstances should it be confuse this field with the interested in field, which means the sexual orientation of the user	```Object {
		  data: [
		    Object {
		      id: string,
		      name: string,
		      category: string,
		      create_time: timestamp
		    }, · · ·]
		}```

likes	All pages (objects) that the user liked. We will also find books here listed in the books field. This is where the Category field has a key role, since all types of objects are thrown into the colloquial one sack	```Object {
		  data: [
		    Object {
		      id: string,
		      name: string,
		      category: string,
		      create_time: timestamp
		    }, · · ·]
		}```

links	Hyperlink published by the user—those published as part of his table	```Object {
		  data: [
		    Link, · · ·
		  ]
		}```

movies	Videos listed in the user's profile as his/her favorite. Same as in the case of books, the category field does not specify the genre of the movie, but always includes the value of "Movie", and the objects mentioned in movies are also found in likes	```Object {
		  data: [
		    Object {
		      id: string,
		      name: string,
		      category: string,
		      create_time: timestamp
		    }, · · ·]
		}```

Table 2 (continued)

Field identifier	Data type	Form (structure)

music	Music listed in the user's profile as his favorite/featured. Similar rules like books and movies	`Object {`
		`  data: [`
		`    Object {`
		`      id: string,`
		`      name: string,`
		`      category: string,`
		`      create_time: timestamp`
		`    }, · · · ]`
		`}`

notes	User notes. Notes are public and have a form like a blog entry	`Object {`
		`  data: [`
		`    Note, · · ·`
		`  ]`
		`}`

outbox	The inbox of the outbox	`Object {`
		`  data: [`
		`    Thread, · · ·`
		`  ]`
		`}`

payments	List of transactions with the application that the user has made. This is one of the fields visible only in the context of the application	`Object {`
		`  data: [`
		`    Object {`
		`      id: string,`
		`      from: integer,`
		`      to: integer,`
		`      status: string,`
		`      application: integer,`
		`      created_time:`
		`timestamp,`
		`      updated_time:`
		`timestamp`
		`    }, · · · ]`
		`}`

(continued)

Table 2 (continued)

Field identifier	Data type	Form (structure)

permissions	A set of permissions granted by the user to our application	```
		Object {
		 data: [
		  Object {
		   type: "permissions",
		   user_about_me:
		boolean,
		    user_likes: boolean,
		.
		.
		.
		  }]
		}
		``
		(the table contains only one object and its keys correspond to the next permissions—those whose value equals 0 (no access) are by default skipped)

photos	Photographs on which the presence of the person has been marked	```
		Object {
		 data: [
		  Photo, · · ·
		]
		}
		``

picture	Current user profile photos	```
		Object {
		 data: [
		  Photo, · · ·
		]
		}
		``

pokes	Ticks (user's "prank")	```
		Object {
		 data: [
		  Object {
		   to: User,
		   from: User,
		   created_time:
		timestamp,
		   type: string
		  }, · · ·]
		}
		``

posts	Posts posted by a given user speaker	```
		Object {
		 data: [
		  Post, · · ·
		]
		}
		``

(continued)

Table 2 (continued)

Field identifier	Data type	Form (structure)

statuses	User status updates. Contains content like posts, however limited to text messages published by a given person (it does not contain, for example, links)	<pre>Object {
		 data: [
		  Status_message, ···
		]
		}</pre>

tagged	Posts in which a given user was mentioned, using the Facebook mechanism of "tagging" people	<pre>Object {
		 data: [
		  Object {
		   id: string,
		   from: User,
		   to: [User, cdots],
		   picture: string,
		   link: string
		   name: string,
		   caption: string,
		   description: string,
		   icon: string,
		   actions: [
		    Object {
		     name: string,
		     link: string
		    }, · · ·
		   ],
		   type: string,
		   application: string,
		   created_time:
		timestamp,
		   updated_time:
		timestamp
		  }, · · ·]
		}</pre>

television	Programs and TV series exchanged in the user profile as his/her favorite/recommended. Similar rules as for books and movies	<pre>Object {
		 data: [
		  Object {
		   id: string,
		   name: string,
		   category: string,
		   create_time: timestamp
		  }, · · ·]
		}</pre>

(continued)

Table 2 (continued)

Field identifier	Data type	Form (structure)
updates	Changes (updates) of the user's inbox. These are de facto messages that were sent by companies and organizations, that is Facebook pages, not actual private users. Sometimes there are messages from users outside your friends	```Object { data: [Object, · · ·] } ``` (the specification only speaks of "message board")
videos	Video files in which the user has been marked ("tagged")	```Object { data: [Video, · · ·] } ```

[a]Facebook allows developers to build the so-called applications that are really all web applications embedded in the Facebook page and operating in its context, communicating with Facebook via the Facebook API

```
node {
        id: integer,
        name: string,
        children: [ node,      ]
}
```

It means a single object consisting of two fields: an integer with id and a text string with the name id, as well as any number of node elements available in the children field. Formal and the definition of the record used here can be found in the Appendix A.

2.3 The Structure of a Single Status on Twitter

Facebook and Twitter services speak by far the most users among those discussed in this work, which is perfectly visible in Tables 2 and 3, as well as in the description of the status object from Twitter, due to its volume limited to the most important elements in this sample of code:

Table 3 Selected user objects in the Twitter API and their character [18]

Field identifier	Data type	Form (structure)
statuses /home_timeline	Returns the statuses seen in the "home" view of the authenticating user, i.e. up to 20 reports published by him or by people he observes	An array of structures representing the status, shown in sample of code in Sect. 2.3
statuses /mentions	Returns up to 20 statuses in which the authenticating user has been mentioned by (@login)	
statuses /public_timeline	It contains the last 20 statuses of all unprotected[a] users	
statuses /retweeted_by_me	The last 20 user authentication messages that are retweets[b]	
statuses /retweets_of_me	Last 20 user authentication messages that have been repeated (retweeted) by others	
statuses /user_timeline	The last 20 messages published by any user—by default, this is the account through which the user authenticates, but We can indicate any other user (unless it is a protected account that did not allow us to access)	
statuses /retweeted_by_user	Equivalent retweeted_by_me with the possibility of indicating any account, with reservations such as for the user_timeline	
statuses /retweeted_to_user	Equivalent retweeted_to_me with the possibility of indicating any account, with reservations such as for the user_timeline	

[a]Users have the option of setting their account as protected, then only those whom they add to friends will have access to their messages
[b]Twitter specific implementation of quoting

```
Object {
  retweet_count: integer,
  possibly_sensitive: boolean,
  geo: object,
  coordinates: object,
  in_reply_to_status_id: integer,
  id_str: string,
  text: string,
  favorited: boolean,
  in_reply_to_status_id_str: string,
  created_at: timestamp,
  in_reply_to_screen_name: string,
  in_reply_to_user_id: integer,
  in_reply_to_user_id_str: string,
  entities: Object {
    user_mentions: [ Object, · · · ],
    hashtags: [ Object, · · · ],
    urls: [ Object, · · · ]
  },
  contributors: Object,
  retweeted: boolean,
  source: string,
  place: Object,
  id: integer,
  user: Object,
  truncated: boolean
}
```

The presence and availability of this information obviously depends on the user's accessibility and the access rights that a given application has for its data. This topic is discussed in more detail in Sect. 2.4. However, the amount of potentially available information turns out to be impressive. Especially Twitter turns out to be able to store and share much more information than it can be seen at first glance through its website. These are not only basic information and the content themselves (as in the Polish Blip website, where besides the message only the time of its publication is available, an indication of the authors—of which we know little (Table 1) and the channel used for communication with the service), and rich information collection about their context and circumstances of sending.

Therefore, please pay attention to what data is available. First, in the similarity of the structures represented by the various Facebook sites, one can notice one of the guiding ideas by which its API was designed (Fig. 6).

Secondly, the information spectrum is wide. We have informative categorizations, such as well-liked books, films, music.[3] It is available, though much less often, about

---

[3]It should be approached from a distance—pages on Facebook are created by users, they are not moderated, and in the category "book" equally will be the actual book, its author, or the page

Fig. 6 Last.fm service cares about the correctness of the information presented about the listened and lost tracks, automatically correcting the tags

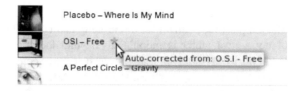

the geographical positions of users, both from the Facebook check-ins mechanism and from the Twitter geo-location data attached to the status. An interesting object is also the network of knowledge and dependence between friends. Finally, text messages are also available, which users of social services "produce" a lot, and there are also dependencies, such as "respond to" relations, as well as the diversification of popularity indicators (like likes and retweets).

Let's also pay attention to the rich information about the interactions between users: they are mentioned likes ("likes" by someone) and retweets ("letting go of status" on Twitter), and mentioning other users: Twitter, Facebook and even Polish Blip, let us clearly indicate in the statement that we are talking about some other user (and create a hyperlink to his profile), informing him about this statement. The number of such references and citations allows us to collect information on the popularity and activity (interactivity) of individual users. Often, people using this (but above all, of course, the number of followers/friends) also publish popular popularity rankings in these types of communities.

Knowing in general how a wide range of data we can potentially have. We still need to divide this category into categories according to the possibilities of using this data—about such categorization is provided in Sect. 2.5. However, before we can think about specific ways to use the collected data (Sect. 3), we must determine whether we have access to them at all and what must happen to obtain it (Sect. 2.4).

2.4 Model of Data Access Authorization

Social websites store huge amounts of information about their users. These are both data allowing their unambiguous identification, and often personal information, which they would not like to share with any stranger. It follows the need to control access to this information so that it is possible to determine which of them is not going to get into the wrong hands. Also, the introduction of access control from the application side allows for a much better protection of the website against activities aimed at helping users or correct functioning of the service itself.

For these reasons, most social networking sites decide to implement certain authorization schemes and limit access to data and methods available. This subsection presents the characteristics of the most common solutions to this problem.

---

entitled "Ladies read." Last.fm considerably cares more about the correctness and form of data in user statistics (Fig. 7 shows an example).

No Authorization
The simplest mechanism for authorizing and controlling access to data is the lack of any control, which is undoubtedly beyond doubt. It should be noted here that this is not an inadvisable approach, even if it may seem so at first glance. Of course, this means very simple (with almost zero effort) to access user data. Therefore, they cannot be sensitive or "sensitive." On the other hand, access is very easy and if the site is focused on openness and publicity, nothing prevents us from using it.

Thus, Twitter does not require client applications to download a user or his statements, unless it is the user who has marked his account as protected. Here the mechanism of protection against the "malicious" application overloading the server and/or trying to retrieve much more data than it needs or "falls out" works. References to API methods of the service are limited in a time unit: up to 350 requests per hour if they are authenticated or up to 150 per hour if not [17].

However, if we decide to authenticate, in exchange for a larger limit of API requests, Twitter gains a bit better control over our access to its data, using the OAuth mechanism.

The Facebook service will give the following user data without any authorization: id, first_name, middle_name, last_name, link, username, gender, locale, name and nickname, login and associated profile address, gender and interface language.

Disadvantages:

- no control over what or who reads users' data,
- no possibility to control the unintended use of user data,
- indicated quantitative access limit in order to protect against excessive server load, especially due to a malfunctioning application,
- in some solutions, granting access consists in giving our login and password to the application, in the belief that the data will not be used in an inappropriate manner.

Advantages:

- simplicity of server implementation,
- simplicity in the implementation of client services,
- lower implementation costs of solutions based on API data.

Last.fm (Old Model)—Validation of the Application Key
Until recently, Last.fm used a very simple verification methodology, which consists only in confirming the access of the whole file (globally) to the user data (all). In other words, once obtaining an API key (a unique string of characters uniquely identifying the application) allowed to use all API methods freely to the hypothetical moment when administrators of Last.fm recognize that the application has exceeded its authority and will receive its access to the services of this site.

This method is just as easy to implement (on the client side) as the lack of self-analysis. In queries, it is enough to additionally send the application key on behalf of which they are called. However, this creates a significant risk of interception of the key—if someone obtains the application identifier, without the knowledge and consent of the owner, invoking API Last.fm methods in his/her application in an inappropriate or disruptive manner, which will block the administrator's key.

Calls to a given key are not particularly tracked or restricted. Of course, the administration can inspect such information, but there is no standard mechanism that would block access using the same key from a new machine until its owner authorizes it. It would not make sense, because it would be contrary to the idea of calling queries directly from desktop applications that are installed on many different users' computers, however, the side effect is the ease with which a key that does not belong to the querying user can be presented in queries. We can test the Last.fm API calls without the slightest problem by pasting the key from the examples in the document, belonging to Last.fm.

In the case of abuse, the application key will be blocked, and the creator will have to apply for a new one and try to prevent malicious abuse on his behalf more effectively in the future.

Disadvantages:

- one-step authorization with a single point of failure (key theft),
- no possibility to deny access from a specific user,
- abuse of the stolen application key is reflected not only on the malicious third party, but also on the creator of the original application.

Advantages:

- simplicity in implementation—low costs,
- the possibility of eliminating access from applications that do not comply with the rules of website use.

OAuth—Last.fm, Facebook, Twitter…
The most popular authentication system currently is OAuth 2.0, the standard published at the end of 2010 [17]. OAuth provides exchanged information on who (application) he asks for access to whom (what user) resources[4] from the specified supplier (owner of the assets). It is therefore a trilateral opera, followed by an agreement between the user, the customer and the resource provider.

(The flow of information between these three sides is illustrated in Fig. 7.)

A: In the first step, the web application sends to the user's browser (and the server of the service we want to use) the client identifier (application) and the URI address to which the user should be redirected after successful authorization.

B: The user authenticates himself to the service that stores his data, which the application wants to access and allows the application for this access—only here the user can enter the login and password for the service. This is important for the user, because he logs into his familiar interface, securely domain and is sure that his credentials will not be stolen by the application.

C: The server responds with an authorization code that is passed back to the application asking for access. This code is not enough to get access, because it could be intercepted on the user's side and it would not give more security than the method validation of the application key.

---

[4]For example, we will gain access to the status published by "protected" users with the user's friends, which we authenticate.

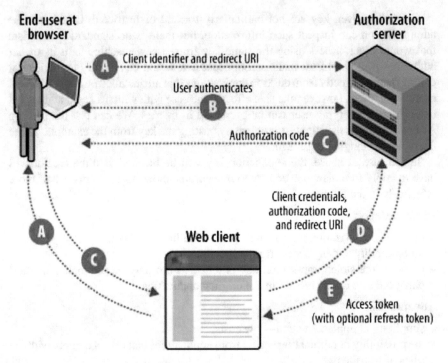

Fig. 7 Information flow in the OAuth 2.0 process from the server point of view; the steps are carried out in the order indicated by subsequent letters; the flow initiates Web client (web application) [17]

D: The obtained authorization code is sent in a direct link between our client application and the server along with the data necessary to authenticate the application itself (the creator of the application usually receives the so-called secret key, acting as its password).

E: Once the application's credentials have been verified, the server issues the final data access code (the so-called token), which can be used in further links between the application and the server (Fig. 8).

This mechanism ensures authentication of both the application and the user, as well as verifying the access rights of the application to its data. What is more, it requires the application to define which specific rights will be required by it, thanks to which the user is presented with precise information on what will be made available (Fig. 7 shows example).

This increases the user's confidence in the given platform. He knows what to expect and how his data is used. It has also been noticed—which seems quite logical—that the more law enforcement exchanges the application in its task, the less users agree to it.[5]

---

[5]"There is a strong inverse correlation between the number of beneficiaries our app is applauding and the number of users who provide it [...], therefore we recommend that our application only expects those rights that it absolutely needs" [14].

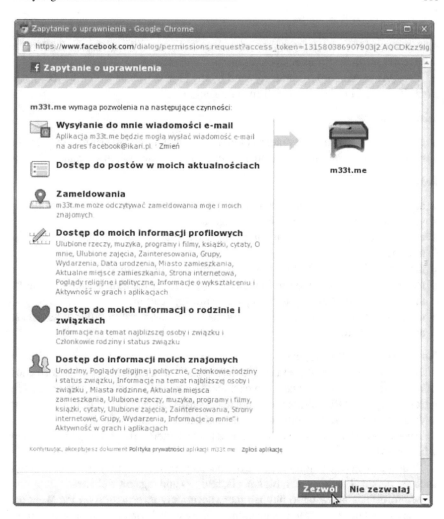

Fig. 8 Screen presented to the Facebook user access rights to his data

Fig. 9 An example graph of knowledge between users, on which we can operate the postgraph library

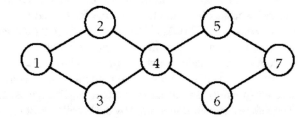

This model uses more and more services, although it is very "fresh" (Working specification [19] comes from the end of July 2011, describing version 2.20 of the standard) OAuth is still intensively developed and refined.

The OAuth standard is currently used by the largest websites on the Web: Facebook, Twitter, Google, Last.fm, Blip and many more…

Disadvantages:

- greater process complexity means higher implementation costs (however, it can be reduced by popularizing the standard causing the creation of appropriate libraries).

Advantages:

- authentication against the content provider of both the user and the application,
- verification of the list of required permissions,
- control over the authorization lying in the user's hands (and not only the administration),
- the user's ability to withdraw from the application of permissions,
- resistance to attacks "man in the middle" (no communication channel is sent all the information necessary to impersonate the application),
- the application does not need to know the user's password in order to be able to act on its behalf.

No Interface for Developers

Although the total lack of API is difficult to classify as a model for the authorization of access to data, it is not wisely to exclude this type of situation in this work. The lack of a typical API interface does not necessarily indicate the lack of information embedded and marked specifically to be read and used automatically.

An example is the so-called microformats, which their official wiki page describes with a pert [20]:

"Microformats are patterns of HTML code, representing widely published things, such as: personal data, events, blog posts, reviews and tags on websites.

Microformats allow us to publish more accurately information on the Web; are the easiest and fastest way to ensure compatibility with devices and mobile programs for information on the website."

With their help, we can download data of various kinds embedded inside "ordinary" websites in a simple way. True for this purpose are robots that use regular expressions or XHTML/HTML parser (provided the correct structure of the site code, it allows for greater flexibility and achieve better results).

Popular information of this type is widely used attribute rel = "nofollow" in carriers on websites, meaning that it should not be considered when analyzing the weights (quality) of websites by Internet search engines. The XFN (XHTML Friends Network) information supported by the WordPress blog platform is also quite widely used [21]—it enriches the hyperlink with information about who the author of the page he is currently browsing to the author of the page he has been contacting (business contact, friend, colleague, nobody), whether we met him personally, whether he is a member of our family or a colleague.

Such techniques make it possible to enrich additional content with minimum workload, and thanks to their standardization, they can be easily read and processed fully automatically. Mining the Social Web [17] gives examples of applications: planning trips thanks to embedded pages on geo microformats, search for recipes (hRecipe), but also—which is of importance to the subject of this work—building information about the network of friends based on XFN microformats.

Facebook also applies microformats—in February 2011, hCalendar support was added and hCard to the information published about it about millions of events [22].

Disadvantages:

- the need to manually extract information from websites,
- only some of them concern the users themselves,
- use field limited to content on websites and language (X) HTML,
- it is not a service as such,
- only public information (no access control).

Advantages:

- low cost of use in already existing websites,
- simplicity of access (microformats do not constitute a new language).

2.5 Types of Available Data

The data collected for the previously discussed methods can be divided due to their form for several groups that will differ in their possible use. The information that can be obtained is varied in terms of precision, unambiguity and—of course—the thematic scope.

For example, once we will have information about the location of the user expressed in precise geographical coordinates, other times it will only be a field with the name of the city in which he is currently staying. Once it is the interest of a given person in a musical performer (which is interesting to us), the other is the "love" of some unidentified side perspective, representing the idea of getting up late or mixing the tea counter-clockwise (which is a bit lower value).

Specific and Unambiguous Data
An example of precise and specific information that is available to us. It will be the date given by Facebook. It is known what this information is, because it is always assigned in one of the birthday or birthday_date[6] fields and after interpreting it

---

[6]The mentioned difference of attribute names shows that despite the efforts of Facebook API creators, the data format is not always completely consistent. This may give the parent some unexpected difficulties when implementing the application. During the implementation of the sample application it turned out; because if the date of birth in a fixed format according to the specification of the User object is in the birthday field, if we use the FQL data access method (p. 66), the field with the same name contains date in a somewhat arbitrary format ("depending on from the language

according to its format, we have unambiguous information (at least within one year) and precise information.

Reliable and "precise" should also be information such as the name and the name of the user, but on the other hand, social websites "do not like" what users like pseudonyms. Popular nicknames users rarely have a field in the profile (not counting login fields), so they are typed by internet users as part of their name. If, therefore, we want to identify users by name and surname, next to the ambiguity, the landscape of popular names (Jan Kowalski, Anna Nowak) or changes related to the change of marital status (change of name or insertion of the member, maiden names), there is also the problem of different ways of giving the same surnames by the user himself. It is worth mentioning here that the newest social website at the time of writing this work, Google Plus, has a very restrictive policy regarding the names given by users and there are cases of account deletion that contained other or fictitious data. This is met with harsh criticism from those Internet users who, even in the circle of close friends, are much better identified by pseudonyms than surnames, so adapting to the "proposed" policy by Google may mean for them the renunciation of their own identity [23].

Unambiguous and precise data are usually (if available) geolocation information, like placing the user at the moment of publishing the given entry. An exemplary status from Twitter [24] includes the term "geo": {"type": "Point", "coordinates": [37.61277829, −122.38639914]}, as well as the following moment: "created_at": "Mon Feb 26 18:05:55 +0000 2007". This gives information about the user's location at different times of time a bit less complete and continuous than a continuous stream of GPS data. However, with very active users willing to publish their items, we can get a detailed history of their location or difference. Similarly, the service offered by Google Latitude [25], although it has emerged from the testing phase, is described as potentially drastically changing at any time.

In the case of data from the GPS system, however, it is worth being prepared for a lot of mistakes: it may be a problem related only to the Google Latitude service, but in the profile of the author of the work, we can see that his current position is "Shanghai, People's Republic of China, "which is certainly not and has not been true at any point in his life.

The smaller the range of values for the variable, and the smaller freedom of filling it with the user, the more certainty can be treated as its value. According to this logic, the most reliable information carries values such as gender or the status of the account verification. The best verification, however, brings practical observations: Facebook administrators, for example, care about the correct list of geographic locations, so although the user enters the name of the city where he lives as a text, thanks to the hints mechanisms he is guided to the correct and verified location 'e free from errors. The situation of tags for music files (song titles, artists' names …) on Last.fm is analogous.

---

setting"), and the field with the expected date is called: birthday date. For the better, the format is different from the previously specified and variable, because it is DD/MM/YYYY when the year of birth is known, when the user does not give, the date is presented in the MM/DD format.

For precise data in the above-mentioned meaning, the fields mentioned in sub-chapter 1 can be considered, regardless of the website they come from, with slightly less-known reasons being given names and surnames, followed by fields such as registered and created at (automatically generated) by the service) and other time-stamp type fields, as well as fields that store logical values (boolean) (verified, subscriber), e-mail (because its verification is a required step when registering in most websites) and all counters (playcount, statuses_count).

Fields such as hometown and location present the previously mentioned problem of correctness and authenticity of the entered data. We can find there both Łódź, which can be translated into a geographical position on the planet Earth, and Hollywood, which is a real location on Earth, but not necessarily for the person it concerns. Finally, we may find a user who decided to put in this field Neverland (or much less value), with which we cannot do anything and not much help.

The significant_other field usually represents a real value representing a person with whom a given user is in any form closer to each other, the values written "for jokes" are, fortunately, a marginal phenomenon.

The other values from Table 1 are usually those described below.

2.6 Groups of Friends, Interpersonal Relationships

Connections between people, such as watching content published by other users, approved knowledge ("friendship") or sharing content published by others (Twitter retweet) are the basic distinguishing feature of a social service.

There are information available in every site (on Facebook we have a friendship, on Twitter and Blip followers) and followed (follow), Last.fm again friends, and in Google Plus—a circle of people) and each of them it easily makes them available in its API (if it exists at all, Google Plus, as of August 20, 2011 still did not have an open interface for programmers, although the appearance of "social" games (we see the results of friends, we can challenge them on duel, etc.) suggests that such an interface already exists somewhere and is tested on the first applications).

Going further, not only lists of friends on websites can interest us. The application here is also microformats, described before. Thanks to them, we can construct a graph of knowledge among Internet users based on XFN tags embedded on websites and typical blogs.

In the case of this type of information, we usually do not have to deal with the problem of heterogeneity—if the information is available, then it is also a precise indication of a specific person, site, account.

Text Information, with Context
Using the data available in social services, especially on Facebook, we have access to a huge number of links between people and "liked" by them.

These are seemingly anonymous objects, because only their ID is known from the machine's point of view, it is known that they have some text name and assigned category. This category is what carries the most information, because it informs about what the object is, embedding it in some context—it becomes a book, a movie, a public figure or a TV series. According to such information, the categories are also placed in the appropriate user resources: music, books, tv, etc. (and others described in the table in Sect. 2.2).

They are therefore data that allow us to create information about the user and his interests. These, in turn, allow finding people with a similar genius to suggest knowledge or allow us to find out (with the help of additional sources or generally understood expert knowledge, that is, credibly introduced knowledge allowing further inference), for example, what genres of music and film genres the general public likes.

Text Information, No Context
In addition to data embedded in certain clearly defined groups, there are also pages (here understood as pages in the context of Facebook) placed in terms of ambiguous terms. Nevertheless, they are available as a kind of user's interest and a positive relationship between him and the content, but he is not the author.

These will be pages like "I'm eternally overwhelmed because at night I'm a superhero" in quite random categories (in FQL referred to as page type), as illustrated by the example page mentioned above—answers a on request FQL select name, type FROM page WHERE page_id = 164290160303599 is:

```
<page>
    <name>I Am Always Tired Because I Become a Superhero at
    Night</name>
    <type>PUBLIC FIGURE</type>
</page>
```

Thus, it is erroneous to inform that the page is about a public figure. Accordingly, a "bent" interpretation could be explained by the fact that comics superheroes are usually publicly known characters, but this is not an explanation for the fact that for programmers, the application introduces only noise to the input data.

Information Completely Free of Context
This group includes, above all, the so-called user statuses, i.e. short text messages published by them in the so-called stream. Seemingly they seem to be useless, but we can use them in two ways. The first one is collecting data about the time of publishing these messages—on this basis new, useful information is obtained (e.g. about the user's activity). The second one, related to such published content more directly, is the analysis of their content in terms of statistics and language.

These issues are described in more detail in the chapter on data interpretation, but here it is worth noting their usefulness, even though it is difficult to embed such content published by users in any unambiguous context, because they are usually downloaded as part of the user's message stream. The name itself indicates—it contains a rich and constantly growing collection of contents different in terms of type.

The usefulness of these user-published statuses is not diminishing due to the limited number of characters imposed by micro-blog platforms (Twitter—140 characters, Blip—160 characters). Sometimes such limitations make even greater care for the brevity of the message and sticking to the main subject, although obviously the abbreviations used by Internet users to fit in the character limit can make it difficult to read the message—not just the machine.

Tags (tags) Entered by Users
A great simplification in the issues discussed in previous section is usually created among micro bloggers to use the so-called hashtags. They are marked by "#" markers, binding statements with some topic or categorizing them and grouping them as statements of a certain kind.

The following tags, for example, are popular on the Blip website (a list of popular tags on the website on 21.08.2011): #blog #drogiblipie #fb #gastrophase #iphone #kot #krakow #lastfm #lol #ludziektorzy #minecraft #mnvr # models #off #oldsql #polonia #rower #slucham #ttdkn #read. The meaning of some of them is obvious often already at the first contact (#blog, #iphone, #fb to facebook, #minecraft is the name of the popular game), some require some knowledge of the idea behind them: #drogiblipie is a tag intended for questions and requests for help (and not for announcements and advertising), for example #drogiblipie where I can buy headphones beats audio? [28]; # people are used to criticize their actions and others' or humorous presentation of them, for example #people who come to the bar at night in pyjamas. Sic! [29].

For automatic processing of messages and attempts to segregate them thematically, the tags *#rower*, *#iphone*, *#models* or *#kot* will be of value—for their actions, the string of characters becomes an opinion, a question or an expression of interest in each topic.

This chapter shows what data are usually available for social services and what security schemes it uses to control access to these data, indicating their advantages and disadvantages. What is particularly important, however, is the presented division of information on account of their basic nature, i.e. their meaning and knowledge of their context. They decide how the collected information will be used. Knowing the meaning of collected information, we can make their interpretation and carry out their analysis. The methods of such analysis and analysis are presented in Sect. 3.

3 Interpretation of Data

When we have a subset of the data mentioned in previous sections, we can start using them. The way in which this can be done depends primarily on the type of information that has been collected. This chapter describes the techniques that can be used, depending on the information contained in the data obtained (for example, certain specific numerical values, information about human knowledge, data on their preferences and interests or the textual statuses published by them) and the way in which the data will be used and the form of possible applications to be obtained.

3.1 Compare of Values

The simplest, yet very interesting for users, the use of data that can be collected is simply to suggest knowledge based on similar values of certain indicators.

Example 1: Let's assume that in the geo table in the database, there are geographical positions of users with coordinates saved in the lat and long fields (respectively width and longitude). We would like to introduce to the user simple suggestions of new friends who are in a similar age and similar geographical location, informing about the distance from a given person at the same time.

For the very determination of points (people) located at the nearest available location (selected user), we can of course calculate the Euclidean distance treat them as co-ordinates on the plane, with special regard to co-ordinates, find in the plane with special regard to the coordinates located near the 180° meridian. For small distances in which it is worth finding proposes, such approximations would not significantly impede the sorting results after the distances.

A much better effect, however, will be obtained using the spherical law of cosines, which for contemporary compilers and interpreters should give satisfactory accuracy at distances of up to 1 m [26], then:

$$d = arccos(\sin(lat_1) \cdot \sin(lat_2) + \cos(lat_1)\cos(lat_2) \cdot \cos(long_2 - long_1)) \cdot R \tag{1}$$

where $R = 6371$ (km), and d is the result (distance) in kilometers.

Example 2: Using the coordinates in the form given in Example 1, we can find the nearest big city near the current location of the user, and then, for example, search for others nearby or simply give this information; in his profile (for example, Now I am visiting: Poznan).

The publicly available Geonames database is perfectly suited for this purpose [30]. If we will imported it into a database, we can build with the following method solution that retrieves information about the nearest cities.

```
public static function nearestCities ($lat, $long,
$limit = 3) {
   $lolat = $lat - 2;
   $hilat = $lat + 2;
   $lolo = $long - 2;
   $hilo = $lońg + 2;
      $cities = self ::
      fetchAll (' SELECT name,
      asciiname , latitude,
      longitude, fcode, country
      FROM geoname WHERE
      fclass=\'P\' AND
      population>0 AND latitude
      BETWEEN $4 AND $5
      AND longitude BETWEEN $6 AND
      $7
      ORDER BY ((latitude-$1) ^ 2 + (longitude-$2) ^
      2) ASC LIMIT $3',
array ($lat, $lońg, $limit , $lolat , $hilat ,
$lolo , $hilo));
return $cities;
}
```

A method code in PHP (and SQL) that retrieves the locations nearest to the given ones.

It is worth noting that this code uses some simplifications to help drastically speed up database searches and sorting results. The trick consists in narrowing the area only to points, whose coordinates do not differ by more than two degrees. This solution and the assumption of indexes in the database table on the fields latitude and longitude allows us to quickly find the answer. Without this limitation, the time would be several hundred times higher due to the need to calculate the square of the Cartesian distance (according to which the results are sorted) for each point in the database, and these are millions.

If an application is created to suggest to the user of potential partners, it is sufficient for selected people from around the same city (or according to the distance, as in Example 1) to apply filtering by age and opposite sex (dates of birth are, as previously discussed, usually easily accessible, instead of the opposite sex criterion, specifically apply the sexes, which the user lists as interested in, this form is, however, much less frequent when applying for private settings).

3.2   Graph of Knowledge

Building a database of dependencies between users, and in the case of Facebook, also between users and websites, give us a very powerful tool.

Six Hand Words, the Shortest Path

There is a theory that two people in the world share the maximum of 5 handprints between friends. In the 1960s he was experimentally examined by the American sociologist Stanley Milgram [27]. Having a sufficiently extensive database about people and their friends, we can therefore be tempted to set the shortest path between them in the graph of knowledge and to present to the user—if we had data about every knowledge between two people, according to this theory, we should never get paths more than 7 vertices. Surprisingly, only the first large Polish social network, now already forgotten by its users, Grono.net, implemented a mechanism showing the shortest path to friends to any person.

However, remember that on Facebook, if we can download a list of friends of one of the users of the application, this operation will fail with the list of friends of any of his friends, unless he is also the user of the application. When attempting such unauthorized access to the list of friends, someone who did not grant permission to do so will receive a message (here with the example numbers of users):

"Can't lookup all friends of 531931138. Can only lookup for the logged in user (1217426661), or friends of the logged in user with the appropriate permission"

Interestingly, we can check the knowledge between two, but specific users of the site—asking for two people who are not familiar with the author of the work (and do not have friends with him), performed in the context of application, in which they did not register (their friends also do not):

SELECT uid1, uid2 FROM friend WHERE uid1=100001854294917 AND uid2=100000828797502[7]

Returns information about the actual existence of knowledge between them:

```
[
  {
    "uid1": "100001854294917",
    "uid2": "100000828797502"
  }
]
```

In order to achieve the desired effect (the shortest path of friends) one can use one of the algorithms to search for the shortest path in the graph, for example, the breadth-searching algorithm. If we decide on such an approach, we face a choice: analysis on the local data base, or interactive polling of the site (for example, Facebook).

Local copy: the speed of action is undoubtedly the advantage of this solution. After creating a copy of the familiarity chart, downloading the list of friends of all users of the application. Disadvantages—we are exposed to false results by incomplete data, because downloading massively, we are not able to go anywhere in the network of knowledge.

---

[7]Query in FQL language.

Table 4 Output of the directed_all_pairs_shortest _paths function from the postgraph package for the graph from Fig. 9

_user1	_user2						
	1	2	3	4	5	6	7
1	0	1	2	3	5	4	5
2	1	0	3	2	4	3	4
3	2	3	0	1	3	2	3
4	3	2	1	0	2	1	2
5	5	4	3	2	0	3	2
6	4	3	2	1	3	0	1
7	5	4	3	2	2	1	0

Interactive action: this approach will require much more time, because it counts with delays at each query, resulting not only from the time needed to generate an answer to the query, but also the delay of the HTTP protocol itself used to transmit atomic information ("whether the edge in the graph exists?") for significant distances, so obtaining the result will be uneconomically time-consuming Advantage: the result will be the most up-to-date and complete.

The ideal situation is the one in which the social service belongs to us, and the number of users is both attractive and computationally acceptable at the same time.

Example 3: If we store information about knowledge between users in the fb_fships table (from Facebook friendships) in the Postgres database, we can use the postgraph library [31], which is a bridge between the Postgres database and the Boost libraries for C++.

It gives us the option of delineating the shortest path between all pair of willows in the graph with one call of function. However, since this function requires the name of the column in which the weight is kept for the edges between the vertices, and for the knowledge graph, it is assumed that the weight is always equal to 1, we need to create an additional view adding the weight column with constant value 1.[8]

```
CREATE VIEW fb_fships_postgraph AS
  SELECT    t2.int_id    AS    _user1,    t3.int_id    AS    _user2,
1::float AS weight
    FROM fb_fships
    LEFT JOIN fb_users t2 ON fb_fships._user1 = t2.uid
    LEFT JOIN fb_users t3 ON fb_fships._user2 = t3.uid;
SELECT                          *                          FROM
directed_all_pairs_shortest_paths('fb_fships_postgraph',
'_user1', '_user2', 'weight');
```

If there is a knowledge graph in the database illustrated in Fig. 4, then the result calling the above-mentioned point is presented in Table 4.

---

[8]For scales different from 1, the Dijkstra algorithm or its modifications are used, whereas for equal to 1, the algorithm will become simply a wide-range algorithm, used in the case of no weighting.

Unfortunately, the use of this package brings some additional requirements to the database design. As the author of the postgraph[9] says, the library cannot cope with the eight-byte bigint fields. So, we cannot use numbers directly from sites like Facebook, because they do not fit in the usual type of integer and must be stored as bigint. It may be necessary to modify the view so that it maps long user numbers to internal ones that fit in the integer type.

The biggest disadvantage of the presented ready solution is the fact that it allows only to specify the length of the shortest path, without giving the vertices on which it passes. They can therefore be used at most to experimentally check this hypothesis.

Proposing Friends
Another application, however, often implemented already in the social network itself, from which we would like to get data, is to suggest people that the user may know, based on many mutual friends or other factors that two people belong to the same social society.

One of the methods of detecting such belonging to a single society is the use of Jaccard's or cosine similarity [32]. For the two vertices in the graph, V {i Vj, the similarity measures mentioned are defined as:

$$Jaccard(v_i, v_j) = \frac{|N_i \cap N_j|}{|N_i \cup N_i|} = \frac{\sum_k A_{ik} A_{jk}}{|N_i| + |N_j| - \sum_k A_{ik} A'_{jk}} \tag{2}$$

$$Cosine(v_i, v_j) = \frac{\sum_k A_{ik} A_{jk}}{\sqrt{\sum_s A_{is}^2 \cdot \sum_{st} A_{it}^2}} = \frac{|N_i \cap N_j|}{\sqrt{|N_i| \cdot |N_j|}}, \tag{3}$$

where N_i is the neighbors of the vertex v_i, A is the neighborhood matrix of the graph, and | * | indicates cardinality. These measures return values from 0 to 1. Naturally, the higher the indication, the more "similar" (in terms of the circles in which it rotates) will be user v_j to user v_i.

There is a problem for them, however, when the two adjacent vertices in the graph get the value 0—this happens for people who, although they know each other, do not have common friends. While the value of the measure makes sense from the point of view of structural equivalence, but if two people know each other, one may suspect that they share some similarity. Therefore, a slightly modified form of the above formulas is used, in which Nv includes the vertex v [32] in this set.

Another modification is the so-called MASI distance (Measuring Agreement on Set-valued Items, measuring the compliance of elements with values being sets [17]), being a modification of Jaccard's measure to reduce distances in situations where the sets overlap:

$$MASI(v_i, v_j) = \frac{|N_i \cap N_j|}{max(|N_i|, |N_j|)} \tag{4}$$

---

[9]From the README.postrgraph file: PostgreSQL allows for 8-byte integers and 4-byte floats. Using those will cause "bad things to happen" (PostgreSQL allows 8-byte integers and 4-byte floating point numbers.) Using them will result in bad things happening.

Proposing Strangers

On the other side, the purpose of our service may be the opposite. Let's assume, hypothetically, that we create a website that helps us find a lover, I will never know which one. Therefore, we are interested in such a group of people with whom we do not have common friends and even friends of our friends.

In practice, this means that we are interested firstly in the longest possible path of mutual friends (subsection Six hand words, the shortest path), and the best place in which the path does not exist. However, these requirements may not be enough.

This topic humorously touched [33] an internet comic created by mathematicians and addressed to mathematicians, the Spiked Math. By looking at the example presented in it, let's discuss possible solutions and complication.

Let's assume that the user would like to keep in touch only with those of the chart in Fig. 10 who do not know each other in order to avoid gossip about their topic. If she meets with Alice and Betty, because they know each other, they will soon find out, "because women like to gossip" [33]. We assume, therefore, that he should not make arrangements with any two who know each other. Therefore, it would be necessary to select the maximum independent subset[10] of the chart of knowledge, which gives the possibility of arranging for six people (for example, Alice, Carol, Debbie, Irena Eve, and Fran).

This, however, may not be enough, it may happen that he will meet with Alice and Carol, and Alice will tell her to his friend Gloria, who will tell Carol. It will then be unmasked. We can limit contacts to one person from every single member of the graph.[11] It will guarantee that the chain of knowledge will not lead to the exchange of rumors about him, allowing to meet up to four people from the graph.

But what if Eve and Fran become friends after the same man would meet both? Author [33] sums it up humoristically with a dialogue: *Mathematics does not work that way! But girls like that.*

Nevertheless, the exploration of completely new acquaintances, from new environments, and those that are interesting for a given user is an interesting issue and worth paying much attention to.

Investigation of the "Suspension" of Towers in the Group

The methods described above are usually based on the tendency of people connecting to groups of friends and local (also around a topic) community. Within a larger graph of knowledge, it is possible to distinguish certain groups of people in which everyone knows each other. From the point of view of graph theory, such a group is called a clique.

Detection of these types of groups can be performed, for example, using the Python networkx library [17]. It is also worth paying attention to those groups, which from the mathematical point of view are not yet a click, because they lack individual connections. We are talking about quasi-clicks [32] or subgroups of γ-dense, where γ fulfills the condition:

---

[10] The maximum subset of vertices in which there is no edge between me and no two.

[11] A coherent sub-graph G is its coherent subgraph not contained in a larger subgraph of the graph G.

Fig. 10 An example of a
graph showing how people
are of the gender we know

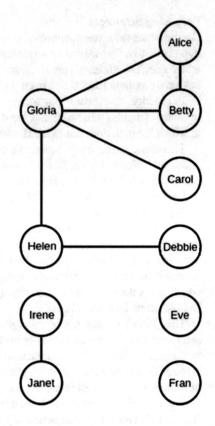

$$\frac{E_s}{V_s(V_s - 1)/2} \geq \gamma \tag{5}$$

For subgraph Gs (Vs, Es).

It is possible to automatically assess the degree of integration of a given group of people who are familiar with it, and also to detect completely closed groups of friends ($\gamma = 1$), or cliques.

The principles described in sub-sections "Proposing friends" and "Investigation of the "suspension" of towers in the group" include mechanisms for detecting the community and clustering (segmentation) of the knowledge graph used for the analysis and visualization of large communities.

3.3 Finding Common Labels (Features)

A great tool for proposing friends to people are their own interests and the compatibility of those with potential new friends. We need some attachment points, saying

Fig. 11 The comparison with the user whose profile we visit is presented by Last.fm to each logged-in user

what a person likes. Popular social networking sites provide exactly the information we need:

- **Facebook**: well-liked websites, placed in separate categories that put them in a certain context, saying that it is a film, a book genre, a musical performer or a field of science.
- **Twitter, Blip**: users often use hashtags, for example *#kot*, so here we have information that sorts the entries thematically in a certain way.
- **Last.fm**: data collected by the website is strictly specialized (listening to music, favorite songs, artists, genres), so we can automatically rely on them similarity measures, without fear of comparing the values coming from all the different worlds. Last.fm also performs such comparisons as shown in Fig. 11.

The methods of calculating the similarity of two users based on the data collected in this way will be very similar to the previously discussed search for people that we know—it is most reasonable to use a measure of similarity, for example expressed in Eq. 4 (MASl distance).

If, therefore, user A has a lot of common interest in books or muse with user B, then it is worth offering them knowledge, indicating the basis of such a proposition (justified settlements are always better taken by the user, and the indication of a common topic can help in making acquaintance).

3.4 Analysis of Publication Time and Message Content

For data such as user statuses and messages published by them, one can use primarily two types of information: circumstances (mainly: time) of publishing these messages and their content.

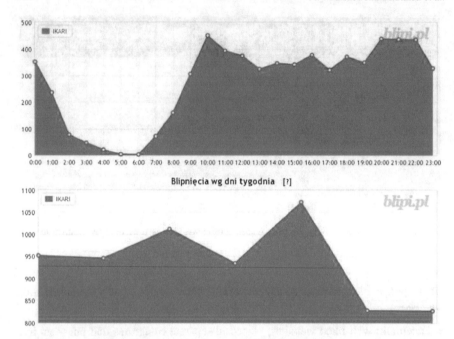

Fig. 12 Statistics generated by the Blipi.pl website based on the statuses published by the user of the Blip.pl website [34]

Figure 12 presents charts generated by the Blipi.pl website based on the data of the selected Blip.pl website user. As we can see, it is a simple statistical analysis consisting in counting messages published in certain specific time intervals. Such a simple analysis gives rather detailed data about the daily rhythm of a given user (we know that he sleeps for the most part between 2 and 6 h) and weekly (we can assume that on Friday and Saturday he rests more actively than with a computer or mobile device).

Such information can then be used to match both proxies and adapt commercial offers to the needs and preferences of a given user, as well as to set a schedule of activities such as online competitions for users of a given website—companies often make mistakes releasing competitions while most of our target audience is at work and cannot participate or engage appropriately.

The second way in which user-published statuses can be used is lexical analysis. By gathering, as in the example above, the statistics for this time the content of the message, we can build an approximate image of it, what the user likes to write about (and therefore what he is interested in).

In practice, the application of this technology can be observed in "side" services focused on services such as *Twitter* or *Blip*. Twitter itself also saw the potential to explore popular expressions in the news and created the Trends panel (trendy), in which it publishes popular in They are mainly mentioned earlier, there are also "ordinary" phrases, which means that every part of the message is taken into account.

The tool created for this purpose is the nltk library (Natural Language Processing Toolkit, a set of tools for processing natural language) for Python—and unfortunately only for him.

Mining the Social Web [17] proposes using TF-IDF (Term Frequency-Inverse Document Frequency—validation of terms—reverse frequency in documents) together with the cosine similarity metric [17]. The basis of TF-IDF is the calculation of two title indicators, term frequency describing the frequency of occurrence of a given term[12] (concept) in the document:

$$tf_{i,j} = \frac{n_{i,j}}{\sum_k n'_{k,j}} \tag{6}$$

where $n_{i,j}$ is the number of occurrences of the term t_i in the document d_j, and $\sum_k n_{k,j}$ in the nominative—the sum of the number of occurrences of all terms in the d_j document; whereas inverse document frequency is defined as:

$$idf_i = log\frac{|D|}{|\{d : t_i \in d\}|} \tag{7}$$

where $|D|$ is the number of documents in the corpus (collection of the texts examined), and $|\{d: t_i \in d\}|$ is the number of documents containing at least one occurrence of a given term. Finally, the TF-IDF value is calculated as [1.40]:

$$(tf - idf)_{i,j} = tf_{i,j} \times idf_i \tag{8}$$

By means of this method, we can calculate for each term-document, how important for a given document is this date. Therefore, for a given concept, it is possible to determine how much a statement is "on the subject" of this concept, but it should be remembered that due to limitations present in (and characteristic of) microblog sites, usually only on other platforms (as it keeps changing example: Facebook does not limit the length of the entry on the user's board) statements will be appropriately "saturated" with the terms to which they refer, because on sites such as Twitter or Blip, users are forced to sit in a tight limit of characters and if only they will they could change the second and subsequent occurrences of a word with a shorter word—he does it.

However, even short messages on these sites carry meaning that can be detected by such methods.

To determine the similarity between entries, when the value of tf-idf is already calculated, the cosine metric is used (as one of many possibilities):

---

[12]Under the concept, the term is understood as a single word (e.g., a "chamber") or a group of words whose meaning is different than individual words forming a group (e.g. "High House") [40].

$$similarity = \cos(\theta) = \frac{\vec{A} \cdot \vec{B}}{\left\|\vec{A}\right\|\left\|\vec{B}\right\|} = \frac{\sum_{i=1}^{n} A_i \times B_i}{\sqrt{\sum_{i=1}^{n}(A_i)^2} \times \sqrt{\sum_{i=1}^{n}(B_i)^2}} \qquad (9)$$

It is calculated, as can be seen, as the cosine of kata θ between the vectors $\vec{A}$ and $\vec{B}$ n of the attributes representing (through the values of A_i ... A_n and analogously B_i ... B_n) two comparable objects. It is therefore scalable to any uncounted number of dimensions (attributes). The result can be interpreted as a similarity, where 1 is identical (vectors do not differ, the angle between them is 0), 0 means total lack of connection, and -1 means the opposite (it is possible to obtain only cases where it allows negative attribute values, otherwise it is not possible to obtain vectors).

For text, vectors are constructed as ordered n-tki values tf-idf for subsequent words, for example $\vec{A} = [\text{tfidfi}_{i,d}, \text{tfidf}_{j,d}]$, where the term t_i is "open", t_j "source", and d_d is a document, for which the vector is designated; then $\vec{A}$ is a vector characteristic of the words open, source[13] in the document d_d.

3.5 Supplementing Data from Various Sources

If the application being built will be based on more than one social network site, we will allow the user to enter, for example, both his Facebook and Last.fm and Twitter accounts, not only the sum of the information that can be collected, but also the product. In other words, there are some common parts between data from several websites, in particular some fields may offer more than accuracy (from one service we know only the year of birth, and the other one can get a full date, d), while others bed and simply broaden the collected data in a given category quantitatively (even if the user's Twitter messages can be extended to those from Facebook services and vice versa). Similarly, one service can only provide approximate and user's location (city), while another—precise coordinates of geographic coordinates (sometimes even with error limits).

On the other hand, one should also take into account the opposite situation—data from two websites concern the same information, but at least they are different or contradictory.

If it differs a, it may be a case where one information is richer than the other in a field whose content is at least somewhat arbitrary: for example, the same user in one site may be called Caesar Pokorski, in another Caesar "ikari" Pokorski, and in another—only ikari, or—even worse—ikari87.

Example 4: In order to detect a situation where one text "expands" another, it
if it is sufficient for it to be considered as the same content but richer one can use one of the algorithms to calculate the distance between text strings. For example, the fuzzystrmatch module [35] for the PostgreSQL database server (most likely we

---

[13]Do not necessarily present side by side!

would like to carry out such operations "as close as possible to the data") has several functions that can help in this:

1. soundex calculating the so-called soundex, the result of a phonetic algorithm for comparing words in English—this is a four-character code representing words that sound similar in a similar or identical way (generally);
2. levenshtein calculates the distance of Levenshtein, also called the distance between two characters—it says how many editions (insertions, deletions or substitutions) should be made on the first subtitle to get the second (its extended variant allows us to set our own cost of cosmic ones) operations);
3. metaphone and dmetaphone (double metaphone) are algorithms with an idea like soundex and generate four-character codes (in this case, they are four letters, for soundex on the other hand—a letter and three digits). Compare their effectiveness for several sample texts that may be an identifier and/or username.

Code below is calculating soundex values in PostgreSQL for several nickname combinations. It compares the values of soundex and difference for several strings, whereby we assume that the "ikari" chains, Caesar Pokorski and Caesar "ikari" Pokorski say to one person, and "anks" and Anna Nowak—the other.

```
SELECT soundex ('Cezar Pokorski'), soundex ('ikari'),
soundex ('Cezar "ikari" Pokorski'), soundex ('anks'),
soundex ('Anna Nowak');
 soundex | soundex | soundex | soundex | soundex
---------+---------+---------+---------+---------
 C261    | I260    | C262    | A520    | A552

SELECT difference ('Cezar Pokorski', 'Cezar "ikari" Pokorski '
),
difference ('Cezar Pokorski', 'Anna Nowak'),
difference ('ikari' , 'Cezar "ikari" Pokorski'),
difference ('ikari', 'Anna Nowak'), difference ('anks', 'Anna
Nowak');
 difference | difference | difference | difference | difference
------------+------------+------------+------------+------------
          3 |          0 |          2 |          0 |          2
```

It should be taken into account that the difference is averse, contrary to its name, to the higher values, the more two chains are like each other. It is defined as the number of compatible items in soundex. Because soundex codes are four-character, it will take values from 0 to 4.

It turns out that such a simple function handles these cases surprisingly well. Codes of sequences Cezar Pokorski and Caesar "ikari" Pokorski differ only by 1, while ikari itself is relatively close to both, which is a good sign when comparing the same pseudonym to the version contained in the surname. The bad sign is when comparing our self with a pseudonym with the name itself—they should have nothing in common, but they contain similar letters—similar soundex codes. On the other hand, in the case of pseudonyms that seem to come in some way from first name or surname (for example, *anks* vel *Anna Nowak*), a high indication of similarity (if we can call the value of 2 or more—50%) correctly suggests the existence of suspicion that the same person. However, since it is assumed that it is known in advance that both accounts describe the same person, such correspondence can be read as a possible alternative spelling (an alternative name) of the name/name of a given user. However, if the same user entered one field "surname" by Cezar Pokorski, in another site by Anna Nowak, the compliance is 0 and we can suspect fraud (unreliability of data).

A not-less-useful distance-editing algorithm, gives the levenshtein function, has a more differentiated value in terms of values, which can be treated as greater precision of the answer (provided well-chosen parameters, usually by trial and error).

Let's compare the names of the same users with his help. The first variant leaves the default cost values of all editing operations equal to 1. The second variant modifies them in such a way that adding new characters is treated best (we want to accept extensions of information about additional words or fuller forms), remove worse, substitute for a completely different sign of the most severe punishment[14]:

```
SELECT
  levenshtein ('Cezar Pokorski', 'Cezar "ikari" Pokorski') AS
lev1,
  levenshtein ('ikari', 'Cezar "ikari" Pokorski') AS lev2,
  levenshtein ('Cezar Pokorski', 'Anna Nowak') AS lev3,
  levenshtein ('ikari' , 'Anna Nowak') AS lev4,
  levenshtein ('ikari' , 'Cezar Pokorski') AS lev5,
  levenshtein ('anks' , 'Anna Nowak') AS lev6;
 lev1 | lev2 | lev3 | lev4 | lev5 | lev6
------+------+------+------+------+------
    8 |   17 |   10 |    9 |   11 |    9
```

```
SELECT
  levenshtein ('Cezar Pokorski', 'Cezar "ikari" Pokorski', 1, 3,
4) AS
lev1,
  levenshtein ('ikari' , 'Cezar " ikari " Pokorski', 1, 3, 4) AS
lev2,
  levenshtein ('Cezar Pokorski', 'Anna Nowak', 1, 3, 4) AS lev3,
  levenshtein ('ikari', 'Anna Nowak', 1, 3, 4) AS lev4,
  levenshtein ('ikari', 'Cezar Pokorski', 1, 3, 4) AS lev5,
  levenshtein ('anks', 'Anna Nowak', 1, 3, 4) AS lev6;
 lev1 | lev2 | lev3 | lev4 | lev5 | lev6
------+------+------+------+------+------
    8 |   17 |   36 |   21 |   17 |   14
```

[14]There is some danger here: if the total penalty for removing the mark and adding a new one will be lower than the penalty for the replacement, it does not matter how much the value of the substitution cost will be determined—instead, separate operations will be calculated Eight and character insertion. It may be difficult to select weights to improve the results, because the distances obtained bed a are identical for insertion, removal and substitution costs equal to 1, 3, 4, respectively, which for 1, 3, 100.

Code is calculating the distance of Levenshtein in PostgreSQL for several combinations a pseudonym and/or name.

It shows that the aforementioned modification of the weights significantly improves the results in the context of the expected use. Without it, ikari was "similarly similar" to Anna Nowak, which is anks. With her, however, the details of Cezar Pokorski always give smaller distances when compared with data also concerning him, than with the name Anna Nowak. If the cost of the operation is too big, as in the transition between Caesar Pokorski and Anna Nowak—it can be suspected that the two values of the field name do not reflect this sameness (in an alternative or more precise way) and do not they will be considered by human to be compatible.

Example 5: A completely different situation will be the one in which we would like to collect information about our favorite music performers, having at our disposal his Facebook account and his *Last.fm* account.

Last.fm provides the `user.getTopArtists` method, which returns the information, for example for a call:

```
http://ws.audioscrobbler.com/2.0/?method=user.gettopartists&user
=ikari87&api_
key=b25b959554ed76058ac220b7b2e0a026
```

We get the answer in code shown below:

```
<lfm status="ok">
  <topartists user="ikari87"type="overall" page="1" perPage="50"
      totalPages="76" total="3790">
   <artist rank="1">
     <name>Placebo</name>
     <playcount>1552</playcount>
     (...)
</artist>
   <artist rank="2">
   <name>Delerium</name>
   <playcount >1351</playcount>
   <mbid>4279aba0-1bde-40a9-8fb2-c63d165dc554</mbid>
   <url>http://www.last.fm/music/Delerium</url>
   (...)
   </artist>
(...)
```

Last.fm's reply to the request for favorite artists user ikari87. The corresponding query in Facebook Query Language (FQL) will, however, look as follows:

```
SELECT page_id, fan_count, name FROM page WHERE page_id IN
(SELECT page_id FROM page_fan WHERE uid= 1217426661)
AND type= 'MUSICIAN /BAND'
```

It will return the answer shown below:

```
        {
            "page_id": 170505350807,
            "fan_count" :2810,
            "name": "DIGIT ALL LOVE"
        },
        {
            "page_id": 55339791254,
            "fan_count": 1583647,
            "name": "Placebo"
        },
        {
            "page_id": 45666355872,
            "fan_count": 938102,
            "name": "Unheilig"
        },
        (...)
```

Facebook response to a query about liked by 1,217,426,661 music performers. This answer gives information about liked contractors that can be used to extend the information previously collected from *Last.fm*. Is this already or is there a certain catch behind it? Well, yes, the information returned is not identical. With *Last.fm*, we will find out which artists the user likes and how much (results s and sorted according to the small number of plays and enriched by the position in the personal ranking), while from *Facebook*, at most, how many people like the performer in general (however, we do not know to what extent it suits the particular user). Nevertheless, we have a solid starting point to broaden information about musical preferences.

3.6   Discovering New Connections, Dependencies, Proposals

Data collected from social networks can be used in many different positively useful ways. The application based on this type of data sets deals with the entire field of computer science referred to as data mining. We can try to discover new connections between facts in the database and on this basis offer new users or use observers in any other way. Examples of problems discovered by the machine may be somewhat trivial, such as: "if we like Britney Spears and we like Shakira, we probably like pop music genre" or "We like Lion King and Pocahontas, we may like Avatar too", but we can also discover more complex and less expected dependencies [36, 37].

Example 6: A simple function given below is detecting dependencies between sites liked by users assuming arbitrarily selected criteria. On this base, we can suggest users who like and discover links between the objects they represent:

Download popular pages (>: min_likes) among users.

```
SELECT _fb_object, COUNT(_fb_user) AS count
FROM fb_likes GROUP BY (_fb_object)
HAVING COUNT(_fb_user) > :min_likes
```

- query retrieving 5 pages liked by the largest
- the number of people who also like the given page, which is to be
- at least 70% of first page fans (as: min_likes)

```
SELECT _fb_object, COUNT(_fb_user) AS count FROM fb_likes
WHERE _fb_user IN
(SELECT _fb_user FROM fb_likes WHERE _fb_object=:page_id)
AND _fb_object <> :page_id GROUP BY _fb_object
HAVING COUNT(_fb_user)> :min_likes ORDER BY count DESC LIMIT 5
```

Code shows SQL queries used to search for links between pages.

In the same way, we can detect correlations between friendships and suggests to users of friends based on those they already know (here are correlations much stronger).

If the aim of application is to propose new knowledge to the user based on how well he/she understands his/her friends (we should use the not-so-well-defined term: how much he likes them), we can try a personalized vector of weights, in which categories the similarity between them and the person most affects relationships, and in what less—as much, of course, as much as it will be possible based on the possessed data. While often the essence of human relations can be elusive, we can, however, hope to make observations like "they understand each other perfectly, because they are interested in similar literature and music." They immediately suggest the use of neural networks for this purpose such a solution would obviously be a much better match between the proposition and the user, but the disadvantage would be slow learning and the need to provide a huge number of examples (in the form of evaluations from the user), and we can never collect such data in the right way [38, 39].

The presented methods give a wide set of tools for analyzing the obtained data. With their help, we can generate a variety of proposals and conclusions. It is worth noting that these methods are usually based primarily on appropriately selected similarities—regardless of whether we are talking about similarity between user's friend groups, his interests, or his personal data [40]. Suggestions-oriented mechanisms for the user will aim to generate results that are close to what is already known about the user, but which he did not report himself. The mechanisms that build the conclusions for the creators of the application will be based on the dependencies discovered based on the similarity between certain groups of information.

4 Conclusion

The presented methods allow us to predict the interests of users, find people and topics that might interest them, discover their communities, and discover more general facts and addictions related not only to Internet users. Decisive disciplines represent objects made accessible through the programming interfaces of the most popular social networks.

Some of the similarly functioning algorithms implement the same social services—Facebook today does not offer friends only based on people we already know. The methods listed in the publication can be identified by cataloging and creating custom maps of user interests. The increase in the number of users of Internet social services is increasing every day. It is possible to expect further progress and development of APIs of different service providers over time. This publication has brought a lot of insight into the use of the interfaces.

References

1. Maciej Laskowski, Problematyka budowy serwisu społecznościowego na przykładzie projektu lokalnego portalu kulturalnego - studium przypadku, 5.2008. http://kis.pwszchelm.pl/publikacje/VII/Laskowski.pdf. Accessed 11 July 2011
2. Google Trends "social media"—search volume index. http://www.google.com/trends?q=social+media&ctab=0&geo=all&date=all&sort=0. Accessed 30 July 2011
3. Erfati A. Wall Street Journal, Google + Pulls in 20 Million in 3 Weeks. http://online.wsj.com/article/SB10001424053111904233404576460394032418286.html. Accessed 31 July 2011
4. Haland L. Time to reach 20 million users (wykres poglądowy zamieszczony przez autora w serwisie Google Plus, wykorzystywany przez elektroniczna prase branzowa tu oryginalne zródło). https://plus.google.com/112418301618963883780/posts/B3s3dd739bG
5. NeXt Up! Research, Social Networking: Facebook. Face of Next Generation Social Networking (v2), 11.10.2009. https://www.sharespost.com/research_report/Facebook-V2.pdf. Accessed 20 March 2011
6. Orenstein D (2000) Computerworld, QuickStudy: Application Programming Interface (API). http://www.computerworld.com/s/article/43487/Application_Programming_Interface. Accessed 31 July 2011
7. Musser J. ProgrammableWeb, Twitter API Traffic is 10x Twitter's Site, 20.09.2007r. http://blog.programmableweb.com/2007/09/10/twitter-api-traffic-is-10x-twitters-site/. Accessed 10 May 2011
8. Blip.pl, Specyfikacja API serwisu Blip.pl. http://blipapi.wikidot.com/
9. NeXt Up!, Social Networking: Twitter. Short and Tweet, 24.09.2010. https://www.sharespost.com/research_report/100924_GSVP_Twitter-V5.pdf. Accessed 30 July 2011
10. Twitter, method reference: GET users/show. https://dev.twitter.com/docs/api/1/get/users/show. Accessed 04 Aug 2011
11. Facebook, API Reference: User. https://developers.facebook.com/docs/reference/api/user/. Accessed 04 Aug 2011
12. Facebook, Graph API. https://developers.facebook.com/docs/reference/api/. Accessed 08 Aug 2011
13. Facebook, API Reference: Authentication. https://developers.facebook.com/docs/authentication/. Accessed 14 Aug 2011
14. Last.fm, user.getInfo—method reference. http://www.lastfm.pl/api/show?service=344. Accessed 04 Aug 2011
15. GNU FM, What is scrobbling?, http://foocorp.org/projects/fm/scrobbling/. Accessed 04 Aug 2011
16. Biernacki A, vbeta.pl, 8.02.2010r, 95% treści w serwisach społeczżnościowych to spam, http://vbeta.pl/2010/02/08/95-tresci-w-serwisach-spolecznosciowych-to-spam. Accessed 05 Aug 2011

17. Russell MA (2011) Mining the social web. O'Reilly Media
18. Twitter, REST API Resources. https://dev.twitter.com/docs/api. Accessed 07 Aug 2011
19. Hammer-Lahav E (ed) Yahoo!, D. Recordon, Facebook, D. Hardt, Microsoft, 25.07.2011. The OAuth 2.0 Authorization Protocol, v2.20 (draft). https://datatracker.ietf.org/doc/draft-ietf-oauth-v2/?include_text=1. Accessed 06 Aug 2011
20. Dłubacz W. Wiki: Mikroformaty. http://microformats.org/wiki/Main_Page-pl. Accessed 14 Aug 2011
21. XHTML Friends Network. http://gmpg.org/xfn/. Accessed Aug 2011
22. Microformats.org, Facebook adds hCalendar support. http://microformats.org/2011/02/17/facebook-adds-hcalendar-hcard. Accessed Aug 2011
23. Watters A. Read Write Web, 12.07.2011, no pseudonyms allowed: is Google Plus's real name policy a good idea? http://www.readwriteweb.com/archives/no_pseudonyms_allowed_is_google_pluss_real_name_po.php. Accessed 19 Aug 2011
24. Strona anonimowego uzytkownika w serwisie Facebook, I Am Always Tired Because I Become a Superhero at Night. http://www.facebook.com/superhero.at.night. Accessed 20 Aug 2011
25. Google, Google Latitude. http://www.google.com/latitude/bZ0. Accessed Aug 2011
26. Veness C. Calculate distance, bearing and more between latitude/longitudepoints. http://www.movable-type.co.uk/scripts/latlong.html. Accessed 20 Aug 2011
27. Andrzej Łodyński, Polityka, 18.03.2000, Sześć uścisków dłoni (artykuł popularnonauko¬wy). http://archiwum.polityka.pl/art/szesc-usciskow-dloni,363689.html. Accessed 10 Aug 2011
28. Uzytkownik „askman" serwisu Blip.pl, przykładowa wypowiedz oznaczona #drogiblipie, http://blip.pl/s/671588629. Accessed 21 Aug 2011
29. Uzytkowniczka „andante" serwisu Blip.pl, przykładowa wypowiedz oznaczona #ludziektórzy. http://blip.pl/s/670986881. Accessed 21 Aug 2011
30. Geonames, baza danych zawierająca lokalizaqe geograficzne. http://www.geonames.org/. Accessed June 2011
31. Keitt T. Postgraph, biblioteka rozszerzająca funkcjonalnosc bazy danych PostgreSQL. https://launchpad.net/postgraph
32. Tang L (Yahoo! Labs), Liu H (Arizona State University) (2010) Community detection and mining in social media. http://www.morganclaypool.com/doi/abs/10.2200/S00298ED1V01Y201009DMK003, cited 21.10.2010
33. Spiked Math, 21.08.2010r, Thedategraph. http://spikedmath.com/287.html. Accessed 21 Aug 2011
34. Blipi.pl, Statystyki użytkownika ikari. http://stats.blipi.pl/ikari. Accessed 23 Aug 2011
35. Biblioteka fuzzystrmatch dla bazy danych PostgreSQL. http://www.postgresql.org/docs/8.3/static/fuzzystrmatch.html. Accessed July 2011
36. Kryvinska N (2012) Building consistent formal specification for the service enterprise agility foundation. Soc Serv Sci J Serv Sci Res Springer 4(2):235–269
37. Kaczor S, Kryvinska N (2013) It is all about services—fundamentals, drivers, and business models. Soc Serv Sci J Serv Sci Res Springer 5(2):125–154
38. Gregus M, Kryvinska N (2015) Service orientation of enterprises—aspects, dimensions, technologies. Comenius University in Bratislava. ISBN: 9788022339780
39. Kryvinska N, Gregus N (2014) SOA and its business value in requirements, features, practices and methodologies. Comenius University in Bratislava. ISBN: 9788022337649
40. Molnár E, Molnár R, Kryvinska N, Greguš M (2014) Web intelligence in practice. Soc Serv Sci J Serv Sci Res Springer 6(1):149–172

Self-similar Teletraffic in a Smart World

Izabella Lokshina, Hua Zhong and Cees J. M. Lanting

Abstract This chapter focuses on building the future smart world by using recent advances in networking science and practice, which range from traffic engineering and control to innovative wireless scenarios and include key challenges arising from the upcoming interconnection of massive numbers of devices, sensors and humans. The authors bridge the gap between performance modeling and real-life operational aspects, as well as they leverage measurement data to provide a better understanding of the wired and wireless networks' operation under realistic conditions. Specifically, the authors examine self-similar teletraffic in communication networks, capable of carrying diverse traffic, including self-similar teletraffic, and supporting diverse levels of Quality of Service (QoS). Advanced self-similar models of sequential and fixed-length sequence generators and efficient algorithms simulating self-similar behavior of teletraffic processes in communication networks are developed and applied. Additionally, buffer overflow simulations are conducted with a finite buffer single server model under self-similar traffic load. The authors confirm this scenario is particularly efficient being used to analyze reliability and performance in communication networks in a smart world, while higher performance networks must be described by lesser buffer overflow probabilities. Numerical examples and simulation results are shown.

Keywords Smart world · Communication networks · Diverse traffic · Quality of service (QoS) · Self-similar teletraffic · Modeling and simulation · Advanced generators of self-similar teletraffic · Self-similar queuing systems · Buffer overflow · Performance analysis

I. Lokshina (✉) · H. Zhong
MMI, SUNY Oneonta, Oneonta, NY, USA
e-mail: Izabella.Lokshina@oneonta.edu

H. Zhong
e-mail: Hua.Zhong@oneonta.edu

C. J. M. Lanting
DATSA Belgium, 3010 Leuven, VBR, Belgium
e-mail: Cees.Lanting@datsaconsulting.com

© This is a U.S. government work and not under copyright protection in the U.S.;
foreign copyright protection may apply 2020
N. Kryvinska and M. Greguš (eds.), *Data-Centric Business and Applications*,
Lecture Notes on Data Engineering and Communications Technologies 30,
https://doi.org/10.1007/978-3-030-19069-9_5

1 Introduction

The growth of broadband networks and the Internet has been exponential in terms of
users and end-user systems as well as in traffic in recent years. In a smart world, communications networks can support a wide range of multimedia applications, such as
audio, video and computer data that differ significantly in their traffic characteristics
and performance requirements [2, 3, 11–13, 18, 21, 23, 25, 27, 31].

One promising goal for telecommunication developers is to build a unified communication network platform capable of carrying diverse traffic, including self-
similar teletraffic, and supporting diverse levels of Quality of Service (QoS), also as
a potential green broadband network solution [16, 17, 24, 26].

Recent studies of communication network traffic in a smart world have shown
that teletraffic, i.e. technical term, identifying all phenomena of control and transport
of information within telecommunications networks, exhibits self-similar properties
over a wide range of time scales [5, 18–20, 28, 32, 33]. Therefore, self-similar queuing
systems, defined as queuing networks with stochastic, long-range dependent self-
similar processes, are appropriate reference models for teletraffic as they provide
capacity to estimate queuing and network performance as well as reliability, allocate
communication and data processing resources, and ensure the QoS [19, 20, 22, 34].

Self-similar teletraffic is observed in Local Area Networks (LANs) and Wide
Area Networks (WANs), where superposition of strictly independent alternating
ON/OFF traffic models, whose ON- or OFF-periods have heavy-tailed distributions with infinite variance, can be used to model aggregate network traffic that
exhibits self-similar, long-range dependent behavior typical for measured Ether-
net LAN traffic over a wide range of time scales. Self-similar teletraffic is also
observed in Asynchronous Transfer Mode (ATM) networks, when arriving at an ATM
buffer, it results in a heavy-tailed buffer occupancy distribution, and a call loss probability decreases with the buffer size, not exponentially, like in traditional Markovian
models, but hyperbolically [18, 20].

Furthermore, self-similar teletraffic is observed in the Internet as many characteristics can be modeled using heavy-tailed distributions, including the distributions
of traffic times, user requests for documents, and file sizes. In IP with TCP networks the transfer of files or messages shows that the reliable transmission and flow
control mechanisms serve to maintain long range dependent structure included by
heavy-tailed file size distributions [4, 18, 31, 34].

Self-similar video traffic provides possibility for developing models for Variable
Bit Rate (VBR) video traffic using heavy-tailed distributions [18, 34]. In a Smart
World, the impact of self-similar models on queuing performance is very significant
and the main trends in such findings are associated with efficient modeling and
simulation of communication network traffic and analyzing queuing models and
protocols in practical traffic scenarios [18–20, 31].

The properties of self-similar teletraffic are very different from properties of traditional models based on Poisson, Markov-modulated Poisson, and related processes.
More specifically, while tails of the queue length distributions in traditional tele-

traffic models decrease exponentially, those of self-similar traffic models decrease much slower. Therefore, the use of traditional models in communication networks characterized by self-similar processes can lead to incorrect conclusions about the performance of analyzed networks. Traditional models can lead to over-estimation of the network performance, insufficient allocation of communication and data processing resources, and consequently difficulties in ensuring the QoS [18, 20, 32].

Self-similarity can be classified into two types: deterministic and stochastic [32, 35]. In the first type, deterministic self-similarity, a mathematical object is assumed to be self-similar if it can be decomposed into smaller copies of itself [29, 34]. That is, deterministic self-similarity is a property, in which the structure of the whole is contained in its parts.

This chapter is focused on stochastic self-similarity. In that case, probabilistic properties of self-similar processes remain unchanged or invariant when the process is viewed at different time scales. This contrasts with Poisson processes that lose their burstiness and flatten out when time scales are changed. However, the time series of self-similar processes exhibit burstiness over a wide range of time scales. Self-similarity can statistically describe teletraffic that is bursty on many time scales [18, 20, 29, 34].

One can distinguish two types of stochastic self-similarity. A continuous-time stochastic process Y_t is strictly self-similar with a self-similarity parameter H ($1/2 < H < 1$), if Y_{ct} and $c^H Y_t$, i.e. a rescaled process with time scale ct, have identical finite-dimensional probability for any positive time stretching factor c [14]. This definition, in a sense of probability distribution, is quite different from that of the second-order self-similar process, observed at the mean, variance and autocorrelation levels [30, 32, 34]. The process X is asymptotically second-order self-similar with $0.5 < H < 1$, if for each k large enough $\rho_k^{(m)} \to \rho_k$, as m $\to \infty$, where $\rho_k = \mathrm{E}\big[(X_i - \mu)(X_{i+k} - \mu)\big]/\sigma^2$.

This chapter takes a different, new approach by using the exact or asymptotic self-similar processes in an interchangeable manner, which refers to the tail behavior of the autocorrelations. Modeling and simulation of self-similar teletraffic is performed with advanced generators of synthetic self-similar sequences, divided into two practical classes: sequential generators and fixed-length sequence generators, developed by Radev and Lokshina [34] and improved by Lokshina [18, 20]. Both classes of self-similar traffic generators are considered in this chapter.

Besides, steady-state simulation of a SSM/M/1/B self-similar queuing system with long-range dependent processes and finite buffer capacity is performed with splitting [21, 22], using a variant of the RESTART/LRE method [19] to accelerate the buffer overflow simulation. Efficient Limited Relative Error (LRE) algorithm accelerating buffer overflow simulation is developed and applied to generate stochastic, long-range dependent self-similar processes with a limited relative error, which is appropriate and particularly efficient being used to evaluate reliability and performance in communication networks in a smart world, while higher performance networks must be described by lesser buffer overflow probabilities [20].

The remaining of this chapter is organized as follows. Section 2 explains unique properties of stochastic, long-range dependent self-similar processes contained within diverse communication network traffic, which do not degenerate with increase of the non-overlapping batch size to infinity in difference with traditional processes that are normally used in network traffic modeling, which do degenerate. Section 3 discusses steady-state simulation of self-similar queuing systems with long-range dependent processes and finite buffer capacity. Section 4 describes Limited Relative Error (LRE) algorithm for accelerated buffer overflow simulation in self-similar queuing systems under long-range dependent traffic load with different buffer sizes. Simulation results and performance analysis are provided in Sect. 5. Sections 6 and 7 discuss sequential generators and fixed-length sequence generators of self-similar network traffic, respectively. Simulation results and performance analysis are provided in Sect. 8. Finally, conclusions are given in Sect. 9, followed by references.

2 Distinctive Properties of Stochastic, Long-Range Dependent Self-similar Processes

Recent studies of communication networks in a smart world have evidently shown that the most striking feature of some second-order self-similar processes is that the accumulative functions of the aggregated processes do not degenerate as the non-overlapping batch size m increasing to infinity. Such processes are known as Long-Range Dependent (LRD) processes [4, 20, 32].

This contrasts with traditional processes used in modeling IP network traffic, all of which include the property that the accumulative functions of their aggregated processes degenerate as the non-overlapping batch size m increasing to infinity, i.e., $\rho_k^{(m)} \to 0$ or $\rho_k^{(m)} = 0(|k| > 0)$, for $m > 1$.

The equivalent definition of long-range dependence is given as (1):

$$\sum_{k=-\infty}^{\infty} \rho_k = \infty. \tag{1}$$

Another definition of LRD is presented as (2):

$$\rho_k \sim L(t)k^{-(2-2H)}, \quad \text{as} \quad k \to \infty, \tag{2}$$

where $\frac{1}{2} < H < 1$ and $L(\cdot)$ slowly varies at infinity, i.e. for all $x > 0$ it could be determined as (3):

$$\lim_{t \to \infty} \frac{L(xt)}{L(t)} = 1. \tag{3}$$

The Hurst parameter H characterizes the relation in (2), which specifies the form of the tail of the accumulative function. One can show that is true for $1/2 < H < 1$, is given in (4):

$$\rho_k = \frac{1}{2}\left[(k+1)^{2H} - 2k^{2H} + (k-1)^{2H}\right]. \tag{4}$$

For $0 < H < 1/2$ the process is Short Range Dependent (SRD) and could be presented as (5):

$$\sum_{k=-\infty}^{\infty} \rho_k = 0. \tag{5}$$

For $H = 1$ all autocorrelation coefficients are equal to one, no matter how far apart in time are the sequences. This case has no practical importance in teletraffic modeling. If $H > 1$, then (6) is true:

$$\rho_k = \begin{cases} 1 & \text{for } k = 0 \\ \frac{1}{2}k^{2H}g(k^{-1}) & \text{for } k > 0 \end{cases} \tag{6}$$

where:

$$g(x) = (1+x)^{2H} - 2 + (1-x)^{2H}. \tag{7}$$

One can see that $g(x) \to \infty$ as $H > 1$. If $0 < H < 1$ and $H \neq 1/2$, then the first non-zero term in the Taylor expansion of $g(x)$ is equal to $2H(2H - 1)x^2$. Therefore, (8) is true:

$$\rho_k/(H(2H-1)k^{2H-2}) \to 1, \quad \text{as } k \to \infty. \tag{8}$$

In the frequency domain, an essentially equivalent definition of LRD for a process X with given spectral density (9):

$$f(\lambda) = \frac{\sigma^2}{2\pi} \sum_{k=-\infty}^{\infty} \rho_k e^{ik\lambda}, \tag{9}$$

is that in the case of LRD processes, this function is required to satisfy the following property (10):

$$f(\lambda) \sim c_{f_1} \lambda^{-\gamma}, \quad \text{as } \lambda \to 0, \tag{10}$$

where c_{f_1} is a positive constant and $0 < \gamma < 1, \gamma = 2H - 1 < 1$.

As a result, LRD manifests itself in the spectral density that obeys a power-law near the origin. This implies that $f(0) = \sum_k \rho_k = \infty$. Consequently, it requires a spectral density, which tends to $+\infty$ as the frequency λ approaches 0.

For a Fractional Gaussian Noise (FGN) process, the spectral density $f(\lambda, H)$ is given by (11):

$$f(\lambda, H) = 2c_f(1 - \cos(\lambda))B(\lambda, H), \tag{11}$$

with $0 < H < 1$ and $-\pi \leq \lambda \leq \pi$, where (12) is true:

$$c_f = \sigma^2(2\pi)^{-1}\sin(\pi H)\Gamma(2H + 1)$$

$$B(\lambda, H) = \sum_{k=-\infty}^{\infty} |2\pi k + \lambda|^{-2H-1}, \tag{12}$$

and $\sigma^2 = \text{Var}[X_k]$ and $\Gamma(\cdot)$ is the gamma function.

The spectral density $f(\lambda, H)$ in (11) obeys a power-law at the origin, as shown in (13):

$$f(\lambda, H) \to c_f\lambda^{1-2H}, \quad \text{as} \quad \lambda \to 0 \tag{13}$$

where $1/2 < H < 1$.

3 Steady-State Simulation of Self-similar Queuing Systems

Let us first think that the first buffer has a finite capacity N. In this case the state space of the driving process (X_t) is finite in $\{0, ..., N\}$. Let us consider Markov additive process (X_t, Z_t). As we have formerly described in [19, 21], there is a significant difference in the queueing performance between traditional models of teletraffic such as Poisson processes and Markovian processes, and those exhibiting self-similar behavior. More specifically, while tails of the queue length distributions in traditional teletraffic models decrease exponentially, those of self-similar traffic models decrease much slower.

Let us consider the potential impact of traffic characteristics, including the effects of self-similar behavior on queuing and network performance, protocol analysis, and network congestion controls. Steady-state simulation of self-similar queueing systems includes generation of self-similar traffic, simulation of queuing process and simulation of overflow probability [19, 21].

This can be demonstrated with buffer overflow simulation in an SSM/M/1/B queueing system (B < ∞, i.e. queuing systems with finite buffer capacity) with self-similar queuing processes. In this case, the difference with an M/M/1/B queuing system is that the arrival rate λ_j into an SSM/M/1/B queuing system is not a constant value. It depends on the sequential number of time-series i, the total number of observation n and the Hurst parameter H, which determine the self-similarity rate. The analyzed SSM/M/1/B queuing system has exponential service times with constant rates $1/\mu$ as shown in Fig. 1.

The flow balance equations are given below [11, 22, 23, 32, 36]:

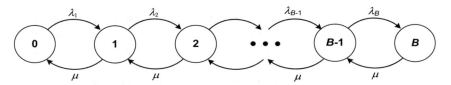

Fig. 1 State transition diagram for a SSM/M/1/B self-similar queuing system

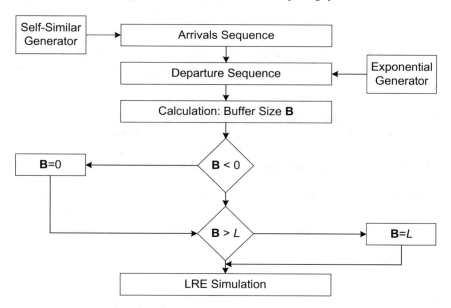

Fig. 2 Steady-state simulation of self-similar queuing process

$$\lambda_j = \lambda(i, n, H); \quad j = 1, 2, \ldots, B$$
$$\lambda_j = 0; \qquad\qquad j \geq B + 1 \tag{14}$$
$$\mu_j = \mu; \qquad\qquad j = 1, 2, \ldots, B + 1$$

This system is stable with a throughput $\rho = \frac{\lambda(i,n,H)}{\mu} < 1$.

Let us consider two separated cases: $\rho = 1$, and $\rho \neq 1$. For $j = 0, 1, 2, \ldots, B$ the distribution of the number of calls in the system is $P_j = \rho^j P_0$, which is determined according to (15):

$$P_j = \frac{\rho^j (1-\rho)}{1-\rho^{B+1}}; \quad \rho \neq 1$$
$$P_j = \frac{1}{B+1}; \qquad \rho = 1 \tag{15}$$

Therefore, the rate at which calls are blocked and lost is λP_B. The self-similar queuing process is described with the steady-state simulation scheme, shown in Fig. 2 [20, 21].

The self-similar traffic can be generated, and the sequence of arrivals is obtained. The fixed length of self-similar traffic is extracted by fixing the number of observations. As the service process is Markovian, the sequence of departures has exponential distribution, generated with inverse transform generator [20, 22, 23, 33].

The next step is buffer size calculation. If the service size is greater than the size of arrivals, than the buffer size $B = 0$, as it is impossible to have negative buffer size. In cases when the buffer is greater than the overflow L, i.e. $B > L$, the traffic is lost, therefore we have assumed that $B = L$.

The simulation is performed with splitting [20, 22, 23] using a variant based on the RESTART method [9, 10], where any chain is split by a fixed factor when it hits a level upward, and one of the copies is tagged as the original for that simulation level.

When any of those copies hits that same level downward, if the copy is the original it just continues its path, otherwise it is killed immediately [20, 22, 23].

This rule applies recursively, and the method is implemented in a depth-first fashion, as follows: whenever there is a split, all the non-original copies are simulated completely, one after the other; then the simulation continues for the original chain [15, 20, 21, 37–39].

The reason for eliminating most of the paths that go downward is to reduce the work. The calculation of the buffer size for all sequences gives possibilities for determining the overflow probability using the RESTART with the Limited Relative Error (LRE) algorithm [20, 21].

4   Limited Relative Error Algorithm for Buffer Overflow Simulation

A standard version of Limited Relative Error algorithm provides possibility to determine the complementary cumulative function of arrivals at single server buffer queues with Markov processes [6, 20, 21].

In order to describe the LRE principles for steady-state simulation in Disctere-Time Markov Chains (DTMC), let us consider a homogeneous two-node Markov chain, which is extended to regular Disctere-Time Markov Chain, consisting of $(k + 1)$ nodes with states, respectively $S_0, S_1, \ldots, S_k$, as shown in Fig. 3.

We obtain the random generated sequence $x_1, x_2, \ldots, x_t, x_{t+1}, \ldots$ for $x = 0, 1, \ldots, k$, for which a transition for state S_j at the time t exists, e.g. $x_t = j$ and there are no constraints to the parameters of the transition probabilities:

$$p_{ij} = P(j|i); \quad (i, j) = 0, 1, \ldots, k; \quad \sum_{j=1}^{k} p_{ij} = 1 \qquad (16)$$

There are no absorbing states S_i at $p_{ii} = 1$ for all stationary probabilities $P_j \, j = 0$, 1, ..., k, which satisfy the constraint condition:

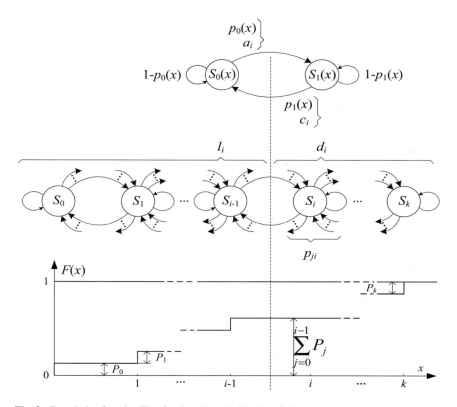

Fig. 3 Cumulative function F(x) for (k + 1)-node Markov chain

$$0 \leq P_j < 1; \quad \sum_{j=0}^{k} P_j = 1. \tag{17}$$

The cumulative distribution $F(x)$ can be presented as:

$$\left. \begin{array}{l} F(x) = F_i; \quad (i - 1) \leq x < i; i = 1, 2, \ldots, k + 1; \\ F_i = \sum_{j=0}^{i-1} P_j; \ F_0 = 0; \quad F_{k+1} = 1; \end{array} \right\} . \tag{18}$$

In order to simulate the $(k + 1)$ nodes of Markov chain, the complementary cumulative distribution $G(x) = 1 - F(x)$ that is more significant, can be determined along with the local correlation coefficient $\rho(x)$ through the limited relative error approach.

After having the two-node Markov chain defined as shown in Fig. 1, with changing the states n times, an estimation of the local correlation coefficient $\hat{\rho}$ can be received, which connects the number of transitions through a dividing line $a_i \approx c_i$, with the total number of observed events $l_i = n - d_i$ ($\beta = 0, 1, \ldots, i - 1$) at left side, and d_i at right side ($\beta = i, i + 2, \ldots, k$).

The value of simulated complementary cumulative distribution $\widehat{G}_i$ can be defined directly by using relative frequency d_i/n, if there is enough number of samples:

$$n \geq 10^3; \quad (l_i, d_i \geq 10^2); \quad (a_i, c_i, l_i - a_i, d_i - c_i) \geq 10. \tag{19}$$

The posterior equations can be used for the complementary function $\widehat{G}(x)$, the average number of generated values of $\hat{\beta}$, the local correlation coefficient $\hat{\rho}(x)$, the correlation coefficient $\mathrm{Cor}[x]$ and the relative error $\mathrm{RE}[x]$:

$$\widehat{G}(x) = \widehat{G}_i = d_i/n \quad \hat{\beta} = \frac{1}{n} \sum_{i=1}^{k} d_i$$

$$\hat{\rho}(x) = \hat{\rho}_i = 1 - \frac{c_i/d_i}{1 - d_i/n} \quad \begin{matrix} i - 1 \leq x < i \\ i = 1, \ldots, k \end{matrix}$$

$$\mathrm{Cor}[x] = \mathrm{Cor}_i = (1 + \hat{\rho}_i)(1 - \hat{\rho}_i) \quad \mathrm{RE}[x]^2 = \mathrm{RE}_i = \frac{1 - d_i/n}{d_i} \cdot \mathrm{Cor}_i. \tag{20}$$

The main advantage of this approach is that the relationships between transitions c_i are obtained with routine statistical calculations.

The necessary total number of simulation trails n is determined with the maximal relative error $\mathrm{RE}_{\max}[x]^2$ and with the less value of the function $G(x)$, presented as $G_{\min} = \widehat{G}_k$ in approximation equation:

$$n = \frac{(1 - G_{\min})}{G_{\min} \cdot \mathrm{RE}_{\max}[x]^2} \approx \frac{\mathrm{Cor}_k}{\widehat{G}_k \cdot \mathrm{RE}_{\max}[x]^2};$$

$$\mathrm{Cor}_k = \frac{1 + \hat{\rho}_k}{1 - \hat{\rho}_k}. \tag{21}$$

This procedure can be described with a standard version of Limited Relarive Error algorithm of random discrete sequences for buffer arrivals.

Algorithm 1. Limited Relative Error of Buffer Overflow Simulation

Step 1: Initialization of minimal and maximal values of the simulation parameter.

Step 2: Estimation and management of the simulation time.

Cycle L_1: Determine the current variable for calculating the Markov chain, e.g. $\omega := \beta$;

Generate a new value for β with given distribution. Increase the number of state h_β;

If the condition $\beta < \omega$ is true, then increase the number of transitions $c_{\beta+1}$ while it reaches the value of c_ω.

Cycle L_2: Determine the total number of events at the left part l_s and at the right part d_s of the Markov chain and number of transitions $a_s := c_s$;

Check on the constraint condition (18) for the index i = s. If the constraint condition (19) is true, then calculate the posterior values of the local correlation coefficient $\hat{\rho}_s$ and relative error RE[x] according to (20);

Calculate whether the relative error RE[x] ≤ REmax[x];

If s < k, than leave the cycle L_2.

If the index s = k is reached, than leave the cycle L_1 and increase the index of the simulation time s: = s + 1;

Step 3: Printing out the experimental results for i = 1, 2, ..., k. The results for the total frequency d_i are determined according to:

$$d_i = \sum_{j=1}^{k} h_j \quad \text{for } i = 0, 1, \ldots, k \quad \text{where} \quad d_0 = n. \tag{22}$$

The values of the complementary function $\widehat{G}_i$, the local correlation coefficient $\hat{\rho}_s$ and the relative error RE[x] are calculated as given in (20).

5 Simulation Results and Performance Analysis

As an example, the overflow probability of an SSM/M/1/B queuing system has been simulated with different characteristics of self-similar arrival processes.

To demonstrate the effects of self-similarity on the buffer overflow probability, the obtained experimental results were compared with the complementary cumulative distribution in a traditional single server finite buffer queue M/M/1/B. The obtained results in logarithmic scale are given in Fig. 4.

To get representative and steady results the sequences of 10 000 observations were used. With the suggested LRE algorithm the values of complementary cumulative function G(x) for different buffer sizes were calculated. The calculations were provided with the step m = 4. One can see in Fig. 4 that the increasing Hurst parameter has led to an insignificant decrease in the overflow probability. For example, for the value of Hurst parameter H = 0.6 the overflow probability is $G(L) = 10.45 * 10^{-2}$, and for H = 0.9 it is $G(L) = 5.6 * 10^{-2}$.

On the other hand, the overflow probability of self-similar queuing system has increased significantly in comparison with the theoretical M/M/1/B queuing system, for which $G(L) = 4.79 * 10^{-5}$.

After that, the simulation was repeated for the SSM/M/1/B queuing system by using self-similar arrival process with H = 0.6 and different buffer sizes. The obtained results for buffer size B = 40, B = 60 and B = 80 are demonstrated in Fig. 5.

One can see that since the buffer size was increased twice, the overflow probability has been changed simply by about two orders of magnitude—from $1.045 * 10^{-1}$ to $6.4 * 10^{-3}$.

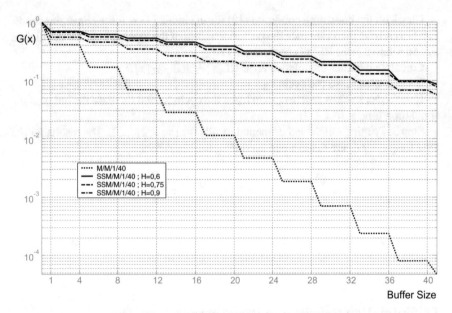

Fig. 4 Buffer overflow probability (L = 41) in SSM/M/1/40 self-similar queuing system

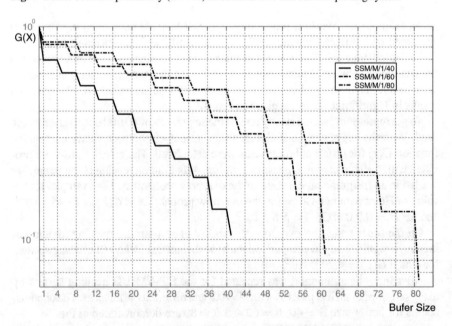

Fig. 5 Buffer overflow probability in SSM/M/1/B self-similar queuing system with different buffer sizes

Finally, it was confirmed that in order to design a finite buffer single server model with long-range dependent self-similar arrival processes, the buffer size has to be increased many times in order to decrease the overflow probability.

6 Sequential Generators of Self-similar Teletraffic

As defined in [19, 34], one can make a sequential Markovian model that imitates a self-similar sequence. However, this approach would have a disadvantage because of connection between the model parameters and its self-similar properties that is difficult to understand.

Markovian models for self-similar traffic require including several control parameters with a wide range of input values. As a result, controlling these values in sequential generators is much more complex than in generators of fixed-length sequences of self-similar processes with a given Hurst parameter [16, 19, 34].

The following efficient candidate sequential generators have been considered, based on:

- Fractal-Binomial-Noise-Drivent Poisson Processes (FBNDP);
- Superposition of Fractal Renewal Processes (SFRP);
- M/G/∞ Processes (MGIP);
- Pareto-Modulated Poisson Processes (PMPP);
- Spatial Renewal and FGN Processes;
- Superposition of Autoregressive Processes (SAP).

For the standard Fractal Renewal Process (FRP), inter-event times are independent random variables [4, 8, 34]. The marginal Probability Density Function (PDF) of such a FRP can be defined as (23):

$$f(t) = \begin{cases} 0, & t \leq A, \\ \delta A^{\delta} t^{-(\delta+1)}, & t > A \end{cases} \tag{23}$$

where $0 < \delta < 2$. Selecting δ in this interval proves far superior to $0 < \delta < 1$ for the same required values that the inter-event time PDF can be further improved [19, 34].

The improved PDF of the FRP decays as a power law, as shown in (24):

$$f(t) = \begin{cases} \delta A^{-1} e^{-\delta t/A}, & 0 < t \leq A, \\ \delta e^{-\delta} A t^{-(\delta+1)}, & t > A, \end{cases} \tag{24}$$

which is continuous for all t, producing smoother spectral density function.

A method, based on the FBNDP processes adds M independent and identically distributed alternating FRP to generate a Fractal Binomial Noise (FBN) process that serves as the rate function for a Poisson process. The FBNDP requires five input parameters to generate self-similar sequences: A, δ, R, Δt, and M. The resulting Hurst parameter H assumes the value $(\alpha + 1)/2$, where α is a shape parameter of beta-distribution. The suggested approach for generating self-similar sequences in a time domain is advancing with the intervals Δt. If S is a simulation clock, which advances in time, and $S^{(j)}$ is the elapsed time of the j-th FRP sequence, then $S^{(j)} = \tau_0^{(j)} + \tau_1^{(j)} + \cdots + \tau_k^{(j)}$ for some k and $j = 1, 2, \ldots, M$, where $\tau_k^{(j)}$ is the inter-arrival time [34].

A method based on the Superposition of the Fractal Renewal Processes (SFRP) uses a group of independent and identical FRP. This method requires three parameters, i.e., α and A from the individual FRPs, and M, which is the number of FRPs superposed. The resulting Hurst parameter H assumes the value $(\alpha + 1)/2$ where α is a shape parameter of beta-distribution [34].

A method based on the Superposition of Autoregressive Processes (SAP) generates asymptotically self-similar sequences when aggregating several independent autoregressive processes. In the simplest case this can be presented as the sum of two autoregressive processes of the first order. The resulting Hurst parameter H is linearly dependent on a shape parameter α_2 of beta-distribution, while α_1 can be selected arbitrary; for example, $\alpha_1 = 1$ in all cases that are considered [34].

Accordingly, the sequences of self-similar pseudo-random numbers $X_0, X_1, \ldots$ have been generated with the algorithms for sequential generators [34], improved by Lokshina [19]. It took 1 min 38 s. on the Centrino-based HP PC (1.76 GHz) to generate the Fractal-Binomial-Noise-Driven Poisson Processes (FBNDP) traffic sequence of 1,048,576 numbers; however, it took only 3.5 s on the Centrino-based HP PC (1.76 GHz) to generate the Superposition of Autoregressive Processes (SAP) traffic sequence of 1,048,576 numbers [19].

7 Fixed-Length Sequence Generators of Self-similar Teletraffic

The most frequently studied discrete-time models of self-similar teletraffic belong to Fractional Autoregressive Integrated Moving-Average (F-ARIMA) and FGN processes. Radev and Lokshina [32, 34] and Lokshina [19] considered the following possible fixed-length sequence generators based on:

- F-ARIMA processes;
- Fast Fourier Transform (FFT);
- FGN and Daubechies Wavelets (DW) [7];
- Random Midpoint Displacement (RMD);
- Successive random additions.

Recent studies of communication networks in a Smart World have evidently shown the use of self-similar time series is very important, since a traditional approach to describing processes by generating calls with standard statistical distributions is not valid for current IP network implementations, where the LRD traffic is observed. Normal self-similar time series cannot describe different applications at once in communication networks. Therefore, a suggestion is made to multiply the values of self-similar time series with those standard statistical distributions, known to be used to adequately describe different applications in communication systems. However, this approach does not work well for the time parameter. Various communication network traffic flows are simulated with different statistical distributions that could indirectly describe the time parameter in frequency domain. The obtained values show self-similar long-range dependent characteristics in wide range of time scales and are appropriate for cluster analysis in frequency domain.

Let us consider F-ARIMA$(0, d, 0)$ method for generating self-similar sequences, where d is the fractional differencing parameter, $0 < d < 1/2$, and let us generate the process $X = \{X_i: i = 0, 1, 2, ..., n\}$ with a normal marginal distribution, the mean of zero and the variance σ_0^2, and the Autocorrelation Function (ACF), $\{\rho_k\}$ ($k = 0, \pm 1, ...$) defined as (25):

$$\rho_k = \gamma_k/\gamma_0 = \frac{\Gamma(1 - d)\Gamma(k + d)}{\Gamma(d)\Gamma(k + 1 - d)}, \tag{25}$$

where:

$$\gamma_k = \sigma_0^2 \frac{(-1)^k \Gamma(1 - 2d)}{\Gamma(k - d + 1)\Gamma(1 - k - d)}. \tag{26}$$

Accordingly, the fixed-length sequence of self-similar pseudo-random numbers has been generated through the following steps of Algorithm 2, developed by Radev and Lokshina [34] and improved by Lokshina [21].

Algorithm 2. Fixed-Length Sequence Generator, based on Fractional Autoregressive Integrated Moving-Average Processes (F-ARIMA)

Step 0: Set $N_0 = 0$ and $D_0 = 1$. X_0, the first pseudo random element in the output self-similar sequence, is generated from the normal distribution $N(0, \sigma_0^2)$, where σ_0^2 is the required variance of the X_i.

Step 1: ($i = 1, ..., n - 1$). Compute mean$_i$ and var$_i$ of X_i recursively, using the following equations:

$$N_i = \rho_i - \sum_{j=1}^{i-1} \phi_{i-1,j}\rho_{i-j}, \quad D_i = D_{i-1} - N_{i-1}^2/D_{i-1}, \quad \phi_{ii} = N_i/D_i,$$

$$\phi_{ij} = \phi_{i-1,j} - \phi_{ii}\phi_{i-1,i-j}, \quad j = 1, \ldots, i - 1,$$

where ϕ_{ij}, $i = 0$, $j = 0, \ldots, n - 1$, is given by:

$$\phi_{ii} = -\binom{i}{j}\frac{(j - d - 1)!(i - d - j)!}{(-d - 1)!(i - d)!}, \quad mean_i = \sum_{j=1}^{i} \phi_{ij}X_{i-j},$$

$$var_i = (1 - \phi_{ii}^2)var_{i-1},$$

Generate X_i from $N(mean_i, var_i)$. Increase i by 1.

Step N: If i = n, then stop.

A self-similar traffic sequence $\{X_1, X_2, \ldots, X_n\}$ has been obtained. It took 4 h, 1 min and 24 s on the Centrino-based HP PC (1.76 GHz) to generate the F-ARIMA traffic sample sequence with 1,048,576 numbers [19]. Therefore, the algorithm, based on the F-ARIMA method, is too rigorous to be used to generate long sample sequences.

In this chapter a new advanced generator of pseudo-random self-similar sequence based on Fractional Gaussian Noise (FGN) and Daubechies Wavelet (DW), called the FGN-DW method, is proposed. The generators based on the DW produce more accurate self-similar sequences than those based on Haar Wavelets (HW) [1, 19, 32, 34]. In that case, not only estimates of H obtained from the DW are closer to the true values than those from the HW, but also variances obtained from the DW are much lower.

The reason is that the DW produce smoother wavelet coefficients that are used in the Discrete Wavelet Transform (DWT), than the HW. The HW are discontinuous, and they do not have good time-frequency localisation properties, since their Fourier transforms decay as $|\lambda|^{-1}$, $\lambda \to \infty$, meaning that resulting decomposition has a poor scale. Therefore, the DW produce more accurate coefficients than the HW.

The suggested approach for generating synthetic self-similar FGN sequences in a time domain is based on a DWT. Wavelets can provide compact representations for a class of FGN processes, because the structure of wavelets naturally matches the self-similar structure of the LRD processes.

This chapter confirms that the FGN-DW method is sufficiently fast being used for practical generation of synthetic self-similar sequences as simulation input data. The wavelet analysis transforms a sequence onto a time-scale grid, where the term scale is used instead of frequency, because the mapping is not directly related to frequency as in the Fourier transform.

The wavelet transform generates the wavelet coefficients $d_x(i, j)$ from a sequence of given numbers. For a LRD process, the variance of the wavelet coefficients at each level i is defined by (27):

$$E[d_x(j, \cdot)]^2 = C P_f 2^{i(2H-1)}, \tag{27}$$

where $C > 0$ and (28) is true:

$$P_f = 2(2\pi)^{1-2H} c_\gamma E(2H - 1) \sin((1 - H)\pi), \tag{28}$$

where $E(\cdot)$ is the Euler function and c_γ is a positive constant. The power parameter P_f that plays a major role in fixing the absolute size of the LRD process and generates effects in the applications is an independent quantitative parameter with the dimensions of variance.

A spectral estimator can be obtained though calculating time average Q_i of the $d_x(i, \cdot)$ at a given scale that could be described as (29):

$$Q_i = \frac{1}{n_i} \sum_{j=1}^{n_i} d_x^2(i, j), \tag{29}$$

where n_i is the number of wavelet coefficients at scale i, i.g. $n_i = 2^{-i}n$, where n is the number of data points. The estimator uses a weighted linear regression as the variance of $\log_2(Q_i)$ vary with $\log_2(2^i)$. An estimated Hurst parameter $\widehat{H}$ could be obtained with the linear regression (20):

$$y_i = \log_2(Q_i) = (2\widehat{H} - 1)i + c + 1/(n_i \ln 2). \tag{30}$$

where c is a constant.

The wavelet transform delivers good resolution in both time and scale, as compared to the Fourier transform, which provides only good frequency resolution.

Accordingly, the fixed-length sequence of self-similar pseudo-random numbers has been generated through the following steps of Algorithm 3, developed in [34] and improved in [19, 21].

Algorithm 3. Fixed-Length Sequence Generator, based on Fractional Gaussian Noise and Daubechies Wavelet (FGN-DW)

Step 1: Given: H, Start for i = 1 and continue until i = n. Calculate a sequence of values $\{f_1, f_2, \dots, f_n\}$ using (31):

$$f(\lambda, H) = c_f |\lambda|^{1-2H} + O(|\lambda|^{\min(3-2H,2)}), \tag{31}$$

where $c_f = \sigma^2 \cdot (2\pi)^{-1} \sin(\pi H) \Gamma(2H + 1)$, $O(\cdot)$ represents the residual error and $f_i = \hat{f}(\pi i/n; H)$, the value of the frequencies f_i corresponds to the spectral density of an FGN process for f_i ranging between π/n and π.

Step 2: Multiply $\{f_i\}$ by realizations of the independent exponential random variable with the mean of one to obtain $\{\hat{f}_i\}$, because the spectral density estimated for a given frequency is distributed asymptotically as the independent exponential random variable with the mean $f(\lambda, H)$.

Step 3: Generate a sequence $\{Y_1, Y_2, \dots, Y_n\}$ of complex numbers such that $|Y_i| = \sqrt{\hat{f}_i}$ and the phase of Y_i is uniformly distributed between 0 and 2π. This random phase technique preserves the spectral density corresponding to $\{\hat{f}_i\}$. It also makes the marginal distribution of the final sequence normal and produces the requirements for FGN.

Step 4: Calculate two synthetic coefficients of orthonormal Daubechies wavelets. The output sequence $\{X_1, X_2, \dots, X_n\}$ representing approximately self-similar FGN process is obtained by applying the inverse Daubechies wavelets transformation of the sequence $\{Y_1, Y_2, \dots, Y_n\}$.

A self-similar traffic sequence $\{X_1, X_2, \dots, X_n\}$ has been obtained. It took only 2 s on Centrino-based HP PC (1.76 GHz) to generate a sequence of 1,048,576 numbers [19, 21].

Analysis of the mean times, required to generate sequence of a given length, confirmed that advanced sequential generators of self-similar teletraffic are more attractive for practical simulation studies of computer networks than the F-ARIMA-based generators, since they are much faster. However, these generators require more input parameters. Selecting appropriate values as input parameters is a problem that remains. Additionally, defining relationship between the Hurst parameter and two shape parameters of beta-distribution in the case of Superposition of Autoregressive Processes (SAP) is also a problem that remains.

8 Simulation Results and Performance Analysis

Let us consider generation of self-similar time series with the help of Algorithm 3 with one million values Hurst parameter $H = 0.5$; time scaling factor equal to 10 and number of wavelet coefficients equal to 10. The obtained results are shown in Fig. 6.

For the generated time series, the values $Q(j_1)$ were calculated as function of j_1, as shown on Fig. 7. When linear regression (30) was implemented, better representation was obtained if to compare to the analytical values, as confirmed in Fig. 8.

The obtained results demonstrate self-similar and long range dependent properties of the generated time series sequence, as established in Figs. 9 and 10.

The Hurst parameter H has been determined with a very high precision that obtained the exact value of 0.504 for the original value of $H = 0.5$. The obtained results were estimated based on correspondence criteria with parameter $Q = 0.16155$,

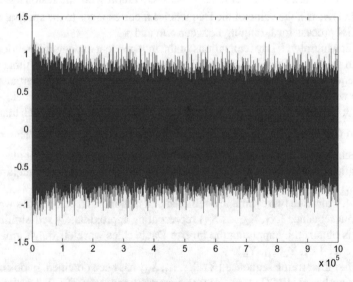

Fig. 6 Self-similar time series with 1,000,000 values

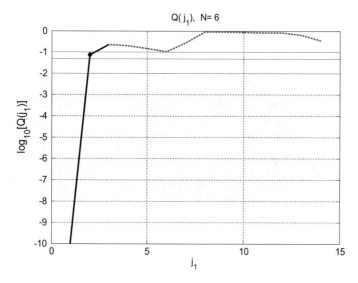

Fig. 7 Comparison of self-similar time series with different scale parameters

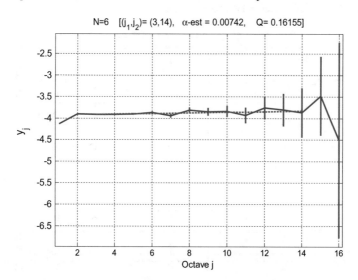

Fig. 8 Comparison between generated time series and analytical values

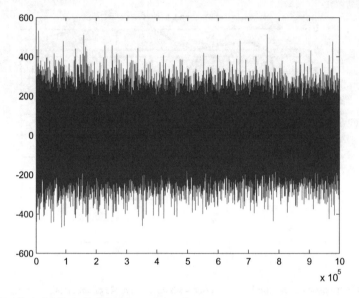

Fig. 9 Product of self-similar and Pareto time series

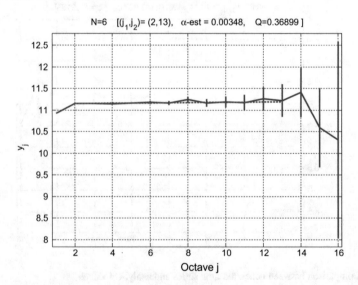

Fig. 10 Comparison between generated time series and analytical values obtained with weighted linear regression

and time series with equivalent length of 1,000,000 values were generated with Pareto distribution and $\alpha = 1.8$. Then, they were multiplied with self-similar time series, generated with the use of Algorithm 3.

The implementation of wavelet criteria provided good results for $Q(j_1)$ as function of j_1, confirmed in Fig. 10, which are very similar to the results presented in Fig. 8. Considering the whole time interval $(j_1, j_2) = [1, 16]$, the optimal values were registered at $(j_1, j_2) = [2, 13]$, in addition to the fact that the correspondence criteria parameter Q increased from 0.16155 to 0.36899.

The linear regression gave an opportunity to define the Hurst parameter with better precision (0.502 vs. 0.504, for the exact value of 0.5). The logarithmic diagram obtained with weighted linear regression is shown in Fig. 10. The confidential intervals are shown with use of vertical lines. They are small for $j_1 < 13$ and obtained with confidential probability of 0.95 after generation of 2^{10} (1,024) samples.

9 Conclusions

This chapter focuses on building the future smart world by using recent advances in networking science and practice, which range from traffic engineering and control to innovative wireless scenarios and include key challenges arising from the upcoming interconnection of massive numbers of devices, sensors and humans. The authors bridge the gap between performance modeling and real-life operational aspects, as well as they leverage measurement data to provide a better understanding of the wired and wireless networks' operation under realistic conditions. Specifically, the authors examine self-similar teletraffic in communication networks, capable of carrying diverse traffic, including self-similar teletraffic, and supporting diverse levels of Quality of Service (QoS).

Recent studies of communication network traffic in a smart world have clearly shown that teletraffic exhibits long-range dependent self-similar properties over a wide range of time scales, which are very different from the properties of traditional models. Therefore, self-similar queuing systems are appropriate reference models for teletraffic as they provide capacity to estimate queuing and network performance and quality, allocate resources, and ensure the Quality of service (QoS).

One of the problems that telecommunications researchers face during simulation is how to generate long synthetic sequential self-similar sequences. To solve the problem three aspects must be considered:

- how accurately self-similar process can be generated;
- how quickly the methods generate long self-similar sequences, and
- how appropriately self-similar processes can be used in sequential simulation.

Advanced self-similar models and efficient algorithms for self-similar teletraffic simulation were developed and applied. Numerical examples and simulations results were shown and analyzed.

Particularly, steady-state simulation of a SSM/M/1/B self-similar queuing system with long-range dependent processes and finite buffer capacity was performed with splitting using a variant of the RESTART/LRE method, in order to accelerate the buffer overflow simulation. The models of the SSM/M/1/40 self-similar queuing system were used with different parameters of long-range dependent self-similar processes and different buffer sizes. Furthermore, efficient algorithms to simulate the buffer overflow probability were developed and applied to generate stochastic, long-range dependent self-similar processes with a limited relative error that are appropriate and particularly efficient being used to evaluate reliability and performance in communication networks in a smart world, while higher performance networks must be described by lesser buffer overflow probabilities.

References

1. Abry P, Flandrin P, Taqqu MS, Veitch D (2000) Wavelets for the analysis, estimation and synthesis of scaling data. In: Self-similar network traffic analysis and performance evaluation. Wiley, pp 39–88
2. Auer L, Kryvinska N, Strauss C (2009) Service-oriented mobility architecture provides highly-configurable wireless services. IEEE, pp 1–1
3. Bashah NSK, Kryvinska N, Do TV (2012) Quality-driven service discovery techniques for open mobile environments and their business applications. J Serv Sci Res 4(1):71–96
4. Bobbio A, Horváth A, Scarpa M, Telek M (2003) Acyclic discrete phase type distributions: properties and a parameter estimation algorithm. Perform Eval 54(1):1–32
5. Boxma O, Cohen J (2000) Self-similar network traffic and performance evaluation. Wiley, New York
6. Bucklew JA (2004) Introduction to rare event simulation. Springer, New York
7. Daubechies I (1992) Ten lectures on wavelets. In: CBMS-NFS regional conference series in applied mathematics, vol 61. SIAM Press, Philadelphia, Pennsylvania
8. Faraj R (2000) Modeling and analysis of self-similar traffic in ATM networks. PhD thesis
9. Georg C, Fuss O (1999) Simulating rare event details of ATM delay time distributions with RESTART/LRE. In: Proceedings of the IEEE international teletraffic congress, ITC16. Elsevier, pp 777–786
10. Georg C, Schreiber F (1996) The RESTART/LRE method for rare event simulation. In: Proceedings of the Winter simulation conference, Coronado, CA, USA, pp 390–397
11. Giambene G (2005) Queueing theory and telecommunications: networks and applications. Springer, New York
12. Kryvinska N, Auer L, Zinterhof P, Strauss C (2008) Architectural model of enterprise multi-service network maintaining mobility. IEEE, pp 1–22
13. Kryvinska N, Strauss C, Collini-Nocker B, Zinterhof P (2011) Enterprise network maintaining mobility—architectural model of services delivery. Int J Pervasive Comput Commun 7:114–131. https://doi.org/10.1108/17427371111146419
14. Kushner H (2001) Heavy traffic analysis of controlled queueing and communications networks. Springer, New York
15. L'Ecuyer P, Demers V, Tuffin B (2006) Splitting for rare-event simulation. In: Proceedings of the 2006 Winter simulation conference, pp 137–148
16. Law AM, Kelton WD (2006) Simulation modeling and analysis, 4th edn. McGraw Hill, New York
17. Lokshina IV (2011) Modeling and simulation of traffic with integrated services at media getaway nodes in next generation networks. Int J Interdiscip Telecommun Netw 3(3):1–14

18. Lokshina IV (2012) Performance evaluation of multi-service UMTS core networks with clustering and neural modelling. Int J Mob Netw Des Innov 4(1):24–31
19. Lokshina IV (2012) Study about effects of self-similar IP network traffic on queuing and network performance. Int J Mob Netw Des Innov 4(2):76–90
20. Lokshina IV (2014) Study on estimating probability of buffer overflow in high-speed communication networks. In: Proceedings of 2014 networking and E-commerce conference (NAEC'2014), Trieste, Italy, pp 306–32
21. Lokshina IV (2015) Study on estimating buffer overflow probabilities in high-speed communication networks. Telecommun Syst (Springer: NY) 62(2):289–302
22. Lokshina IV, Bartolacci MR (2012) Accelerated rare event simulation with Markov chain modelling in wireless communication networks. Int J Mob Netw Des Innov 4(4):185–191
23. Lokshina IV, Bartolacci MR (2012) Buffer overflow simulation in queuing networks with finite buffer capacity. Int J Mob Netw Des Innov 5(3):144–151
24. Lokshina IV, Lanting CJM (2018) A qualitative evaluation of IoT-driven eHealth: knowledge management, business models and opportunities, deployment and evolution. In: Hawaii international conference on system sciences (HICSS-51), pp 4123–4132
25. Lokshina I, Zhong H (2017) Analysis of turbo code behavior with extrinsic information transfer charts in high-speed wireless data services. Int J Interdiscip Telecommun Netw 9(4):26–36
26. Lokshina IV, Durkin BJ, Lanting CJM (2017) Data analysis services related to the IoT and Big Data: potential business opportunities for third parties. In: Hawaii international conference on system sciences (HICSS-50), pp 4402–4411
27. Lokshina IV, Durkin BJ, Lanting CJM (2017) Data analysis services related to the IoT and Big Data: strategic implications and business opportunities for third parties. Int J Interdiscip Telecommun Netw 9(2):37–56
28. Marie RR, Blackledge JM, Bez HE (2007) Characterization of internet traffic using a fractal model. In: Proceedings of the conference on signal processing, pattern recognition and applications (SPPRA 2007), pp 156–167
29. Park K, Willinger W (2000) Self-similar network traffic and performance evaluation. Wiley, New York
30. Popescu A (2001) Traffic self-similarity. In: Proceedings of the IEEE international conference on telecommunications (ICT), pp 20–24
31. Radev D, Lokshina I (2006) Performance analysis of mobile communication networks with clustering and neural modeling. Int J Mob Netw Des Innov 1(3/4):188–196
32. Radev D, Lokshina I (2007) Modeling and simulation of self-similar teletraffic. In: Proceedings of the industrial simulation conference 2007, Delft, The Netherlands, pp 355–359
33. Radev D, Lokshina I (2009) Modeling and simulation of self-similar wireless IP network traffic. In: Proceedings of the IEEE wireless telecommunications symposium (WTS) 2009, Prague, Czech Republic, pp 1–6
34. Radev D, Lokshina I (2010) Advanced models and algorithms for self-similar network traffic simulation and performance analysis. J Electr Eng 61(6):341–349
35. Stathis C, Maglaris B (2000) Modeling the self-similar behaviour of network traffic. J Comput Netw 34(1):37–47
36. Trivedi K (2001) Probability and statistics with reliability, queueing, and computer science applications. Willey, New York
37. Villen-Altamirano J (1998) RESTART method for the case where rare events can occur in retrials from any threshold. Int J Electron Commun 52(3):83–190
38. Villen-Altamirano M, Villen-Altamirano J (2006) On the efficiency of RESTART for multidimensional systems. ACM Trans Model Comput Simul 16(3):251–279
39. Villen-Altamirano M, Villen-Altamirano J (2011) The rare event simulation method RESTART: efficiency analysis and guidelines for its application. Network performance engineering: a handbook on convergent multi-service networks and next generation internet. Springer, Berlin, Heidelberg, pp 509–547

Izabella Lokshina Ph.D. is Professor of MIS and chair of Management, Marketing and Information Systems Department at SUNY Oneonta, Oneonta, NY, 13820, USA. Her main research interests are intelligent information systems and communication networks, as well as complex system modeling.

Hua Zhong Ph.D. is Associate Professor of Management in Management, Marketing and Information Systems Department at SUNY Oneonta, Oneonta, NY, 13820, USA. His main research interests are supply chain modeling, scheduling, queuing systems and simulation.

Cees J. M. Lanting Ph.D. is Senior Consultant at DATSA Belgium, Leuven, VBR, 3010, Belgium. He heads the IoT Development of Electronics and Applications (IDEA) and IoT Data Acquisition (IDA) labs. His main research interests are smart communications and IoT.

Assessment of the Formation of Administration Systems in the Enterprise Management

Nataliia Ortynska, Oleg Kuzmin, Vadym Ovcharuk
and Volodymyr Zhezhukha

Abstract The paper develops a method for assessing the level of administration systems formation in the enterprise management. This method is based on the multi-criteria approach and the use of the trapezoidal membership function of confidentiality, availability and integrity parameters of the administration systems in certain fuzzy term sets, which enables the adoption of managerial decisions to ensure a higher level of compliance of such systems with the parameters "as is" and "to be". It describes peculiarities of diagnosing the actual state of administration systems in the enterprise management as the preconditions for assessing their level of formation. The article substantiates expediency of isolating confidentiality, availability and integrity as grounds for identifying gaps between the "as is" and "to be" of administration systems in the enterprise management. It suggests using fuzzy logic tools and mathematical apparatus when assessing the level of formation of administration systems. Method for assessing the level of such formation has been applied in practice in the activities of a number of companies.

Keywords Administration · Documentation · Information · Enterprise · Reengineering · Management

N. Ortynska · O. Kuzmin · V. Ovcharuk · V. Zhezhukha (✉)
Lviv Polytechnic National University, Bandera Street, 12, Lviv, Ukraine
e-mail: volodymyr.y.zhezhukha@lpnu.ua

N. Ortynska
e-mail: nataliia.v.ortynska@lpnu.ua

O. Kuzmin
e-mail: oleh.y.kuzmin@lpnu.ua

V. Ovcharuk
e-mail: vadym.v.ovcharuk@lpnu.ua

© Springer Nature Switzerland AG 2020   161
N. Kryvinska and M. Greguš (eds.), *Data-Centric Business and Applications*,
Lecture Notes on Data Engineering and Communications Technologies 30,
https://doi.org/10.1007/978-3-030-19069-9_6

1 Introduction

A typical feature of management processes at enterprises in the present conditions is the shift of attention from solving local problems related to partial objects (individual business processes, organizational management structure, personnel incentives, marketing and logistics, foreign economic activity, etc.) to complex diagnostics of all production and economic activities of the company and adoption of such managerial decisions that will make it possible to achieve a complex effect. Obviously, such a change in priorities determines changes in the roles and significance of the administration systems at the enterprises.

Review and summarization of literature sources makes it possible to conclude that today in theory and practice, effective approaches have been formed both for decision-making in administration systems and for performing all kinds of work in them. More and more company executives are becoming aware of the importance and necessity of such systems as an important factor in gaining competitive advantages in the domestic and foreign markets. At the same time, the scale and diversity of areas of management activity in organizations determines the complexity of multivector diagnostics of administration systems, in particular, from the point of view of their formation.

Identification of the level of formation of administration systems necessitates the solution of a wide range of practical tasks, in particular:

– Diagnosing all processes in each administration system regardless of its level;
– Establishment of the order for the execution of works in the administration system, as well as study of the key parameters of the relevant processes;
– Thorough description of information and documentary data streams in administration systems, etc.

Results of assessing the level of formation of administration systems in the enterprise management will also facilitate the possibility of improving the regulation of relevant processes in such systems. It means that functions will be identified within the administrative processes and will be allocated to concrete executors. Methodological tools can, in particular, be known in the theory and practice of the match matrix "function-link" (functions in the administration system allocated to a certain link of the organizational management structure).

Results of assessing the level of administration systems formation in the enterprise management also provide the possibility of forming an information base for the reengineering of such systems, i.e., their partial or complex reorganization. This area of using assessment results is important given that today both general management systems and administration systems are influenced by dynamic factors of the internal and external environment. In addition, the state of uncertainty and low predictability of the business environment of the companies' activities should be discussed.

2 Model Preparing

2.1 Diagnosing the Actual State of Administration Systems in the Enterprise Management as the Precondition for Assessing Their Level of Formation

The study of the theory and practice makes it possible to conclude that the assessment of the level of formation of administration systems in the enterprise management should be based on a thorough analysis of the state of such systems in a particular organization. At the same time, in this case it is necessary not only to take into account the internal variables, but also to take into account the current global trends in the administration processes. Responsible actors should determine which of these tendencies have a significant impact on the company's administration systems, which will affect it in the short term, and which are highly unlikely to be typical of the administration. Moreover, results of the performed studies give us an opportunity to specify the subject of the analytical study of the actual state of the administration system, which includes:

- Compliance of actual information and document flows in the administration system with the expectations of stakeholders;
- Provision of the administration system with the necessary resource support, first of all, material and financial one (in this case, special attention should be paid to the adequacy of the level of competence of subjects in such systems);
- Compliance of the existing regulations concerning the processes in the administration systems with the requirements of completeness, systemicity and real compliance in the entity's practical activity;
- Compliance of the administration systems and the processes taking place in them with the strategic and tactical goals of the company, as well as corresponding stage of its life cycle;
- Search of competitive advantages of the enterprise in the context of its administration systems;
- Diagnosis of the company's management decisions made over the recent period of time about the administration processes at different levels of the management organizational structure, etc.

Diagnosis of the actual state of the administration systems in the enterprise management as a precondition for assessing their level of formation is complicated in many organizations because such systems may have different levels of complexity or, for example, be non-linear, disproportional, bureaucratic, non-structurally decentralized and have a high risk level in general. Taking into account the research findings [1, pp. 110–111], it is worth noting that an analyst can effectively study such systems during the diagnosis of the actual state of administration systems "only if they are relatively simple for research, determined, linear, sufficiently democratic, and selective". In addition, one should agree with the opinion of these authors on the importance of analyzing administration systems "from the methodological perspective (through the lens of the researcher) and improving the research procedures" [1, p. 111], the functioning and development of these systems in general, that is, "as objects the researcher focuses on".

Assessment of the level of administration systems formation in the enterprise management from strategic or tactical perspectives provides the possibility of automation of relevant processes (for example, information-document flows, algorithms of actions, etc.) (Fig. 1). At the same time, it's important to understand all the processes in such systems, to manage them at a high level, to regulate them, to control them, etc.

Results of assessing the level of administration systems formation in the enterprise management also provide the possibility of forming an information base for the reengineering of such systems, i.e., their partial or complex reorganization. This area of using assessment results is important given that today both general management systems and administration systems are influenced by dynamic factors of the internal and external environment. In addition, the state of uncertainty and low predictability of the business environment of the companies' activities should be discussed. Therefore, taking into account the definition of the concept of reengineering offered by its author M. Hammer, it should be noted that it is through reengineering that there is a "fundamental rethinking and rigorous redesign" of the administration systems of companies that aim to "dramatically improve their performance, such as cost, quality, service, speed".

As you know, this approach is quite often considered in the practice of engineering and reengineering of business processes, when based on the results of the analysis of the situation on the basis of "as is" principle the corresponding "as is" description is generated and the "as is" model is built.

In the analyzed context, it should be noted that at a certain stage of its life cycle each enterprise faces urgent task of reorganizing existing administration systems based on the use of reengineering tools. Quite often, formation of completely different necessary and desirable parameters of such systems is conditioned by changes in the strategy of its activity. As the review and summary of literature sources shows,

Fig. 1 Key areas of use of the results of assessing the level of administration systems formation in the enterprise management. *Note* Singled out by the authors

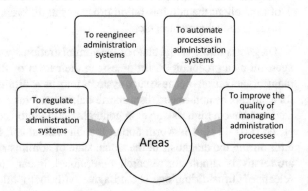

it is due to reengineering in the administration systems that the following tasks are solved:

- Direction of administration processes to achieve the entity's goals;
- Elimination of duplicate functions in administration systems, duplicate information flows, documentation, indicators of performance evaluation, etc.;
- Ensuring efficient management of administration systems taking into account the needs of the "owners" of the relevant processes and the established management nodes, as well as planning and reporting indicators;
- Bridging gaps in the administration systems, in particular, at the interface between different subjects of such systems;
- Optimization of the resource support necessary for the effective functioning of the administration systems;
- Reasonable distribution of work between officials and units that perform various tasks within the administration system;
- Increase in personal responsibility of the subjects for the efficiency of building and using administration systems;
- Increase in the level of efficiency and completeness of information and documentary support in the administration systems necessary to ensure the targeted influence of the managing system on the managed system at all stages of the management technology;
- Development of cross-cutting administration systems, etc.

The above has made it possible to conclude that the key task of diagnosing the actual state of the administration systems in the enterprise management as a precondition for assessing their level of formation is to establish a situation for each administration system based on the well-known "as is" principle. As a result, in the long run, we can talk about the characteristics of these systems from the "to be" standpoint (Fig. 2).

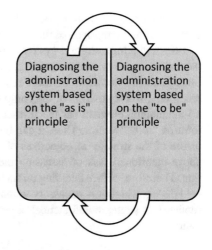

Fig. 2 Vectors of diagnosing the actual state of administration systems in the enterprise management as the preconditions for assessing their level of formation. *Note* Developed by the authors

Diagnosing the administration system based on the "as is" principle

Diagnosing the administration system based on the "to be" principle

Thereafter, the situation is described based on the "as needed" principle and the model of the same name is formed. As a result, the action plan is developed and implemented, and the changes managed until the situation is "as needed" [2–6].

The use of this approach to identify the "as is" and "to be" situation as the key idea of assessing the level of formation of administration systems in enterprise management involves establishing a level of such compliance. Consequently, we can conclude that each individual administration system will be considered to have been formed with a certain level if its actual state "as is" corresponds to a certain "to be" level. In this case, the methodological perspective involves combining the assessment methodology and administration system into a single whole. Such a methodology should take into account the assessment subjects, the basic principles of their work, technologies, approaches, tools, information and other resource support. It is equally important to take into account the subjective interpretation of the assessment results of the administration system formation level, which can lead to conflicts of interest.

Study of theory and practice makes it possible to conclude that the same assessment subjects play a key role when choosing a methodology for assessing the level of administration systems formation: their assessment attitude, as well as the tasks that they must solve at the same time. Availability of appropriate resource support also significantly influences the process of such assessment.

Identifying "as is" and "to be" situation as the key idea of assessing the level of formation of administration systems in the enterprise management necessitates the description of the entity's activities in general, its areas of activity, functions, organizational management structure (organizational models), distribution of responsibility among staff members, etc. This applies not only to each individual administration system, but also to the company as a whole.

In the analyzed context, it should be noted that the "as is" and "to be" approach when assessing the level of formation of administration systems should be taken into account from the point of view of the possibility of such systems to promote the implementation of the company's strategy and goals achievement. In other words, such systems can be considered to be formed at a high level if they provide the opportunity to solve both the global tasks of the company and achieve the goals of each individual employee of the enterprise.

Effective administration systems are aimed at contributing to the strategic and tactical financial or non-financial goals of the organization in every possible way. Such goals are known to be more related to the market and are aimed at strengthening the market position of the company. Availability of strategy and goals is also a fundamental element for the further development of the administration systems towards "to be" vector. Thus, it can be concluded that the absence or unstructured nature of the strategy or objectives of the entity will not enable the solution of the above-mentioned task of assessing the level of the administration system formation. Therefore, before deciding on such an assessment, a strategic session should be held during which it is necessary to realize the mission of the enterprise, to develop (update) its strategic and tactical objectives, and to identify the ways of achieving them.

The study of theory and practice, as well as the results of the studies carried out help us conclude that the assessment of the level of administration systems formation in the enterprise management is interdisciplinary; this is due to the multidisciplinarity of administration systems as such. In this context, it is important to establish effective interdisciplinary communications. Thus, solution of a certain task within the assessment gives an opportunity for the exchange of ideas, problems, methodologies, tools, approaches, etc. among the employees and units. On the whole, it should be noted that the openness of the administration systems highlights the need for continuous monitoring of the discrepancies between the "as is" and "to be" principles with regard to each system. Using questionnaires, observations, surveys and other tools, it is possible to carefully diagnose not only this discrepancy, but also the key causes. Quality of the administration systems diagnostics on the basis of "as is" principle will determine the quality of the identified problems, contradictions, differences, and will establish specific steps to improve the efficiency of this area.

2.2 Parameters for Assessing the Level of Administration Systems Formation in the Enterprise Management

Assessment of the level of administration systems formation in the enterprise management from the "as is" standpoint should be carried out not only taking into account the processes implemented in such systems, but also management processes in administration systems implemented by the administration subjects (vertical, horizontal, those involving interaction between the administration subject and administration object, etc.). Such an assessment requires a balanced combination of the strengths of the company's administration processes, as well as the real situation in the business environment, embodied in the assessment of the challenges of the environment, in the selection of key areas for the development of administration systems, in determining the common points of contact between the interests of the employee and the employer.

In assessing the level of administration systems formation in the enterprise management, it is also necessary to determine the areas in the context of which the gap between the "as is" and "to be" of these systems should be identified. The study of theory and practice, as well as the results of the research allow us to talk about the appropriateness of taking into account such areas as the confidentiality of the administration system, its integrity and availability (Fig. 3).

Appropriateness of this approach is primarily explained by the identification of key threats (substantiated in the literature) of the documentation and information support for the processes of targeted influence of the managing subsystem on the managed subsystem at all stages of management technology. As, for example, Dosmukhamedov [7, pp. 141–143] notes, these key threats are the threats of confidentiality, integrity and availability. Taking into account the research results of this author, the area of confidentiality of administration systems in the enterprise management,

Fig. 3 Areas of
identification of gaps
between the "as is" and "to
be" parameters of the
administration system in
enterprise management. *Note*
Singled out by the authors

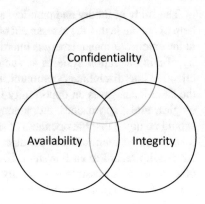

taking into account European integration processes, should reflect the protection of
such systems from the point of view of unauthorized access. The area of integrity
of the administration systems reflects the ability of such systems to maintain the
preset appearance and quality, while the area of the availability of the administration
systems should reflect the ability of such systems to generate necessary information
for the targeted influence of the managing subsystem on a managed subsystem at
all stages of management technology and in a timely manner. Moreover, it is not
just about the current information, but also about retrospective and forward looking
information.

2.3 *Methodical Tools for Assessing the Level of Administration Systems Formation in the Enterprise Management*

Fuzzy logic, in particular, its mathematical apparatus, which is thoroughly described
in the works [8–16] is proposed as methodical tools for solving the defined task of
assessing the level of administration systems formation in the enterprise management.
As the review and summary of literary sources suggests, the fuzzy sets theory tool is
actively used to solve various tasks not only in the economy, but also in healthcare,
politics, IT, administrative management, environmental protection, etc. Application
variation and relative ease of use in solving various problems provides the opportunity
to combine multidirectional economic phenomena, parameters and processes into
a single whole. At the same time, it is not required to take into account precise
and unambiguous statements at each stage of economic processes simulation. Other
advantages of using mathematical apparatus of the fuzzy sets theory in assessing
the level of administration systems formation in the enterprise management are as
follows:

– Sufficient number of distribution functions suitable for research and diagnostics;

- Possibility of comparative diagnosis of various administration systems with a given level of probability;
- Possibility to take into account both quantitative and qualitative features in the assessment;
- Possibility to take into account incoming data with fuzzy boundaries in the assessment;
- Optimization of resources, especially time resources, to determine the level of accuracy of input assessment parameters;
- Possibility to take into account various linguistic variables in the assessment;
- Immediacy of performing assessment procedures and the possibility to compare certain parameters with a certain level of accuracy;
- Analysis of fuzzy input information which sometimes cannot be quantified at all;
- Obtaining assessments under the conditions of limited information support;
- Possibility to substantiate probabilistic hypotheses from among a number of assumptions within the assessment;
- Possibility to take into account the so-called estimated "corridor" of expert values;
- Possibility of fuzzy formalization of criteria for assessment and comparison;
- Possibility to take into account not only assessment values, but also the level of their reliability and distribution, etc.

The author of the theory of fuzzy sets Zadeh [17, p. 338] emphasized the possibility of broad application of his concept of fuzzy logic according to which it became possible to operate judgments "partial truth" or "incomplete truth", which cannot be taken into account within the classic Boolean Logic.

The study of theory and practice makes it possible to conclude that the use of mathematical apparatus of the theory of fuzzy sets when assessing the level of administration systems formation in the enterprise management is based on a long history of successful experience. Moreover, it is worth considering the ideas of Dorofeyeva [18, p. 134] and quote the statement that the process of such assessment includes three blocks of problems of an unspecified nature—uncertainties (Fig. 4).

For practical reasons, these problems of uncertain nature in the assessment of the level of administration systems formation in the enterprise management are difficult to integrate into classic mathematical problems, in view of the need to minimize this uncertainty. On the other hand, as we know, most mathematical models are not capable of providing unambiguous conclusions when assessing various economic phenomena and processes. Thus, when evaluating the formation of administration systems in the enterprise management, it is necessary to abandon a clear mathematical description of the problem because the it is qualitative parameters of the assessment that are determining: the "as is" and "to be" parameters, as well as the availability, integrity and confidentiality as guidelines for identifying gaps between the "as is" and "to be" parameters of the administration systems.

As Nedosekin [19, p. 47] notes, when making a choice within the fuzzy set theory of subjective probabilities, the well-known James Gibbs principle is often used. It states that "from among all probabilistic distributions consistent with the source information about the uncertainty of a corresponding indicator, it is recommended

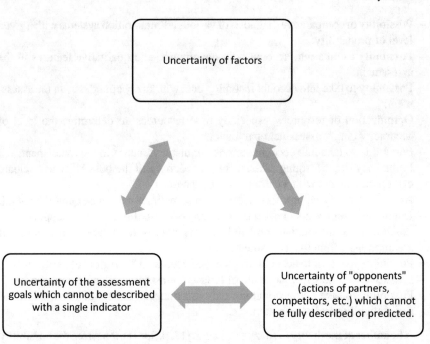

Fig. 4 Three blocks of uncertainties in the assessment of the level of administration systems formation in the enterprise management. *Note* Taking into account [18, p. 134]

to choose the one with the highest entropy". In this case, the Gaussian membership function μ defined on the site [0; 1], which is known to have its maximum and distribution width, is used.

Taking into account the terminology of the theory of fuzzy sets, when determining the level of formation of administration systems in the enterprise management for a certain fuzzy set N, we establish the membership function μ_N. Therefore, $\mu_N(x)$ is a number in the interval [0; 1] which has to identify the level of affiliation of an element x to a certain fuzzy set N. Obviously, the value of the membership function μ_N is determined by the expert method.

The study of theory and practice, as well as the results of the summary of literature sources on the problem of using the theory of fuzzy sets to solve various scientific problems [9, 10, 12, 13, 20–22] make it possible to identify the main stages in the assessment of the level of formation of administration systems in the enterprise management presented below. Thus, first of all, it is necessary to identify the parameters of such a level, which, as noted above, should include confidentiality, availability and integrity of the administration system. Each of these parameters will be further described by linguistic variables (terms), and the weighting factors u will be established in such a way that:

$$\sum_{i=1}^{A} u_i = 1, \tag{1}$$

where A is the number of parameters for assessing the level of administration system formation u.

Study of the theory and practice of applying fuzzy sets theory to solving various tasks of economic nature allows us to conclude that different approaches are proposed to the formation of term sets (fuzzy intervals) and identification of corresponding linguistic variables. In particular, when assessing the level of administration systems formation in the enterprise management, rectangular, trapezoidal, triangular and other functions $L - R$ may be taken into account. At the same time, in the analyzed context, it is most expedient to apply trapezoidal membership functions as those that have become the most widely used in practice. They, as noted in [20, p. 10], have the following mathematical expression:

$$\mu_N(x) = \begin{cases} 0, & x < u_1; \\ \frac{x-u_1}{u_2-u_1} & u_1 \le x < u_2; \\ 1 & u_2 \le x \le u_3; \\ \frac{u_4-x}{u_4-u_3} & u_3 < x \le u_4, \end{cases} \tag{2}$$

where u_1, u_2, u_3, u_4—is the value of the trapezoidal function $L - R$. It should also be taken into account that $u_1 \le u_2 \le u_3 \le u_4$ u_1 is left zero value of the function; u_2 i u_3 is peak of the trapeze which reflects the most realistic expert assessment of a particular element affiliation to a certain term-set; u_4 is right zero value of the function.

Graphical representation of trapezoidal function $L - R$ proposed for use in the assessment of administration system formation in the enterprise management is shown in Fig. 5.

Thus, taking into account the foregoing, the trapezoidal term sets for the parameters of assessing the level of administration systems formation in the enterprise management (confidentiality of administration systems, their integrity and availability) will have the following linguistic interpretation:

- u_1—negative assessment;
- u_2, u_3—the most realistic assessment;
- u_4—positive assessment;

In the future, when assessing the level of administration systems formation in the enterprise management, phasing process should be carried out, i.e. it is necessary to define fuzzy trapezoidal values of the assessment parameters in the form of corresponding linguistic variables:

$$K_{sa} = (u_1, u_2, u_3, u_4);$$
$$C_{sa} = (u_1, u_2, u_3, u_4);$$

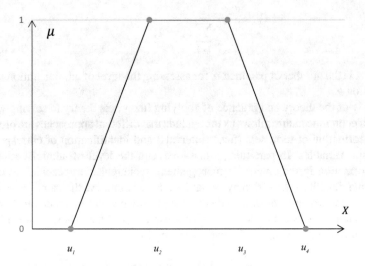

Fig. 5 Trapezoidal function $L - R$ proposed for use in the assessment of administration system formation in the enterprise management. *Note* On the basis of [23, p. 110]

$$D_{sa} = (u_1, \dot{u}_2, u_3, u_4),$$

where K_{sa}, C_{sa}, D_{sa}—confidentiality, integrity and availability of administration systems in the enterprise management, respectively, taking into account the "as is" and "to be" parameters.

Analysis of the content of confidentiality, integrity and availability of administration systems in enterprise management makes it possible to conclude that they do not foresee intermediate values of parameters, as is the case when many economic tasks are solved. This significantly simplifies calculations.

It is also important to emphasize that the content and nature of the proposed parameters for assessing the level of administration systems formation in the enterprise management are similar, given that the trapezoidal scale for these parameters is common (Table 1).

Table 1 Proposed scale of the trapezoidal function $L - R$ for assessing confidentiality, integrity and availability of administration systems in the enterprise management, respectively, taking into account the "as is" and "to be" parameters

Linguistic variables	Very low	Low	Average	High	Very high
Trapezoidal scales of assessments	(0; 0; 0.1; 0.3)	(0.1; 0.3; 0.3; 0.5)	(0.3; 0.5; 0.5; 0.7)	(0.5; 0.7; 0.7; 0.9)	(0.7; 0.9; 1.0; 1.0)

Note Developed on the basis of [20, p.16]

The content of confidentiality, integrity and availability parameters of administration systems in enterprise management makes it possible to conclude that they will not provide for the valuation of fuzzy trapezoidal numbers. At the same time, the presence of three assessment parameters necessitates their aggregation taking into account the weighting factors for obtaining the final trapezoidal number. To do this, we have to use the formula:

$$X = (\sum_{i=1}^{n} x_{i1}u_i; \sum_{i=1}^{n} x_{i2}u_i; \sum_{i=1}^{n} x_{i3}u_i; \sum_{i=1}^{n} x_{i4}u_i) = (u_1; u_2; u_3; u_4), \qquad (3)$$

where x_i—weighting factors of the parameters for assessing the level of administration systems formation in the enterprise management.

The process of dephasing the final result of the assessment (obtaining an integral estimate) of the level of administration systems formation in the enterprise management can be carried out using various known theoretical and practical approaches, in particular [9, 13, 20, 24]:

– Chiu-Park method;
– Dubois-Prade method;
– Chang's method;
– Center of gravity method;
– Kaufmann-Gupta method;
– Jane's method, etc.

Regardless of the chosen approach to dephasing the final result of assessing the level of formation of administration systems in the enterprise management, calculations should be completed with substantiated conclusions, which should result in approved management decisions. In view of this, the assessment process itself should not be interpreted as a self-sufficient process the purpose of which is to obtain qualitative or quantitative assessments; it is an auxiliary tool to form proposals for improving administration systems and thereby ensure a more effective impact of the managing system on the managed system at all stages of management technology at each level of the organizational structure.

2.4 Testing of the Method for Assessing the Level of Administration Systems Formation in the Enterprise Management

Practical application of the improved method for assessing the level of administration systems formation in the enterprise management is carried out in the activities of a number of business entities. As an example, let's look at the relevant calculations regarding the formation of individual administration systems at PJSC "Ukrelektroaparat", one of the largest machine-building enterprises in Ukraine which carries out

its activities in the field of electrical engineering. The company has been successfully producing power, traction transformers, transformer substations, as well as
other electrical equipment (reactors for electric locomotives and electric trains, special low-power transformers, gas control units for block and cabinet construction,
etc.) for more than 60 years. The technological processes of manufacturing most
types of enterprise's products is characterized by a full production cycle, ranging
from metalworking to paintwork of finished parts.

An expert team of 5 people was formed to assess the formation of administration
systems of PJSC "Ukrelektroaparat". It consisted of the co-author of the work, 2
representatives of the institutional level of this enterprise's management, as well
as 2 external experts. The experts largely worked using the brainstorming method
in conjunction with the collective "notebook" method. The expert team decided to
assess the formation of the administration system of PJSC "Ukrelektroaparat" in
terms of the processes of supply, production and sales, each of which occupies a
significant place in the production and economic activities of the company.

Based on the results of discussions, parameters for assessing the formation level of
the administration systems in the supply, production and sales at PJSC "Ukrelektroaparat" included confidentiality of the administration systems (K_{sa}), their integrity
(C_{sa}) and availability (D_{sa}). These parameters were considered taking into account
the "as is" and "to be" parameters for each of the analyzed administration systems.
Fuzzy trapezoidal numbers for such parameters were identified taking into account
the linguistic variables given in Table 1 and their quantitative estimates. For this
purpose, members of the expert team developed a questionnaire, which, in addition
to direct members of the team, was also sent to 15 employees of PJSC "Ukrelektroaparat": 5 for each of the analyzed areas (supply, production, sales). It should be
noted that the established concordance coefficient W was 0.95 at $\chi^2 = 168.8$ (table
value $\chi^2 = 16.9$). Such calculations make it possible to make a conclusion on the
consistency of the statements of experts when working with questionnaires.

Summary of assessments for each of the parameters for assessing the formation level of the administration systems in the supply, production and sales at PJSC
"Ukrelektroaparat" made it possible to determine the corresponding fuzzy trapezoidal numbers, namely:

(a) For the administration system in the supply:

$$K_{sa} = (0.1; 0.3; 0.3; 0.5);$$
$$C_{sa} = (0.3; 0.5; 0.5; 0.7);$$
$$D_{sa} = (0.3; 0.5; 0.5; 0.7);$$

(b) For the administration system in the production:

$$K_{sa} = (0.3; 0.5; 0.5; 0.7);$$
$$C_{sa} = (0.3; 0.5; 0.5; 0.7);$$
$$D_{sa} = (0.5; 0.7; 0.7; 0.9);$$

(c) For the administration system in the (b) For the administration system in the production:

$$K_{sa} = (0.1; 0.3; 0.3; 0.5);$$
$$C_{sa} = (0.1; 0.3; 0.3; 0.5);$$
$$D_{sa} = (0.3; 0.5; 0.5; 0.7).$$

The use of Thurstone Matrix made it possible for the expert team to establish the weighting factors for each of the analyzed parameters for assessing the formation level of the administration systems at PJSC "Ukrelektroaparat", namely:

– For the parameter of administration system confidentiality (K_{sa})—$\alpha = 0.27$;
– For the parameter of administration system integrity (C_{sa})—$\alpha = 0.31$;
– For the parameter of administration system availability (D_{sa})—$\alpha = 0.42$.

As a result, the expert team calculated the final fuzzy trapezoidal numbers for the parameters for assessing the level of formation of administration systems in the supply, production and sales at PJSC "Ukrelektroaparat", namely:

(a) For the administration system in the supply:

$$X = (0.1 \times 0.27 + 0.3 \times 0.31 + 0.3 \times 0.42; 0.3 \times 0.27 + 0.5 \times 0.31 + 0.5$$
$$\times 0.42; 0.3 \times 0.27 + 0.5 \times 0.31 + 0.5 \times 0.42; 0.5 \times 0.27 + 0.7$$
$$\times 0.31 + 0.7 \times 0.42)$$
$$= (0.25; 0.45; 0.45; 0.65);$$

(b) For the administration system in the production:

$$X = (0.3 \times 0.27 + 0.3 \times 0.31 + 0.5 \times 0.42; 0.5 \times 0.27 + 0.5 \times 0.31 + 0.7$$
$$\times 0.42; 0.5 \times 0.27 + 0.5 \times 0.31 + 0.7 \times 0.42; 0.7 \times 0.27 + 0.7$$
$$\times 0.31 + 0.9 \times 0.42)$$
$$= (0.38; 0.58; 0.58; 0.78);$$

(c) For the administration system in the (b) For the administration system in the production:

$$X = (0.1 \times 0.27 + 0.1 \times 0.31 + 0.3 \times 0.42; 0.3 \times 0.27 + 0.3 \times 0.31 + 0.5$$
$$\times 0.42; 0.3 \times 0.27 + 0.3 \times 0.31 + 0.5 \times 0.42; 0.5 \times 0.27 + 0.5$$
$$\times 0.31 + 0.7 \times 0.42)$$
$$= (0.18; 0.38; 0.38; 0.58).$$

The process of dephasing of the final result of the assessment (obtaining an integral assessment) of the level of formation of the administration systems in the supply, production and sales at PJSC "Ukrelektroaparat" is carried out using the formula:

$$y = \frac{\int_{min}^{max} u \times \varphi(u)dx}{\int_{min}^{max} \varphi(u)dx} = \frac{(u_3^2 + u_4^2 + u_3 \times u_4 - u_1^2 - u_2^2 - u_1 \times u_2)}{3 \times (u_4 + u_3 - u_1 - u_2)}. \quad (4)$$

Thus, the following results were obtained:

(a) For the administration system in the supply:

$$y = \frac{(0.45^2 + 0.65^2 + 0.45 \times 0.65 - 0.25^2 - 0.45^2 - 0.25 \times 0.45)}{3 \times (0.65 + 0.45 - 0.25 - 0.45)}$$

$$= \frac{0.54}{1.2} = 0.45;$$

(b) For the administration system in the production:

$$y = \frac{(0.58^2 + 0.78^2 + 0.58 \times 0.78 - 0.38^2 - 0.58^2 - 0.38 \times 0.58)}{3 \times (0.78 + 0.58 - 0.38 - 0.58)}$$

$$= \frac{0.696}{1.2} = 0.58;$$

(c) For the administration system in the (b) For the administration system in the production:

$$y = \frac{(0.38^2 + 0.58^2 + 0.38 \times 0.58 - 0.18^2 - 0.38^2 - 0.18 \times 0.38)}{3 \times (0.58 + 0.38 - 0.18 - 0.38)}$$

$$= \frac{0.456}{1.2} = 0.38.$$

Calculation results make it possible to conclude that the level of formation of the administration systems at PJSC "Ukrelektroaparat" in the supply, production and sales, taking into account the parameters "as is" and "to be", are 45%, 58% and 38%, respectively (Fig. 6). The most challenging for the analyzed company was the sphere of administration in the sphere of sales, which has also affected in many respects the deterioration of the financial results of its activities in recent years.

When assessing the level of formation of the administration systems at PJSC "Ukrelektroaparat" in the supply, production and sales, the expert team interpreted the assessment results using the well-known theory and practice of the universal Harrington scale, which is known to be widely used in different scientific studies.

Taking into account the intervals of the universal Harrington scale given in Table 2, the level of formation of the administration systems at PJSC "Ukrelektroaparat" in the supply, production and sales was determined by members of the expert team as average.

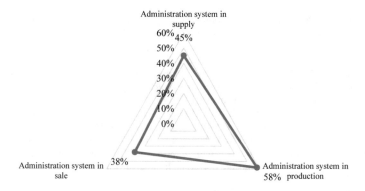

Fig. 6 The level of formation of administration systems at PJSC "Ukrelektroaparat". *Note* Calculated by the authors

Table 2 Quantitative and linguistic assessments according to the Harrington scale

Linguistic assessments	Score intervals	Average scores
Excellent (very high)	0.8–1	0.9
Good (high)	0.63–0.8	0.71
Satisfactory (average)	0.37–0.63	0.5
Bad (low)	0.2–0.37	0.28
Very bad (very low)	0–0.2	0.1

Note On the basis of [24, pp. 263–264; 25, p. 277]

3 Conclusion

Thus, the use of the fuzzy sets theory when assessing the formation of administration systems in enterprise management allows us to obtain the resulting distribution of a combination of fuzzy sets for their further diagnosis, integration, comparison, etc., for the adoption of appropriate management decisions to ensure a higher level of compliance of the administration system with the "as is" and "to be" parameters.

References

1. Khorol'tseva YeB, Fedorova AV (2013) Issledovaniye sistem upravleniya. Vestnik PAGS 110–117
2. Taraskina YuV (2010) Model' razrabotki proyekta reinzhiniringa biznes-protsessov neftyanykh kompaniy. Vestnik AGTU. Seriya: Ekonomika 1:203–210
3. Cherep AV, Potopa KL, Tkachenko OV (2009) Reinzhynirynh – filosofiya upravlinnya pidpryyemstvom kharchovoyi promyslovosti. Kondor, Kyyiv
4. Hakim A, Gheitasi M, Soltani F (2016) Fuzzy model on selecting processes in business process reengineering. Bus Process Manag J 22(6):1118–1138

5. Kryvinska N (2012) Building consistent formal specification for the service enterprise agility foundation. Soc Serv Sci J Serv Sci Res Springer 4(2):235–269
6. Kaczor S, Kryvinska N (2013) It is all about services—fundamentals, drivers, and business models. Soc Serv Sci J Serv Sci Res Springer 5(2):125–154
7. Dosmukhamedov BR (2009) Analiz ugroz informatsii sistem elektronnogo dokumentooborota. Vestnik Astrakhanskogo gosudarstvennogo tekhnicheskogo universiteta. Seriya: Upravleniye, vychislitel'naya tekhnika i informatika 2:140–143
8. Alshibani A, Hassanain MA (2018) Estimating facilities maintenance cost using post-occupancy evaluation and fuzzy set theory. J Qual Maint Eng 24(4):449–467
9. Asiain MJ, Bustince H, Mesiar R, Kolesarova A, Takac Z (2018) Negations with respect to admissible orders in the interval-valued fuzzy set theory. IEEE Trans Fuzzy Syst 26(2):556–568
10. Deschrijver G (2007) Arithmetic operators in interval-valued fuzzy set theory. Inf Sci 177(14):2906–2924
11. Deschrijver G, Kerre EE (2007) On the position of intuitionistic fuzzy set theory in the framework of theories modelling imprecision. Inf Sci 177(8):1860–1866
12. Salah A, Moselhi O (2015) Contingency modelling for construction projects using fuzzy-set theory. Eng Constr Arch Manag 22(2):214–241
13. Simić D, Simić S, Kovačević I, Svirčević V (2017) 50 years of fuzzy set theory and models for supplier assessment and selection: a literature review. J Appl Log 24:85–96
14. Gregus M, Kryvinska N (2015) Service orientation of enterprises—aspects, dimensions, technologies. Comenius University in Bratislava. ISBN: 9788022339780
15. Kryvinska N, Gregus M (2014) SOA and its business value in requirements, features, practices and methodologies. Comenius University in Bratislava. ISBN: 9788022337649
16. Molnár E, Molnár R, Kryvinska N, Greguš M (2014) Web intelligence in practice. Soc Serv Sci J Serv Sci Res Springer 6(1):149–172
17. Zadeh LA (1965) Fuzzy sets. Inf Control 8(3):338–353
18. Dorofeyeva VV (2011) Ispol'zovaniye teorii nechetkikh mnozhestv dlya formalizatsii pokazatelya finansovoy privlekatel'nosti predpriyatiya. Izvestiya IGEA 6:s.133–135
19. Nedosekin AO (2000) Primeneniye teorii nechetkikh mnozhestv k zadacham upravleniya finansami. Audit i finansovyy analiz 2:45–54
20. Anshin VM, Demkin IV, Tsar'kov IN, Nikonov IM (2008) Primeneniye teorii nechotkikh mnozhestv k zadache formirovaniya portfelya proyektov. Problemy analiza riska 3(5):8–21
21. Lam CY, Tai K (2018) Modeling infrastructure interdependencies by integrating network and fuzzy set theory. Int J Crit Infrastruct Prot 22:51–61
22. Zimmermann H (2010) Fuzzy set theory. Wiley Interdiscip Rev 2:317–332
23. Molokanova VM (2012) Otsinyuvannya yakisnykh pokaznykiv portfelyu proektiv za dopomohoyu teoriyi nechitkykh mnozhyn. Upravlinnya proektamy ta rozvytok vyrobnytstva 3:106–114
24. Didyk AM (2016) Polivektornyy rozvytok pidpryyemstv: sotsial'no-ekonomichni ta rehulyatyvni aspekty. Halyts'ka vydavnycha spilka, L'viv
25. Nechyporuk OV (2007) Metodyka reytynhovoyi otsinky investytsiynoyi pryvablyvosti promyslovykh pidpryyemstv. Kommunal'noe khozyaystvo horodov 75:275–284

Franchising and Customer Experience Management in Telecommunication Industry Phenomenon—Literature Review (2011–2018)

Bianka Chorvátová and Peter Štarchoň

Abstract Franchising is a phenomenon largely discussed for about 50 years now. Its types and techniques are similar all over the world, but cultural and legislative differences are specific for every country. Nowadays, the phenomenon of franchising is expanding also in telecommunications companies. They change from state controlled monopolies and start to penetrate also the private market. As the competition is growing, customer experience management is more important to focus on than ever before. This literature review provide the latest literature available in a specific database that discuss the connection between franchising, telecommunications and customer experience management. A classification of 48 up-to-date articles and publications is made with a focus on its research area and researched problems.

Keywords Franchising in telecommunications · Telecommunication franchise opportunities · Data mining in CEM · CEM in telecommunication · Franchise management

1 Introduction

Franchising can be defined as a relationship, where a franchisor (a person or company that grants the license to a 3rd party for the conducting of a business beneath their marks) gives the franchisee (a person or company which is granted the license to try the business beneath the trademark and brand by the franchisor) the right to use its trademark or trade-name. By paying a fee and percentage of sales, franchisee gets an access to name recognition, products, design, business information, employee training and ongoing help in promotion and product updates [1].

B. Chorvátová (✉) · P. Štarchoň
Faculty of Management, Comenius University, Bratislava, Slovakia
e-mail: bianka.chorvatova@fm.uniba.sk

P. Štarchoň
e-mail: peter.starchon@fm.uniba.sk

© Springer Nature Switzerland AG 2020
N. Kryvinska and M. Greguš (eds.), *Data-Centric Business and Applications*,
Lecture Notes on Data Engineering and Communications Technologies 30,
https://doi.org/10.1007/978-3-030-19069-9_7

Franchising is an expanding business model which allows the distribution of goods and services via licensing. In franchising, franchisor specify the product and services which will be offered to and then by the franchisees, additionally they are given complete support. Business Franchising Format is the most recognizable by the common person. In this an exceedingly large business format, the franchisor provides to the franchisee not simply its brand, product and services, but a complete system for operation of the business. The franchisee usually receives web site choice and development support, operational manuals, training, complete standards, internal control, promoting strategy and business consultative support from the franchisor [2]. The use of franchising is linked to the enlargement of the church associate (an early central government management), it dates way back in the Middle Ages. Some have say that it came as a method or style because of the Roman Empire that has given the need of enormous territorial controls, in addition to the shortage of contemporary transportation and communication at the time, there's affordable basis for this assumption [3].

CEM is a management discipline that uses the foremost relevant insights regarding the client to drive the correct actions across acceptable domain of the business and measures the outcomes of these actions to refine each insights and actions within the future [4]. CEM (also referred to as CXM) is completely different than CRM (Customer Relationship Management). CRM programs are available on the market for about twenty years. With CRM, the main target was on internal processes to raise and support customers, reduce cost and increase satisfaction. It had been inside-out approach to manage a relationship with customers. CEM, on the opposite hand, takes an "outside-in" approach. CEM is proactive, even antecedent to CRM. CEM proactively initiates contact, delivers content, pre-emptively solves issues and understands client preferences, typically before the client raise it or maybe is aware of them. For a distinct segment (customer focused) market reminiscent of telcos, CEM is more necessary than it had been a decade ago. Operators are measured by providing new plans, products, features, technologies and everything is simply a click away. CEM groups will target the best client expertise to customers such as the churn rate to be as low as attainable. It's a notable and undeniable fact that a brand new client prices 5 times more than to retain an existing client. Another necessary issue is penetration of social media in client perception, concerning the network. "*A single unsatisfied customer can become more influential than the largest of companies and can spread enough negative sentiments which can impact their businesses both in short and long terms. Telecom operators generally have low Net Promoter Score (NPS) score according to research firm Temkin. This essentially suggests that only a small percentage of CSP's customers have positive sentiments about their network and services. These unsatisfied customers are more likely to churn and even while they remain as customers they are more likely to complain about the services being offered, making calls to customer care, enquiring about inflated bills etc. thereby increasing effective cost per customer in customer operations. An effective CEM program should address these issues.*" [5].

An information system is a tool for gathering and using information in the management practice of any systems, fulfilling the functions of collecting, processing, distributing, storing and using information and becoming an essential part of the management processes [6]. The evolution of IT has enabled assortment and storage of giant amounts of information and data. The dimensions of databases these days will vary from almost nothing to terabytes. It's virtually not possible to investigate these volumes of data using traditional methods. For this reason data processing has attracted an excellent deal of attention. We can search and explore data, generate hypotheses, and learn from them. Data mining is the process of mechanically discovering helpful information in massive information repositories. It's supported by inexpensive data storage, affordable processing power, data availability, and many commercial data mining tools available [7]. *"Since the goal of data mining is extracting meaningful patterns and relationships from large data sets, data mining can redefine and improve customer relationships. Fortunately, telecommunications companies know more about their customers than anyone else. They know who their customers are, they can easily keep track of their customers' activities. A huge amount of data generated by telecommunications companies cannot be analyzed in a traditional manner, by using manual data analysis. That is why different data mining techniques ought to be applied. Data mining helps a business understand its customers better."* [6].

The aim of this article is to find up-to-date literature sources that can be used for further research about franchising and customer experience management in telecommunication industry. A table of 48 publications was made to create a functional database for following review. These publications were obtained by qualitative research—comparison of publications, by research, analysis and data synthesis, we were able to deduce conclusions useful for further analysis of available literature about customer relationship management in telecommunication industry and franchising as a business growth strategy.

2 Classification

To store and analyze a literature database, we used the Zotero program. It allows us to sort the data by author, journal, year or keywords, which will come in handy later during the process of creating a final classification table for this literature review. To focus on franchising in telecommunication industry, we set for following keywords for our research:

- franchising in telecommunications,
- co-opetition in telecommunications,
- strategic alliances in telecommunications,
- telecommunication franchise opportunities,
- data mining in CEM,
- CEM in telecommunication,

Table 1 Number of publications for each keyword in specific time period

Keywords	1900–2018	1900–1960	1961–1980	1981–1990	1991–2000	2001–2010	2011–2018
Franchising in telecomm	16,400	53	714	2200	7860	15,000	13,800
Co-opetition in telecomm	9520	1	3	3	326	3640	5500
Strategic alliances in telecomm	37,200	257	1130	4460	16,600	20,900	17,300
Telecommunication franchise opportunities	16,200	28	382	1340	5600	14,000	13,500
Data mining in CEM	18,400	439	699	780	2070	8230	16,100
CEM in telecomm	5990	7	40	83	467	1990	3210
Franchise management	121,000	9300	9720	13,200	24,300	48,200	23,700
Franchise CEM	3780	228	120	108	400	1220	1670
Data mining in franchising	17,400	1720	2020	2620	6030	15,300	16,600
Strategic partnership with franchisor	17,600	71	493	1690	7500	16,100	17,000
Number of publications	263,490	12,104	15,321	26,484	71,153	144,580	128,380

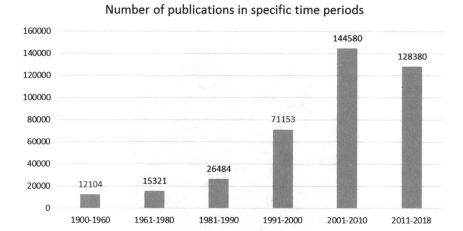

Graph 1 Number of publications in specific time period

- franchise management,
- franchise CEM,
- data mining in franchising,
- strategic partnership with franchisor.

Telecommunications and customer experience management is important in this research, but franchising is the core. According to Michael Seid, "*…the evolution of modern franchising, created by the innovative companies and the pioneers that have led them, is an exciting tale in itself. The future, energized by still unimagined new concepts, new business techniques and international expansion, promises to add still more dynamic chapters to the continuing and growing adventure of franchising.*" [3].

The chosen database for research is Google Scholar, and we focus only on full preview publications. In Table 1 we can see the number of publications (articles, books,…) for specific time periods.

As we can see, our chosen keywords started to be researched more than hundred years ago but only in 2001–2018 academics and other experts started to publish more and more publications related to franchising and franchising in telecommunication industry.

The general growth can be seen in Graph 1.

As we can see in Graph 2, the more general keywords are, the more publications we can find. Which means that we can find some irrelevant publications to our specific research if we choose to use only general keywords. Specific keywords can help with finding more accurate publications to our research. After period-by-period literature research we decided to choose publications from 2011 to 2018 for further research, as we find them the most relevant and up to date. Even publications from earlier years are published as new editions and new information about franchising and cus-

Number of publications 1900-2018

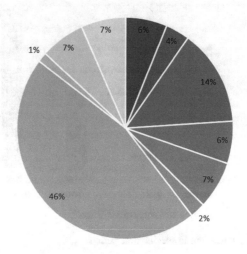

■ franchising in telecommunications   ■ co-opetition in telecommunications

■ strategic alliances in telecommunications   ■ telecommunication franchise opportunities

■ data mining in CEM   ■ CEM in telecommunication

■ franchise management   ■ franchise CEM

■ data mining in franchising   ■ strategic partnership with franchisor

Graph 2 The percentage of each keyword-chosen publications in research

tomer experience management in telecommunication industry practice is included. We collected 50 publications, from which 48 had PDF preview. As we mentioned before, we have chosen papers with full PDF preview only. Some publications were duplicates—they contained several of our chosen keywords. In the end, we managed to create a database of 48 publications.

3 Literature Review

3.1 Authors, Titles, Topics, Journals and Research Areas

In Table 2 we can see all chosen papers for this research. Among the authors, we can see some interesting names. Such as Benjamin W. Cramer, who is a telecommunications policy and media law expert, Johannes Bauer, who is a professor of Michigan State University and an expert in communications, governance and telecommunication, or Kendal H. Tyre, who is a franchise companies counsellor. As we can see,

Table 2 Chosen papers

	Authors	Year	Title/Researched subject	Research area	Methodology
1	Ajao, Blessing Funke; Oyebisi, Timothy; Aderemi, Helen; Jegede, Oluseye	2015	Status and impact of strategic technology alliances among telecommunications firms in Nigeria	Technology, Business	Quantitative (survey), descriptive
2	ZHANG, Zhi-yuan; AN, Li-na	2011	Review and prospect of study on franchise of public utilities	Law, Business	Qualitative, analytical
3	Ali, Christopher	2017	Regulatory (de) convergence: localism, federalism, and nationalism in American telecommunications policy	Law	Qualitative, analytical
4	Cramer, Benjamin W.	2016	Right way wrong way: the fading legal justifications for telecommunications infrastructure rights-of-way	Law, Communication	Qualitative, analytical
5	Parker, James G.	2011	Statewide cable franchising: expand nationwide or cut the cord	Law, Technology	Qualitative, analytical
6	Bergholz, Lindsey	2018	National Association of Telecommunications Officers & Advisors v. FCC	Law, Technology	Qualitative, analytical

(continued)

Table 2 (continued)

	Authors	Year	Title/Researched subject	Research area	Methodology
7	Neocleous, Andreas; Co, L. L. C.; Bello, Gallardo; Bonequi y García, S. C.; Gilfillan, Bowman; Chajec, Don-Siemion; Advisors, Zyto Legal; Harney, Coulson; Phillips, Davies Ward; Vineberg, L. L. P.	2015	The international comparative legal guide to: telecoms, media & internet laws & regulations 2016	Law, Marketing, Technology	Qualitative, analytical
8	Snyder, Thomas W.	2011	Putting a price on dirt: the need for better-defined limits on government fees for use of the public right-of-way under section 253 of the Telecommunications Act of 1996	Finance, Law, Technology	Qualitative, analytical
9	Tyre, Kendal H.	2013	Franchising in Africa	Law	Qualitative, analytical
10	Winter, Sidney G.; Szulanski, Gabriel; Ringov, Dimo; Jensen, Robert J.	2012	Reproducing knowledge: inaccurate replication and failure in franchise organizations	Business	Quantitative, descriptive
11	Johnson, Derek	2013	Media franchising: creative license and collaboration in the culture industries	Business, Marketing	Qualitative, analytical
12	Yadon, Robert E.; Umansky, Barry D.; Coomes, Paul A.	2014	Telecom regulatory review: Kentucky 2014	Law, Business	Qualitative, analytical

(continued)

Table 2 (continued)

	Authors	Year	Title/Researched subject	Research area	Methodology
13	Hulsink, Willem	2012	Privatisation and liberalisation in European telecommunications: comparing Britain, the Netherlands and France	Law, Business	Qualitative, analytical
14	Latipulhayat, Atip	2014	Telecommunications licensing regime: a new method of state control after privatisation of telecommunications	Law, Business	Qualitative, analytical
15	FEE, FRANCHISE		Utilities	Law	–
16	Gaskill, Frank J.	2011	Conditions favoring the globalization of the franchising format: a conceptual model	Law, Business	Qualitative, analytical
17	Sherman, Andrew	2011	Franchising and licensing: two powerful ways to grow your business in any economy	Business, Finance	Qualitative, analytical
18	Chen, Yi-Min; Su, Chien-Tai	2014	Brand equity heterogeneity among strategic groups in service franchising	Business, Marketing	Quantitative, descriptive
19	Dana, Léo-Paul; Granata, Julien; Lasch, Frank; Carnaby, Alan	2013	The evolution of co-opetition in the Waipara wine cluster of New Zealand	Business, Marketing, Agriculture	Qualitative, descriptive

(continued)

Table 2 (continued)

	Authors	Year	Title/Researched subject	Research area	Methodology
20	Hajdul, Marcin	2014	Virtual co-opetition in transport-T-Scale platform case study	Logistics	Qualitative, analytical
21	Ilvonen, Ilona; Vuori, Vilma	2013	Risks and benefits of knowledge sharing in co-opetitive knowledge networks	IT, Communication	Qualitative, analytical
22	Evens, Tom	2014	Co-opetition of TV broadcasters in online video markets: a winning strategy?	Marketing, Business, Technology	Qualitative, analytical
23	Kang, Jin-Su; Lee, Hong-Yuh; Tsai, Julio	2011	An analysis of interdependencies in mobile communications technology: the case of WiMAX and the development of a market assessment model	Technology, Marketing, Business	Qualitative, analytical
24	Bauer, Johannes M.	2014	Platforms, systems competition, and innovation: reassessing the foundations of communications policy	Law, Technology, Communication	Qualitative, analytical
25	Ahokangas, Petri; Horneman, Kari; Posti, Harri; Matinmikko, Marja; Hanninen, Tuomo; Yrjola, Seppo; Goncalves, Vania	2014	Defining "co-primary spectrum sharing"—a new business opportunity for MNOs?	Technology, Business	Qualitative, analytical

(continued)

Table 2 (continued)

	Authors	Year	Title/Researched subject	Research area	Methodology
26	Kossyva, Dorothea I.; Galanis, Konstantinos V.; Sarri, Katerina K.; Georgopoulos, Nikolaos B.	2014	Adopting an information security management system in a co-opetition strategy context	Technology, Innovation, Law	Qualitative, analytical
27	Hung, Chia-Liang; Chou, Jerome Chih-Lung; Ding, Chung-Ming	2012	Enhancing mobile satisfaction through integration of usability and flow	Technology	Quantitative, descriptive
28	Daidj, Nabyla	2011	Media convergence and business ecosystems	Business, Innovation, Technology	Qualitative, analytical
29	Gomes, Emanuel; Weber, Yaakov; Brown, Chris; Tarba, Shlomo Yedidia	2011	Mergers, acquisitions and strategic alliances: understanding the process	Business	Qualitative, analytical
30	Bengtsson, Maria; Kock, Sören	2014	Coopetition—Quo vadis? Past accomplishments and future challenges	HR, Law	Qualitative, analytical
31	Kwak, Jooyoung; Lee, Heejin; Chung, Do Bum	2012	The evolution of alliance structure in China's mobile telecommunication industry and implications for international standardization	Business, Law	Qualitative, analytical
32	Meier, Matthias	2011	Knowledge management in strategic alliances: a review of empirical evidence	Business, Technology	Qualitative, analytical

(continued)

Table 2 (continued)

	Authors	Year	Title/Researched subject	Research area	Methodology
33	Bigliardi, Barbara; Ivo Dormio, Alberto; Galati, Francesco	2012	The adoption of open innovation within the telecommunication industry	Innovation, IT, Technology	Quantitative, descriptive
34	Fernandez, Anne-Sophie; Le Roy, Frédéric; Gnyawali, Devi R.	2014	Sources and management of tension in co-opetition case evidence from telecommunications satellites manufacturing in Europe	Business, Management	Qualitative, analytical
35	Li, Lee; Qian, Gongming; Qian, Zhengming	2013	Do partners in international strategic alliances share resources, costs, and risks?	Technology, Business	Qualitative, analytical
36	McCarter, Matthew W.; Mahoney, Joseph T.; Northcraft, Gregory B.	2011	Testing the waters: using collective real options to manage the social dilemma of strategic alliances	Business, Management	Qualitative, analytical
37	Gerybadze, Alexander	2011	Strategic alliances and process redesign: effective management and restructuring of cooperative projects and networks	Business, Law	Qualitative, analytical
38	de Reuver, Mark; Verschuur, Edgar; Nikayin, Fatemeh; Cerpa, Narciso; Bouwman, Harry	2015	Collective action for mobile payment platforms: a case study on collaboration issues between banks and telecom operators	Management, Business	Qualitative, analytical

(continued)

Table 2 (continued)

	Authors	Year	Title/Researched subject	Research area	Methodology
39	Yang, Haibin; Zheng, Yanfeng; Zhao, Xia	2014	Exploration or exploitation? Small firms' alliance strategies with large firms	Business, Management	Qualitative, analytical
40	Dodgson, Mark	2018	Technological collaboration in industry: strategy, policy and internationalization in innovation	Business, Innovation	Qualitative, analytical
41	Pellicelli, Anna Claudia	2012	Strategic alliances	Business	Qualitative, analytical
42	Clifton, Judith; Comín, Francisco; Díaz-Fuentes, Daniel	2011	From national monopoly to multinational corporation: how regulation shaped the road towards telecommunications internationalisation	Law, Business	Qualitative, analytical
43	Serrat, Olivier	2017	Learning in strategic alliances	Business, Technology	Qualitative, analytical
44	Fang, Eric	2011	The effect of strategic alliance knowledge complementarity on new product innovativeness in China	Business, Innovation, Technology	Qualitative, analytical
45	Arasa, Robert; Gathinji, Loice	2014	The relationship between competitive strategies and firm performance: a case of mobile telecommunication companies in Kenya	Management, Business	Qualitative, analytical

(continued)

Table 2 (continued)

	Authors	Year	Title/Researched subject	Research area	Methodology
46	Kim, Kyung Kyu; Ryoo, Sung Yul; Jung, Myung Dug	2011	Inter-organizational information systems visibility in buyer–supplier relationships: the case of telecommunication equipment component manufacturing industry	Logistics, IT	Quantitative, descriptive
47	Gomes, Emanuel; Cohen, Marcel; Mellahi, Kamel	2013	When two African cultures collide: a study of interactions between managers in a strategic alliance between two African organizations	HR, Management	Qualitative, analytical
48	Ajao, Blessing Funke; Oyebisi, Timothy; Aderemi, Helen; Jegede, Oluseye	2015	Status and impact of strategic technology alliances among telecommunications firms in Nigeria	Business, Law, Management	Qualitative, analytical

this range of expertise among authors can bring us a broad spectrum of views and knowledge that can draw relevant conclusions.

Titles vary from specific topic descriptions, such as Collective action for mobile payment platforms: A case study on collaboration issues between banks and telecom operators and Right way wrong way: The fading legal justifications for telecommunications infrastructure rights-of-way or general topics which cover detailed information addressed in their abstracts.

For example, the article Media Convergence and Business Ecosystems by Nabyla Daidj doesn't seem to cover our research topic—franchising and customer experience management. But as the abstract follows *"Because the markets in which Apple, Google and Microsoft compete are characterized by rapid technological advances, their ability to compete successfully is dependent on their strategies to ensure the launch of competitive products, services and technologies. … Paper focuses on convergence and links with the reconfiguration of value chains in the "new media" sector and diversification strategies adopted by the three companies. As these organizations are made up of different business units, a question arises to how resources and competencies are to be allocated across these businesses. Performance and profitability are determined by an organization's resources and competencies. The different modes of growth (strategic alliances, partnerships, mergers & acquisitions) and in particular, the emergence of business ecosystems will be analysed."* [8], we can now clearly see that in this article we can find information about franchising—as one of the mode of growth.

In Graph 3 we can see which journals are our chosen article published in, and we can see that Elsevier is, for our research topic, the most popular one.

There is one article that has no journal or website, as it is the ORDINANCE #3, 2012—*an ordinance granting a nonexclusive franchise to RT Communications, Inc., to construct, acquire, operate, maintain and supply high speed internet access to the town and inhabitants thereof and to use the streets, roads, alleys and other public places within the town.* A detailed review of articles can be found in Table 2, which contains chosen papers for this research.

For further research, we also split all chosen papers into following fields/research areas—technology, business, law, communication, marketing, finance, agriculture, IT, innovation, management, HR, logistics [9]. In Graph 4 we can see how many articles contribute in which research area. As we can see, the main research area is Business, then Law and Technology. This is common for various reasons. First, franchising and telecommunications are mostly business related topics and depend on technology (as well as innovation, it is interesting that only 5 articles contribute to innovation in franchising or telecommunications topics). They also create new laws and regulations as the market evolve. Management and Marketing do hand in hand when it comes to business, also financial strategies and goals are important not only for franchising in telecommunications in general, but also in customer experience management—as employees selling telecommunication products and services are paid by the amount of products and services they sell.

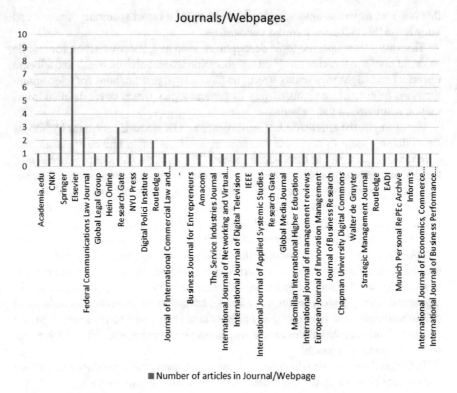

Graph 3 Journals

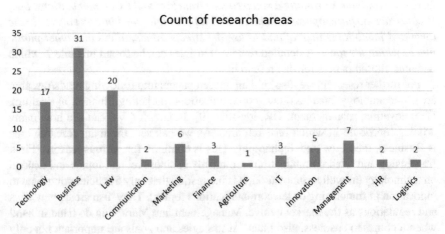

Graph 4 Count of research areas

3.2 Methodologies and Final Literature Classification

Within every major methodology, there are numerous styles how to understand and handle an issue. They give a framework or philosophy for the study, and are different than the particular strategies used for detailed research. For instance, a case study style focuses on exploring and describing a selected instance, person or cluster. A research worker might use observations, interviews or self-reports from the topic to create an entire image. This image, or case, provides a close example of a development which will then be generalized for common public [10].

"*Methodology is the systematic, theoretical analysis of the methods applied to a field of study. It comprises the theoretical analysis of the body of methods and principles associated with a branch of knowledge. Typically, it encompasses concepts such as paradigm, theoretical model, phases and quantitative or qualitative techniques.*" [11]. Certain methodology does not provide solutions—it offers support for understanding which method (best practice) we should apply to a case. Methodology can be also defined as:

- "*the analysis of the principles of methods, rules, and postulates employed by a discipline*" [12];
- "*the systematic study of methods that are, can be, or have been applied within a discipline*" [10];
- "*the study or description of methods*" [13].

Before starting any scientific research, you want to decide which methodology to use. This can guide your study, assist you to settle on the simplest way to gather information and aid in your analysis. Researchers use 3 primary methodology types: qualitative, quantitative and mixed strategies. Inside these broad classes, a lot of specific strategies incorporate number of choices, case studies, self-reporting and surveys [10]. According to Erica Loop, *qualitative research seeks to explore a specific phenomenon, not prove a prediction, according to "Qualitative Research Methods: A Data Collector's Field Guide," published by Family Health International. Often used in the social sciences and education, qualitative methodologies use interviews, focus groups and observations to collect data. Qualitative methods provide rich, contextual explorations of the topic that are often personally or culturally meaningful* [10]. On contrary, *quantitative research is more objective than qualitative methods. In this type of methodology, the researcher crafts a hypothesis and then tests it through structured means. Instead of exploring or describing a phenomena, quantitative methods deal with facts and statistics. This type of research is often used in science or medicine* [10]. Mixed strategies mix qualitative and quantitative analysis. This kind of methodology uses many completely different measures that embrace each discourse understanding like interviews or observations together with facts or statistics. Applying mixed strategies will facilitate the research worker to investigate a subject on multiple levels, gaining completely different views and a comprehensive view of the topic. A mixed methodology meshes over one philosophical perspective, giving the combination of various theories and concepts [10].

We decided to search for analytical or descriptive research as well. *Descriptive research includes surveys and fact-finding enquiries of different kinds. The major purpose of descriptive research is description of the state of affairs as it exists at present. In social science and business research we quite often use the term Ex post facto research for descriptive research studies. The main characteristic of this method is that the researcher has no control over the variables; he can only report what has happened or what is happening. Most ex post facto research projects are used for descriptive studies in which the researcher seeks to measure such items as, for example frequency of shopping, preferences of people, or similar data. Ex post facto studies also include attempts by researchers to discover causes even when they cannot control the variables. The methods of research utilized in descriptive research are survey methods of all kinds, including comparative and co-relational methods. In analytical research, on the other hand, the researcher has to use facts or information already available, and analyse these to make a critical evaluation of the material. This type of research is appropriate for the present study since. It is Ex post facto, there is no control over variables in both the very dynamic economies, it is an attempt to discover cause and therefore compare* [14].

Literature investigation and data mining system varies for every topic. As we found out, we made a few literature investigation attempts that were a failure. But through considerate work we managed to create a database of 48 papers that seem to be the most relevant to the topic. We marked each publication according to its methodology, results can be seen in Table 2.

In this research, we used all four methodologies discussed above; qualitative, quantitative, descriptive and analytical. If a mixed methodology was used, we identified the one that outweighed the other by number of methods used in a specific article.

In Graph 5, we can see how many publications used descriptive approach to the number of publications that used analytical approach, and how many publications used qualitative research to the number of publications that used quantitative research. We assume that qualitative and analytical methodology is more appropriate for our research, as we use best practice, laws and general information available from "real life". New information can be obtained by quantitative and description methodology, that can be used further in following research.

Our final classification consists 48 articles and the classification consists of Year (as up to date data are required for topic such as franchising and telecommunication), Author (various authors from various countries and with various occupation give us the almost-perfect selection of different articles), Title (the more specific title the better, but we can find more detailed information in Researched subject column), and we added a new category—Researched subject (to add a little bit of extra information about an article needed if the title is not specific enough or if we want to compare several articles at glance).

This way we can find in the table all the information needed and if we, or anyone else, wanted to add some other literature to this review, it would be easy and understandable for them to see how to contribute. Final classification table (year 2011–2018) can be seen in Table 3.

Table 3 Final classification

	Year	Authors	Title	Research subject
1	2015	Ajao, Blessing Funke; Oyebisi, Timothy; Aderemi, Helen; Jegede, Oluseye	Status and impact of strategic technology alliances among telecommunications firms in Nigeria	Benefits of strategic technology alliance and benefits to service providers and subscribers
2	2011	ZHANG, Zhi-yuan; AN, Li-na	Review and prospect of study on franchise of public utilities	Franchising, franchise reform, legal issues concerning franchising
3	2017	Ali, Christopher	Regulatory (de) convergence: localism, federalism, and nationalism in American Telecommunications Policy	Telecommunications Act, regulatory capture, policy failure
4	2016	Cramer, Benjamin W.	Right way wrong way: the fading legal justifications for telecommunications infrastructure rights-of-way	Negative impact of telecommunications
5	2011	Parker, James G.	Statewide cable franchising: expand nationwide or cut the cord	TV regulation, franchising legislation
6	2018	Bergholz, Lindsey	National Association of Telecommunications Officers & Advisors v. FCC	Franchising authorities regulating cable rates, effective competition
7	2015	Neocleous, Andreas; Co, L. L. C.; Bello, Gallardo; Bonequi y García, S. C.; Gilfillan, Bowman; Chajec, Don-Siemion; Advisors; Zyto Legal; Harney, Coulson; Phillips, Davies Ward; Vineberg, L. L. P.	The international comparative legal guide to: telecoms, media & internet laws & regulations 2016	Competition and foreign investment in telco, legislation of media
8	2011	Snyder, Thomas W.	Putting a price on dirt: the need for better-defined limits on government fees for use of the public right-of-way under section 253 of the Telecommunications Act of 1996	Telecommunications Act 1996, telecommunications as monopoly industries, fees and revenues

(continued)

Table 3 (continued)

	Year	Authors	Title	Research subject
9	2013	Tyre, Kendal H.	Franchising in Africa	Franchising in Africa, franchise legislation, law and regulation in African franchising sector
10	2012	Winter, Sidney G.; Szulanski, Gabriel; Ringov, Dimo; Jensen, Robert J.	Reproducing knowledge: inaccurate replication and failure in franchise organizations	Expanding by replication versus local adaptation
11	2013	Johnson, Derek	Media franchising: creative license and collaboration in the culture industries	Franchising, from ownership to partnership, production and marketing
12	2014	Yadon, Robert E.; Umansky, Barry D.; Coomes, Paul A.	Telecom regulatory review: Kentucky 2014	Regulation, competition in telecommunications industry
13	2012	Hulsink, Willem	Privatisation and liberalisation in European telecommunications: comparing Britain, the Netherlands and France	Sector-specific study of comparative telecommunications regimes
14	2014	Latipulhayat, Atip	Telecommunications licensing regime: a new method of state control after privatisation of telecommunications	Change from state regulated telecommunications to private sector participation
15		FEE, FRANCHISE	Utilities	Ordinance granting a nonexclusive franchise access
16	2011	Gaskill, Frank J.	Conditions favoring the globalization of the franchising format: a conceptual model	Global growth of franchising, rise of middle classes, franchising provides secure environment in countries with very different legal traditions
17	2011	Sherman, Andrew	Franchising and licensing: two powerful ways to grow your business in any economy	Building a franchising program, financial strategies

(continued)

Table 3 (continued)

	Year	Authors	Title	Research subject
18	2014	Chen, Yi-Min; Su, Chien-Tai	Brand equity heterogeneity among strategic groups in service franchising	Franchisor x franchisee relationship and brand identity
19	2013	Dana, Léo-Paul; Granata, Julien; Lasch, Frank; Carnaby, Alan	The evolution of co-opetition in the Waipara wine cluster of New Zealand	Competition in agriculture, evolution of agriculture cluster
20	2014	Hajdul, Marcin	Virtual co-opetition in transport-T-Scale platform case study	Coopetition in transportation, environmental issues
21	2013	Ilvonen, Ilona; Vuori, Vilma	Risks and benefits of knowledge sharing in co-opetitive knowledge networks	Knowledge sharing in coopetition
22	2014	Evens, Tom	Co-opetition of TV broadcasters in online video markets: a winning strategy?	Coopetition in TV broadcasts against YouTube/Netflix to reduce cost and reach scale in marketplace
23	2011	Kang, Jin-Su; Lee, Hong-Yuh; Tsai, Julio	An analysis of interdependencies in mobile communications technology: the case of WiMAX and the development of a market assessment model	Prospective 4G technology and its issues with stakeholders
24	2014	Bauer, Johannes M.	Platforms, systems competition, and innovation: reassessing the foundations of communications policy	Effects of policy on communications sector, dynamic competition
25	2014	Ahokangas, Petri; Horneman, Kari; Posti, Harri; Matinmikko, Marja; Hanninen, Tuomo; Yrjola, Seppo; Goncalves, Vania	Defining "co-primary spectrum sharing"—a new business opportunity for MNOs?	Licensed sharing with mobile network operators
26	2014	Kossyva, Dorothea I.; Galanis, Konstantinos V.; Sarri, Katerina K.; Georgopoulos, Nikolaos B.	Adopting an information security management system in a co-opetition strategy context	ISO standards when implementing coopetition strategies

(continued)

Table 3 (continued)

	Year	Authors	Title	Research subject
27	2012	Hung, Chia-Liang; Chou, Jerome Chih-Lung; Ding, Chung-Ming	Enhancing mobile satisfaction through integration of usability and flow	User mobile satisfaction with playing mobile games
28	2011	Daidj, Nabyla	Media convergence and business ecosystems	Different growth strategies within reconfiguration of value chains in new media
29	2011	Gomes, Emanuel; Weber, Yaakov; Brown, Chris; Tarba, Shlomo Yedidia	Mergers, acquisitions and strategic alliances: understanding the process	Strategic alliances, integration management
30	2014	Bengtsson, Maria; Kock, Sören	Coopetition—Quo vadis? Past accomplishments and future challenges	Lack of unified definitions in coopetition
31	2012	Kwak, Jooyoung; Lee, Heejin; Chung, Do Bum	The evolution of alliance structure in China's mobile telecommunication industry and implications for international standardization	Alliance formation in telecommunications in China, 3G a 4G
32	2011	Meier, Matthias	Knowledge management in strategic alliances: a review of empirical evidence	Knowledge management in alliances, current state of the art in empirical research
33	2012	Bigliardi, Barbara; Ivo Dormio, Alberto; Galati, Francesco	The adoption of open innovation within the telecommunication industry	ICT and different ways to manage the open innovation processes
34	2014	Fernandez, Anne-Sophie; Le Roy, Frédéric; Gnyawali, Devi R.	Sources and management of tension in co-opetition case evidence from telecommunications satellites manufacturing in Europe	Separation and integration of competition and cooperation in order to manage tension in coopetition
35	2013	Li; Lee; Qian, Gongming; Qian, Zhengming	Do partners in international strategic alliances share resources, costs, and risks?	Risks, costs, resources sharing in strategic alliances – low-tech industries (tenable), high-tech industries (untenable)

(continued)

Table 3 (continued)

	Year	Authors	Title	Research subject
36	2011	McCarter, Matthew W.; Mahoney, Joseph T.; Northcraft, Gregory B.	Testing the waters: using collective real options to manage the social dilemma of strategic alliances	Managing social uncertainty in strategic alliances by collective real options, resource investment
37	2011	Gerybadze, Alexander	Strategic alliances and process redesign: effective management and restructuring of cooperative projects and networks	Formal and informal cooperation, OECD and inter-firm agreements
38	2015	de Reuver, Mark; Verschuur, Edgar; Nikayin, Fatemeh; Cerpa, Narciso; Bouwman, Harry	Collective action for mobile payment platforms: a case study on collaboration issues between banks and telecom operators	Mobile payments in Western societies, collaboration and competitions between banks and operators
39	2014	Yang, Haibin; Zheng, Yanfeng; Zhao, Xia	Exploration or exploitation? Small firms' alliance strategies with large firms	How is small firms' valuation affected by large firms, alliance governance
40	2018	Dodgson, Mark	Technological collaboration in industry: strategy, policy and internationalization in innovation	The use of IT and management skills in future vertical and horizontal collaboration to support innovation
41	2012	Pellicelli, Anna Claudia	Strategic alliances	Goals of alliances, alliances being common nowadays, types and management of alliances
42	2011	Clifton, Judith; Comín, Francisco; Díaz-Fuentes, Daniel	From national monopoly to multinational corporation: how regulation shaped the road towards telecommunications internationalisation	International expanding of former national telecommunications monopolies, regulation
43	2017	Serrat, Olivier	Learning in strategic alliances	Collaborative advantage of strategic alliances and managing the partnership

(continued)

Table 3 (continued)

	Year	Authors	Title	Research subject
44	2011	Fang, Eric	The effect of strategic alliance knowledge complementarity on new product innovativeness in China	Cultural and other business difficulties in strategic alliances
45	2014	Arasa, Robert; Gathinji, Loice	The relationship between competitive strategies and firm performance: a case of mobile telecommunication companies in Kenya	Competitive strategies and firm performance, product differentiation and low cost leadership
46	2011	Kim, Kyung Kyu; Ryoo, Sung Yul; Jung, Myung Dug	Inter-organizational information systems visibility in buyer–supplier relationships: the case of telecommunication equipment component manufacturing industry	Supply chain management of producers of telecom equipment
47	2013	Gomes, Emanuel; Cohen, Marcel; Mellahi, Kamel	When two African cultures collide: a study of interactions between managers in a strategic alliance between two African organizations	Human resource management in international joint venture
48	2015	Ajao, Blessing Funke; Oyebisi, Timothy; Aderemi, Helen; Jegede, Oluseye	Status and impact of strategic technology alliances among telecommunications firms in Nigeria	Improvement of services rendered to subscribers, GSM network operators

Graph 5 Count of methodologies used in articles

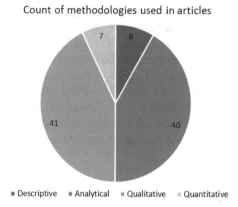

Count of methodologies used in articles

7

8

41

40

■ Descriptive   ■ Analytical   ■ Qualitative   ■ Quantitative

4 Conclusion

Literature research done in this article shows the number of 48 up-to-date articles and publications that discuss the connection between franchising, telecommunications and customer experience management. These articles were published in 2011–2018 period, which means they are the most relevant to this topic when it comes to new information and innovation process. There can be found more articles later this year (2018) that can provide in depth information about our discussed topic for the year 2018. Some articles may not be completely connected with our researched topic, but they help us understand the keywords and the "real-life" use of franchising in telecommunications and how customer experience management shapes telecommunication industry. Therefore with further research we could suggest better solutions when it comes to this matter. The final classification for this research can be found in Table 3.

We believe that having a literature database with the year of publication, author, title and research subject, it is easy for anyone using this database to find the most relevant information for their research or if they just seek more information. Types of methodologies used in articles are important but if the research subject is of no interest for a researcher or reader, he or she doesn't need the information about methodology used in a certain publication. Therefore, we decided to not include this information in our final classification.

We used two methodologies for this paper's research and literature review—qualitative and analytical. We were searching for up-to-date articles that would provide an in depth information about telecommunication industry, franchising and customer management. We have chosen qualitative research as we wanted to develop ideas for potential quantitative research. Also, we wanted to uncover the trends in franchising in telecommunication industry, as well as trends in customer experience management that seems to get more attention than customer relationship management nowadays.

By this research we found out that the most discussed areas were Telecommunications Act 1996, fees and revenues, monopoly and oligopoly in telecommunication industry, rules and relationships in franchising, and how customer experience management is used to pay employees by what they sell and that customers can be seen as threat on social media. The most interesting articles about franchising in telecommunications industry can be found from Taiwan and African countries, such as Nigeria and Kenya. It's because of their potential and large number of middle class citizens (therefore they are able to afford computers and Wi-Fi, mobile phones and mobile internet). These countries focus on technology, innovation and development, so it's interesting for other companies to see how they incorporate franchising and other strategic alliances in telecommunications industry in daily use.

References

1. Business Dictionary (2018) What is franchising? In: BusinessDictionary.com [online]. http://www.businessdictionary.com/definition/franchising.html
2. International Franchise Association (2018) What is a franchise. In: franchise.org [online]. https://www.franchise.org/what-is-a-franchise
3. Seid MH (2018) Where it all began. The evolution of franchising. In: franchise-chat.com [online]. http://www.franchise-chat.com/resources/where_it_all_began_the_evolution_of_franchising.htm
4. Devan A (2015) What is customer experience management (CEM) in a telecom context.... In: linkedin.com [online]. https://www.linkedin.com/pulse/what-customer-experience-management-cem-telecom-context-arumugam
5. Kochhar R (2015) Opinion: why CEM has emerged as top priority for telcos. In: Telecom Tech News [online]. https://www.telecomstechnews.com/news/2015/dec/16/opinion-why-cem-has-emerged-top-priority-telcos/
6. Ćamilović D (2008) Data mining and CRM in telecommunications. In: sjm06.com [online]. http://www.sjm06.com/sjmissn1452-4864/3_1_2008_may_1-125/3_1_61-72.pdf
7. Ristvej J, Zagorecki A (2011) Information systems for crisis management—current applications and future directions. In: Komunikácie—communications—scientific letters of the University of Žilina, vol 13, Iss 2, pp 59–63. ISSN: 1335-4205
8. Daidj N (2012) Media convergence and business ecosystems. In: globalmediajournal.com [online]. http://www.globalmediajournal.com/open-access/media-convergence-and-business-ecosystems.php?aid=35304
9. Bauer C, Strauss C (2016) Location-based advertising on mobile devices: a literature re-view and analysis. Manag Rev Q(MRQ) Springer 66(3):159–194
10. Loop E (2018) Different types of methodologies I synonym. In: classroom.synonym.com [online]. https://classroom.synonym.com/different-types-methodologies-7459438.html
11. Irny SI, Rose AA (2005) Designing a strategic information systems planning methodology for malaysian institutes of higher learning (isp-ipta). Issues Inf Syst 6(1)
12. Merriam-Webster (2018) Definition of methodology. In: merriam-webster.com [online]. https://www.merriam-webster.com/dictionary/methodology. Accessed 12 July 2018
13. Baskerville R. Risk analysis as a source of professional knowledge. Comput Secur 10(8):749–764
14. Kumar Jharoti A (2015) Research methodology. In: shodhganga.inflibnet.ac.in [online]. http://shodhganga.inflibnet.ac.in/bitstream/10603/73387/13/13_chapter4.pdf

Conceptualization of Predictive Analytics by Literature Review

Katarína Močarníková and Michal Greguš

Abstract Predictive Analytics, together with Big Data Analytics, learning algorithms, and machine learning are the most advanced technical innovations of this time. The notion of predictive analytics was introduced in the 20th century and become more and more expanded and applied in many fields like healthcare, business, supply chain management, telecommunications, and many others. The aim of this paper is a detailed literature analysis on Predictive Analytics, mainly in articles published in relevant journals during the selected time period, from 2010 till now. Various databases were used in order to find the most relevant articles for this topic. Articles were systematically analyzed regarding the author (authors), year of publication, the area of research, output, and journal, where the article was published. The main contribution of this article is evidence of the most relevant articles related to Predictive analytics, which can be used for every reader and also an overview, where, or in which fields Predictive analytics is applied and how was used during last years in various researches.

Keywords Predictive analytics · Predictive modelling · Big data · Literature review

1 Introduction

Analysis of data is a process, in which we are searching for data, then cleaning them, transforming and modeling them in order to highlight the important content for our goal and conclude a conclusion [8]. Predicting is a process of forecasting events that have not yet occurred.

K. Močarníková (✉) · M. Greguš
Comenius University in Bratislava, Bratislava, Slovakia
e-mail: katarina.mocarnikova@fm.uniba.sk

M. Greguš
e-mail: michal.gregusml@fm.uniba.sk

© Springer Nature Switzerland AG 2020
N. Kryvinska and M. Greguš (eds.), *Data-Centric Business and Applications*,
Lecture Notes on Data Engineering and Communications Technologies 30,
https://doi.org/10.1007/978-3-030-19069-9_8

Predictive analytics is a branch of advanced analytics. Predictive analytics is used for making predictions about future events which are currently unknown. In predictive analytics, mathematical formulas are applied to data and decision for a given situation or problem should be discovered [33].

Throughout the years, Predictive analytics become very common notion and was applied in many fields. The field in which predictive analytics is applied the most is medicine. Pioneers in this area were Klindworth, who was trying to, using predictive analytics, identify fraud in Medicare program [23] and Gotz with his platform for intelligent care delivery [11]. Predictive analytics is also used in predicting heart-related issues [13], predicting chronic kidney diseases [9], diabetes [28] or probability of readmission [36] or even in predicting Parkinson's disease using video games [32].

Medicine is probably one of the most useful fields, where predictive analytics has the biggest application. However, there are also other important fields, where predictive analytics can be applied in order to make our lives better and easier. For example, predictive analytics can be applied into traffic in order to create and implement intelligent transportation systems, which can significantly reduce the traffic congestion [40] or for predicting the development of traffic situations under unusual conditions from real movement data [1].

Very common type of data analytics is business intelligence, which covers data analysis of business information [8]. We can consider business intelligence as predictive analytics implemented in business, in order to help e.g. managers or sales representatives to make a better decision on business issues. Predictive analytics is more and more frequently used in business to predict customer churn or customer attrition [35].

Education is an area, where predictive analytics can be applied too. It can help students to choose a suitable course, predict their performance on that course or advise them future career path [39], or even predict factors of their attrition in the first year of study [16].

This literature review was mainly inspired by review developed by Kryvinska [25]. We were also inspired by other articles from Kryvinska [20, 26, 27] or Gregus [12].

2 Literature Analysis on Predictive Analytics

An extensive literature review of articles related to predictive analytics published as articles in journals was performed. This review was performed in various databases, e.g. IEEE Xplore, SAGE, SpringerLink, ScienceDirect (Elsevier) and Google Scholar. In order to find the most relevant articles, cross-referenced techniques were used. It has to be noted that the mentioned analysis may not be fully exhaustive, but the most relevant articles starting in 2010 were selected for this analysis. 120 articles were selected and our selection has following structure: author/authors, year of publication, title, field or discipline related to that article, name of the journal and output from article. Selected articles were analyzed chronologically in order to describe their process of evolution. All selected articles with the structure mentioned above are shown in next chapter followed by sub-chapters with detailed analysis.

2.1 Literature Analysis of Predictive Analytics (2010–2018)

See Table 1.

2.2 Author-Year Evolution

First two columns of Table 1 contain information about the author(s) and the year of publication. A deeper analysis of the number of publications per year and authors who published during these years will be described in following subheads.

Evolution of the Number of Published Articles Starting 2010. Our research was focused on articles published in journals starting in 2010. First articles related to predictive analytics were published in 2010. Pioneers in this area were Balkan and Demirkan with their article about using information services in advanced marketing and how to refine predictive models and implement them into corporate strategy [2]. Other authors were Gotovac and Kraljević, who were trying to define a predictive model able to predict prepaid churn in telecommunications [24].

As we can see in Fig. 1, publications related to Predictive Analytics have a rising trend, except the year 2015, when the number of published articles decreased. A number of articles published in 2018 seem to be lower too, but it's because 2018 is not over and there is still a few months for publishing articles. We can expect the growth of articles published in the next years. In 2011, the number of publication decreased and only two articles were published. It was for the first time when Predictive Analytics was applied in fields like medicine when Jacob [19] published their article about using predictive models in predicting the risk of developing breast cancer. Like for medicine, also for automobile industry was the year 2011 time, when Predictive Analytics step into and Hanumanthappa and Sarakuty published their article about the implementation of predictive analytics into automotive marker in order to predict future sales based on historical data [15].

The year 2012 brought us new fields, where predictive analytics was implemented, like insurance [7], HR [21] or finance [5]. In the second half of 2013, Dharmaji and Sridar started with applying predictive analytics into business with their article 'A Comparative Study on How Big Data is Scaling Business Intelligence and Analytics', in which they were studying the emergence of big data and their role in advanced analytics, where techniques like predictive analytics can help to understand the current state of business and help to make productive and persuasive decisions [38]. Gaughan, together with Gold and McClaren contributed in 2013 with the article to the field of entertainment, in which they were writing about predicting Academy awards (Oscar) [10]. In 2014, it was medicine and healthcare area which were dominant. Interesting contributors to this area were Ajourlou with Shams and Yang with predicting readmissions among patients with heart failure [37] or Ng and co-authors with modeling platform for health records [34]. In 2015, the number of articles published decreased rapidly and medicine and health care weren't the only dominant area. In this year, stock price prediction related articles started to be published from Li and his team [31] or Zhang with his team [42]. In 2016, the number of published

Table 1 Literature review of selected articles

	Author(s)	Year	Title
1	Bihani and Patil	2014	A comparative study of data analysis techniques
2	Dharmaji and Sridhar	2013	A comparative study on how big data is scaling business intelligence and analytics
3	Piri et al.	2017	A data analytics approach to building a clinical decision support system for diabetic retinopathy: developing and deploying a model ensemble
4	Kumar and Garg	2018	A hybrid deep learning model for predictive analytics
5	Poornima and Pushpalatha	2016	A journey from big data towards prescriptive analytics
6	Janiesch and Matzner and Schwegmann	2013	A method and tool for predictive event—driven process analytics
7	Ajorlou and Shams and Yang	2014	A predictive analytics approach to reducing 30-day avoidable readmissions among patients with heart failure, acute myocardial infarction, pneumonia, or COPD
8	Wang	2013	A proactive complex event processing method for intelligent transportation systems
9	Renjith	2015	An integrated framework to recommend personalized retention actions to control B2C E-commerce customer churn
10	Dhall and Solanki	2017	An IoT based predictive connected car maintenance approach
11	Upendran et al.	2016	Application of predictive analytics in intelligent course recommendation
12	Klindworth et al.	2012	Assessment of predictive modeling for identifying fraud within the medicare program
13	Paulus and Ward	2013	Augmenting austrian flood management practices through geospatial predictive analytics: a study in carinthia

(continued)

Table 1 (continued)

	Author(s)	Year	Title
14	Fard et al.	2018	Automated robot-assisted surgical skill evaluation: predictive analytics approach
15	Harvey and Luckman	2014	Beyond demographics: predicting student attrition within the bachelor of arts degree
16	Baig and Jabeen	2016	Big data analytics for behavior monitoring of students
17	Irani and Shah and Sharif	2017	Big data in an HR context: exploring organizational change readiness, employee attitudes and behaviors
18	Koseleva and Ropaite	2017	Big data in building energy efficiency: understanding of big data and main challenges
19	Rumsfeld and Shah	2017	Big data in cardiology
	Field	Journal	Output
1 cont	Research	IJETTCS	Comparative study of few of the data analysis techniques was provided
2 cont	Business	IJERSTE	Through better analysis of the large volumes of data, there is the potential for making faster advances
3 cont	Medicine	Decision Support Systems	Developed system provides several important practical implications
4 cont	Mathematics, IT	International Journal of Research in Advent Technology	The application of the model is proposed
5 cont	Research	Journal of Engineering and Applied Sciences	This paper helps to understand its basics
6 cont	Research	Wirtschaftsinformatik	Review of the literature revealed a lack of methodological
7 cont	Healthcare	Health Care Management Science	Results shows improved discrimination power compared to the literature
8 cont	Traffic	Lecture Notes on Information Theory	Simulation experiments show that this method can reduce the traffic congestion
9 cont	Business	International Journal of Engineering Trends and Technology	This framework helps sensing the potential customer attrition

(continued)

Table 1 (continued)

	Field	Journal	Output
10 cont	Automobil industry	International Journal of Interactive Multimedia and Artificial Intelligence	High level architecture of how 'connected cars' can be implemented is presented
11 cont	Education	Procedia Computer Science	Proposed system is likely to benefit course choice as well as career options
12 cont	Medicine, Security	Health Management, Policy and Innovation	National Medicare fraud and abuse estimates based on the predictive model were presented
13 cont	Geology	Natural Hazards and Earth System Sciences	The use of this technology led to a complementary summary of cultural influence on risk perception
14 cont	Medicine	The International Journal of Medical Robotics and Computer Assisted Surgery	Proposed framework is able to classify surgeon's expertise with 82.3% accuracy
15 cont	Education	The International Journal of the First Year in Higher Education	Course preference and first year educational performance were correlated as predictors of attrition
16 cont	Education	Procedia Computer Science	Arguing that enough data is available in a university environment that can be harnessed by Big Data model
17 cont	HR	Journal of Business Research	Set of practical and management implications and recommendations for future research in the area
18 cont	Energetics	Procedia Engineering	Three main problems with Big data in energy field were marked
19 cont	Medicine	European Heart Journal	There needs to be a shift in focus from what big data might do for cardiovascular care to proving what it can do
	Author(s)	Year	Title
20	Gaudi et al.	2016	Case Study: IBM watson analytics cloud platform as analytics-as-a-service system for heart failure early detection
21	Doctor and Iqbal and Mahmud	2016	Cloud enabled data analytics and visualization framework for health-shocks prediction

(continued)

Table 1 (continued)

	Author(s)	Year	Title
22	Lewis et al.	2018	Combining hospital and general practice data to predict the risk of hospitalisation in the Australian context
23	Bansal and Rangra	2014	Comparative study of data mining tools
24	Basha et al.	2017	Comparative study on performance analysis of time series predictive models
25	Amarasingham et al.	2016	Consensus statement on electronic health predictive analytics: a guiding framework to address challenges
26	Schoenherr and Speier-Pero	2015	Data science, predictive analytics, and big data in supply chain management: current state and future potential
27	Lismont et al.	2017	Defining analytics maturity indicators: a survey approach
28	Balamurugan and Devi and Kris	2016	Developing a modified logistic regression model for diabetes mellitus and identifying the important factors of type II DM
29	Scheer et al.	2017	Development of a preoperative predictive model for major complications following adult spinal deformity surgery
30	Jacob and Ramani	2011	Discovery of knowledge patterns in clinical data through data mining algorithms: multi-class categorization of breast tissue data
33	Zhang et al.	2015	Dynamic business network analysis for correlated stock price movement prediction
34	Ou et al.	2016	Dynamic cost forecasting model based on extreme learning machine—a case study in steel plant
35	Bhadoria and Wazurkar	2018	Effective modelling for predictive analytics in data science
36	Goar and Sarangdevot and Singla	2018	Effectual implementation of emotions mining and predictive analytics from twitter social media
37	Bhalla	2012	Enhancement in predictive model for insurance underwriting
38	Halim et al.	2018	Evaluating predictive analytics model performance accuracy for network selection mechanism
39	Marchevsky and Walts and Wick	2017	Evidence-based pathology in its second decade: toward probabilistic cognitive computing

(continued)

Table 1 (continued)

	Author(s)	Year	Title
40	Andrienko and Andrienko and Rinzivillo	2015	Exploiting spatial abstraction in predictive analytics of vehicle traffic
	Field	Journal	Output
20 cont	Medicine	Future Internet	Obtained results are comparable to results from literature
21 cont	Healthcare	Future Generation Computer Systems	Proposed system has high performance in predicting health-shock
22 cont	Medicine	International Journal of Integrated Care	Predictive models were implemented and had positive predictive value
23 cont	Research	IJARCSSE	Tools (Rapid Miner, Orange, Knime, Weka, Keel and R) were described and compared
24 cont	Mathematics	International Journal of Grid and Distributed Computing	ARIMA technique was found as technique with better performance
25 cont	Healthcare	eGEMs: The Journal of Electronic Health Data and Methods	List of recommendations for framework was created
26 cont	SCM	Journal of Business Logistics	Article brought insight into SCM predictive analytics
27 cont	Business	International Journal of Information Management	Analytics is nowadays more commonly applied
28 cont	Medicine	Indian Journal of Science and Technology	Logistic model is built from the sigmoid function using the regression coefficients, produces high accuracy
29 cont	Medicine, Healthcare	Journal of Neurosurgery	Successful model predicting major intra or perioperative complications after ADS surgery was built
30 cont	Medicine	International Journal of Computer Applications	Results indicate the level of accuracy of the algorithms for detecting the breast cancer
31 cont	Economy	IEEE Intelligent Systems	Prediction based on article summaries outperform prediction based on full-length articles
32 cont	SCM	IJMUE	LDA outperforms QDA and their performance was compared

(continued)

Table 1 (continued)

	Field	Journal	Output
33 cont	Economy	IEEE Intelligent Systems	Designed model outperforms the best baseline model
34 cont	Economy	Computers & Industrial Engineering	Forecasting model can offer an accurate result of raw material price
35 cont	Research, IT	Journal of Fundamental and Applied Sciences	Algorithm for create and segregate data based on different collect was proposed
36 cont	Social media	IJCBR	Automatic sentiment extractor from a tweet was built
37 cont	Insurance	International Journal of Computer Science & Engineering Technology	Enhancing the underwriting algorithms helps the concentrate more on the ones which involve higher degree of risk
38 cont	Telecommunication	Journal of Fundamental and Applied Sciences	The most relevant predictive model, decision tree, was observed
39 cont	Medicine	Human Pathology	IBM Watson provides results much more rapidly
40 cont	Traffic	International Journal of Geo-Information	Potential of using real-time movement data for prediction of development of traffic situations
	Author(s)	Year	Title
41	Doyle et al.	2014	Forecasting significant societal events using the embers streaming predictive analytics system
42	Cao and Sharma and Wang	2016	From knowledge sharing to firm performance: A predictive model comparison
43	Kapoor and Sherif	2012	Global human resources (HR) information systems
44	Alharthi	2018	Healthcare predictive analytics: an overview with a focus on Saudi Arabia
45	Prabavathi and Shanthipriya	2018	Healthcare predictive analytics
46	Denley	2014	How predictive analytics and choice architecture can improve student success
47	Lama and Mishra and Pal	2016	Human resource predictive analytics (HRPA) for HR management in organizations
48	Argentinis and Chen and Weber	2016	IBM watson: how cognitive computing can be applied to big data challenges in life sciences research

(continued)

Table 1 (continued)

	Author(s)	Year	Title
49	Sun et al.	2014	iCARE: a framework for big data-based banking customer analytics
50	Gotz et al.	2012	ICDA: a platform for intelligent care delivery analytics
51	Jeble et al.	2018	Impact of big data & predictive analytics capability on supply chain sustainability
52	Bansal and Goel and Sharma	2017	Improved K-mean clustering algorithm for prediction analysis using classification technique in data mining
53	Benoit and Van den Poel	2012	Improving customer retention in financial services using kinship network information
54	Taber et al.	2015	Inclusion of dynamic clinical data improves the predictive performance of a 30-day readmission risk model in kidney transplantation
55	Ghasemaghaei and Hassanein and Turel	2017	Increasing firm agility through the use of data analytics: the role of fit
56	Balkan and Demirkan	2010	Information services for advanced marketing
57	Danforth and Geece	2017	Instagram photos reveal predictive markers of depression
58	Chen	2017	Integrated and intelligent manufacturing: perspectives and enablers
59	Sudhir and Sundaram	2017	Is predictive intelligence going to be the 5 th P of marketing? a study into insight of future marketing
60	Litsey and Mauldin	2018	Knowing what the patron wants: using predictive analytics to transform library decision making
	Field	Journal	Output
41 cont	Social	Big Data	Architecture of EMBERS was described
42 cont	Business	Journal of Business Research	Results indicate that in the best performing model, innovation and IC simultaneously mediate the relationship between KS and FP
43 cont	HR	Kybernetes	By applying analytical techniques on the global database, HRs professionals will get intelligent business insight

(continued)

Table 1 (continued)

	Field	Journal	Output
44 cont	Healthcare	IRJET	There is a real and pressing need to digitize health records in Saudi Arabia
45 cont	Healthcare	Journal of Infection and Public Health	Algorithms for evaluation were compared and algorithm with better predictions was identified
46 cont	Education	Research & Practice in Assessment	Course recommendation system was examined and results showed its impact on student success
47 cont	HR	IJSTR	HRPA has potential to achieve 100% accuracy in decision making for HR.
48 cont	Research	Clinical Therapeutics	Current pilot projects are beginning to yield insight into whether
49 cont	Banking	IBM Journal of Research and Development	The advantages of the iCARE framework have been confirmed in a real case study of a bank in southeast China
50 cont	Medicine, Healthcare	AMIA Annual Symposium Proceedings	ICDA provides a universal standards-based analytics environment
51 cont	Business/SCM	International Journal of Logistics Management	Potential of predictive analytics in achieving desired goal was explained
52 cont	IT	International Journal of Computer Applications	Modification of K-mean clustering was proposed
53 cont	Finance, CRM	Expert Systems with Applications	Predictive power of the churn model can be improved by adding the social network (SNA-) based variables
54 cont	Medicine	Transplantation	Modeling clinical data outperformed models utilizing immutable data in predicting 30DRA
55 cont	Business	Decision Support Systems	Understanding of the impacts of data analytics use on firm agility
56 cont	Marketing	International Journal of Data Analysis and Information Systems	This tutorial is a high level summary of sequence of information services for advanced marketing

(continued)

Table 1 (continued)

	Field	Journal	Output
57 cont	Psychology	EPJ Data Science	Proposed model outperformed general practitioners' average
58 cont	Manufacturing	Engineering	A new industrial revolution is on the horizon
59 cont	Marketing	Asia Pacific Journal of Research	Predictive Intelligence with its tools will help the marketers for personalizing the customer's behavior
60 cont	Services	The Journal of Academic Librarianship	Description of tool able to monitor library behavior and provide description of library behavior
	Author(s)	Year	Title
61	Bell and Jalali and Olabode	2012	Leveraging cloud computing to address public health disparities: an analysis of the SPHPS
62	Levin et al.	2018	Machine-learning-based electronic triage more accurately differentiates patients with respect to clinical outcomes compared with the emergency severity index
63	Martens et al.	2016	Mining massive fine-grained behavior data to improve predictive analytics
64	Gotovac and Kraljević	2010	Modeling data mining applications for prediction of prepaid churn in telecommunication services
65	Berecibar et al.	2016	Online state of health estimation on NMC cells based on predictive analytics
66	Balaji and Caytiles and Iyengar	2017	Optimal predictive analytics of pima diabetics using deep learning
67	Ng et al.	2014	PARAMO: A PARAllel predictive MOdeling platform for healthcare analytic research using electronic health records
68	Liu et al.	2013	Parkinson's disease predictive analytics through a pad game based on personal data
69	Balaji Prabhu and Dakshayini	2018	Performance analysis of the regression and time series predictive models using parallel implementation for agricultural data
70	Danner et al.	2017	Physiologically-based, predictive analytics using the heart-rate-to-systolic-ratio significantly improves the timeliness and accuracy of sepsis prediction compared to SIRS

(continued)

Table 1 (continued)

	Author(s)	Year	Title
71	Nickerson and Rogers	2014	Political campaigns and big data
72	Gerber	2014	Predicting crime using Twitter and kernel density estimation
73	Tsakalidis et al.	2015	Predicting elections for multiple countries using Twitter and polls
74	Homer et al.	2017	Predicting falls in people aged 65 years and older from insurance claims
75	Thornton et al.	2013	Predicting healthcare fraud in medicaid: a multidimensional data model and analysis techniques for fraud detection
76	Hanumanthappa and Sarakutty	2011	Predicting the future of car manufacturing industry using data mining techniques
77	Ichikawa and Oyama and Shimoda	2018	Prediction models to identify individuals at risk of metabolic syndrome who are unlikely to participate in a health intervention program
78	Khan and Quadri	2016	Prediction of angiographic disease status using rule based data mining techniques
79	Taylor et al.	2016	Prediction of in-hospital mortality in emergency department patients with sepsis: a local big data–driven, machine learning approach
	Field	Journal	Output
61 cont	Healthcare	Online Journal of Public Health Informatics	SPHPS can be a hub of Population health record for reducing health disparities
62 cont	Healthcare	Annals of Emergency Medicine	E-triage predictions demonstrated improved identification of clinical patient outcomes
63 cont	Banking	MIS Quarterly	The use of behavioral similarity for predictive modeling
64 cont	Telecommunication	Journal for Control, Measurement, Electronics, Computing and Communications	More complexity is brought into modeling by a lower data amount available for prepaid users
65 cont	Power sources	Journal of Power Sources	This work allows a deep comparison of the different estimation techniques

(continued)

Table 1 (continued)

	Field	Journal	Output
66 cont	Medicine, Healthcare	International Journal of Database Theory and Application	The comparison shows that the proposed model is definitely more effective than the rough set theory model
67 cont	Healthcare	Journal of Biomedical Informatics	Efficient parallel predictive modeling platform can be developed for EHR data
68 cont	Medicine, Healthcare	International Journal of Information Technology	The implemented game can detect possible PD symptoms at an early stage
69 cont	Agriculture	Procedia Computer Science	Predictive model with the best performance results was selected
70 cont	Medicine, Healthcare	The American Journal of Surgery	Physiologically-based predictive analytics improved the accuracy and expediency of sepsis
71 cont	Politics	Journal of Economic Perspectives	Evolution and utility of data in political campaign were described
72 cont	Social media, Security	Decision Support Systems	This research has implications for decision makers concerned with geographic spaces occupied by Twitter-using individuals
73 cont	Social science	IEEE Intelligent Systems	Better results than several past works and the commercial baseline were achieved
74 cont	Healthcare	The American Journal of Medicine	Individuals in this large cohort at high risk of falls could be readily identified up to 2 years in advance
75 cont	Healthcare	Procedia Technology	The paper contributes to the literature by analysis techniques useful at each level for fraud detection
76 cont	Automobil industry	International Journal on Information Technology	This paper introduces the application of data mining technology in the car manufacturing unit
77 cont	Healthcare	International Journal of Medical Informatics	Predictive models outperformed existing conventional methods
78 cont	Medicine	Biological Forum—An International Journal	Summary of various data mining techniques that can help in prediction for early medical diagnosis
79 cont	Medicine, Healthcare	Academic Emergency Medicine	Machine learning approach outperformed existing CDRs for predicting in-hospital mortality of emergency department patients with sepsis

(continued)

Table 1 (continued)

	Author(s)	Year	Title
80	Priyadarshini and Rathi and Rastogi	2010	Predictive analysis for customer relationship management
81	Jeyanthi and Prasad and Radhakrishnan	2010	Predictive analytics using genetic algorithm for efficient supply chain inventory optimization
82	Hashimzade and Myles and Rablen	2016	Predictive analytics and the targeting of audits
83	Kui and Wang and Zhu	2013	Predictive analytics by using bayesian model averaging for large-scale internet of things
84	Akbar et al.	2017	Predictive analytics for complex IoT data streams
85	Adeduro and Oladapo and Omotosho	2018	Predictive analytics for increased loyalty and customer retention in telecommunication industry
86	Boukenze and Haqiq and Mousannif	2016	Predictive analytics in healthcare system using data mining techniques
87	Higdon et al.	2013	Predictive analytics in healthcare: medications as a predictor of medical complexity
88	Archana and Kumari and Malisetty	2017	Predictive analytics in HR management
89	May and Shannon and Vargas	2016	Predictive analytics model for healthcare planning and scheduling
90	Rachuri and Shin and Woo	2014	Predictive analytics model for power consumption in manufacturing
91	Cheng and Yang	2015	Predictive Analytics on CSI 300 index based on ARIMA and RBF-ANN combined model
92	Meena and Revathi	2013	Predictive analytics on healthcare: a survey
93	Dinov et al.	2016	Predictive big data analytics: a study of parkinson's disease using large, complex, heterogeneous, incongruent, multi-source and incomplete observations

(continued)

Table 1 (continued)

	Author(s)	Year	Title
94	Gao and Chen and Wang	2018	Predictive complex event processing based on evolving Bayesian networks
95	Mishra et al.	2010	Predictive data mining: promising future and applications
96	Raju	2018	Predictive healthcare informatics using deep learning—a big data approach
97	Lee et al.	2013	Predictive manufacturing system—trends of next-generation production systems
98	Kumar et al.	2015	Predictive methodology for diabetic data analysis in big data
99	Shameer et al.	2017	Predictive modeling of hospital readmission rates using electronic medical record-wide machine learning: a case-study using mount sinai heart failure cohort
100	Norton	2013	Predictive policing the future of law enforcement in the trinidad and tobago police service
	Field	Journal	Output
80 cont	CRM	JDCTA	Approach which let a shopkeeper to provide the best configuration of laptops
81 cont	SCM	International Journal of Computer Science and Network Security	An innovative and efficient methodology for precise determination of the most probable excess stock level was proposed
82 cont	Audits	Journal of Economic Behavior & Organization	The use of predictive analytics yields a significant increase in revenue over a random audit strategy
83 cont	Mathematics, IT	International Journal of Distributed Sensor Networks	Proposed prediction analytic method has better accuracy compared to traditional methods
84 cont	Transportation	IEEE Internet of Things Journal	Proposed architecture was implemented using open source components
85 cont	Telecommunication	International Journal of Computer Applications	Proposed model can predict customer retention with really high accuracy

Table 1 (continued)

	Field	Journal	Output
86 cont	Medicine, Healthcare	Journal of Computer Science & Information Technology	Used classifier proved its performance in predicting with best results in terms of accuracy and minimum execution time
87 cont	Medicine, Heatlhcare	Big data	Screens for predicting the complexity of patients were implemented
88 cont	HR	Indian Journal of Public Health Research & Development	Key areas are identified from which the predictive analytics can create the values for HR perspective
89 cont	Healthcare	European Journal of Operational Research	Model produces globally optimal estimates
90 cont	Industries	Procedia CIRP	Design of a big data analytics model was presented
91 cont	Mathematics	Journal of Mathematical Finance	Combined model with multiple input indicators is of higher precision
92 cont	Healthcare	International Journal of Science and Research	Pre-implementation of the predictive analytic tool
93 cont	Medicine	PloS one	Proposed model outperform model-based techniques
94 cont	Transportation	Pattern Recognition Letters	Proposed method is effective for predictive complex event processing
95 cont	Research	IJCCT	An overview of some of the notable techniques for prediction was presented
96 cont	Healthcare	IJSRCSEIT	Survey to help the researchers to propose the best framework for medical analysis
97 cont	Manufacturing	IFAC Proceedings Volumes	Predictive manufacturing system is presented as the next transformation of manufacturing evolution
98 cont	Medicine, Healthcare	Procedia Computer Science	Big Data Analytics in Hadoop's implementation provides systematic way for achieving better outcomes
99 cont	Medicine	Biocomputing	A data-driven predictive model is developed to predict readmission rates in heart failure patients
100 cont	Security	International Journal of Computer Applications	Predictive policing is slowing making way to the forefront of strategic law enforcement

(continued)

Table 1 (continued)

	Author(s)	Year	Title
101	Bos and Lauter and Naehrig	2014	Private predictive analysis on encrypted medical data
102	Stefanovic	2014	Proactive supply chain performance management with predictive analytics
103	He et al.	2018	Random forest as a predictive analytics alternative to regression in institutional research
104	Cuccaro-Alamin et al.	2017	Risk assessment and decision making in child protective services: predictive risk modeling in context
105	Giannetti and Ransing	2016	Risk based uncertainty quantification to improve robustness of manufacturing operations
106	Hashimzade and Myles	2017	Risk-based audits in a behavioral model
107	Kumar and Prakash	2014	Role of big data and analytics in smart cities
108	Poucke et al.	2016	Scalable predictive analysis in critically Ill patients using a visual open data analysis platform
109	Lee and Kao and Yang	2014	Service innovation and smart analytics for Industry 4.0 and big data environment
110	Óskarsdóttir et al.	2017	Social network analytics for churn prediction in telco: model building, evaluation and network architecture
111	Nithya	2016	Study on predictive analytics practices in health care system
112	Babu and Jayanthi and Rao	2017	Survey on clinical prediction models for diabetes prediction
113	Cohen et al.	2014	The legal and ethical concerns that arise from using complex predictive analytics in health care
114	Gaughan and Gold and McClaren	2013	The lessons Oscar taught US: data science and media & entertainment
115	Lu et al.	2017	The state-of-the-art in predictive visual analytics
116	Rudin and Tulabandhula	2014	Tire changes, fresh air, and yellow flags: challenges in predictive analytics for professional racing

(continued)

Table 1 (continued)

	Author(s)	Year	Title
117	Modu et al.	2017	Towards a predictive analytics-based intelligent malaria outbreak warning system
118	Leon-Jimenez and Fernandez-Granero and Sanchez-Morillo	2016	Use of predictive algorithms in-home monitoring of chronic obstructive pulmonary disease and asthma: a systematic review
119	Coraddu et al.	2017	Vessels fuel consumption forecast and trim optimisation: a data analytics perspective
120	Adjekum and Ienca and Vayena	2017	What is trust? ethics and risk governance in precision medicine and predictive analytics
	Field	Journal	Output
101 cont	Medicine	Journal of Biomedical Informatics	Implementation of a cloud service for performing private predictive analysis
102 cont	SCM	The ScientificWorld Journal	Described models give very accurate KPI projections and provide valuable insights into newly emerging trends
103 cont	Research, IT	Practical Assessment, Research & Evaluation	It was argued that random forest is a valuable tool for predictive analytics tasks
104 cont	Services	Children and Youth Services Review	Predictive analytics offer much promise to the field of child protection
105 cont	Industries	Computers & Industrial Engineering	Novel algorithm was proposed and described
106 cont	Audits	Public Finance Review	Analysis shows that predictive analytics are successful in raising compliance
107 cont	Research	International Journal of Science and Research	Big data analytics will help in analyzing and predicting information from smart devices
108 cont	Medicine, IT	PloS one	The major benefit from the proposed integration is seamless manipulation and data extraction
109 cont	Manufacturing	Procedia CIRP	Prognostics-monitoring system is a trend of the smart manufacturing environment

(continued)

Table 1 (continued)

	Field	Journal	Output
110 cont	Telecommunication	Expert Systems with Applications	The best configuration is a non-relational learner enriched with network variables, without collective inference
111 cont	Healthcare	IJETTCS	Swarm optimization algorithms can be implemented to solve many optimization problems
112 cont	Healthcare	Journal of Big Data	Detail description of predictive modeling is presented
113 cont	Healthcare	HEALTH AFFAIRS	It is essential that predictive analytics models be constantly evaluated, updated, reimplemented, and reevaluated
114 cont	Media and Entertainment	Big data	Illustrations of some of the strengths of data and gave a few considerations for future entertainment data projects
115 cont	Research	Journal of the Eurographics Association	Summarization of the state-of-the-art in predictive visual analytics
116 cont	Racing	Big Data	Validity of designed prediction models was demonstrated
117 cont	Healthcare	Applied Sciences	Deployed system will help in healthcare take precautions in time and utilize their resources in case of emergency
118 cont	Medicine	Chronic Respiratory Disease	Development of predictive models with clinically useful levels of accuracy, has not yet been achieved
119 cont	Economy	Ocean Engineering	New strategy for the optimisation of the trim of a vessel is proposed
120 cont	Medicine	OMICS: A Journal of Integrative Biology	Proposing a 'points to consider' on how best to enhance trust in precision medicine

articles increased rapidly again and almost half of the papers were from the area of medicine and healthcare. Newcomers for this year were areas of audits [18] and power sources [6]. Last year, in 2017, we had the highest number of articles published so far and we can expect that this growing trend will continue and at the end of 2018, number of publications will be higher than in 2017.

Authors of Selected Articles. Many authors contributed to predictive analytics topic since 2010 to date. Most of them have only one article in our literature review but few of them contributed more to this area than the others. Author, who contributed to this field the most was Yongheng Wang, who was part of three articles published since 2010. Two articles were published in 2013, one was written by him only and was related to intelligent transportation systems [40] and second article in this year was written by him together with co-authors Kui and Zhu. The second article was

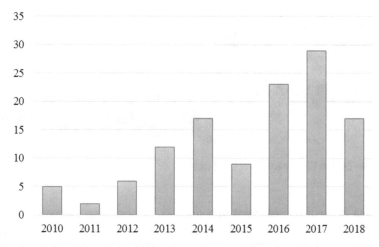

Fig. 1 Evolution of the number of published articles starting 2010

from different area and authors were proposing the use of dynamic Bayesian model averaging to develop a high-accuracy prediction analytic method for large-scale IoT application in it [43]. The last article, in which this author cooperated was published this year and was related to intelligent transportation systems again [41]. Basha and his team published two articles in two different journals in 2017. The first article was a comparative study in which they were comparing the performance of forecasting techniques [3] and in the second article, they were again comparing predictive techniques and doing experiments in the supply chain risk management area [4]. Another author with more than one article was Nigar Hashimzade, who contributed in the field of audits [18] and in 2017 with his second article related to audits too [17]. Lee contributed to our analysis with two articles too. The first article from him and his team was published in 2013 and were about predictive analytics in manufacturing, predictive manufacturing [30]. A year later, they published another article from the manufacturing area about smart predictive informatics tools [29].

2.3 Titles

When we were analyzing titles of selected articles, we could see that there are various articles from various fields or areas, but it wasn't clear, which area is the dominant one. After creating 'word cloud', we can clearly identify which area stands out and which, on the contrary, is not very well represented (Fig. 2).

We created 'word cloud' from all words that were in titles of our selected articles and as we can see, words 'predictive' and 'analytics' were the most common words used in titles of our publications. Other common words were data, prediction, healthcare, model or analysis. Topics or researched subjects were mainly the imple-

Fig. 2 Words from titles of articles in word cloud

mentation of predictive analytics in some area, or predicting some event, mainly in healthcare, or studying predictive models. Challenges or problems were mainly related to doing a research, comparing multiple predictive techniques and their performance, designing a new predictive model or developing predictive models.

2.4 Output

The output from our articles was mostly design of the new predictive model, implementation of the predictive model, review of the literature, a set of practical advice or implications, or summarization of article's problem. This corresponds with the challenge and research methodology of an article.

2.5 Publications

In this section, we were analyzing journals, in which selected articles were published. We created a table, in which we can see the list of all journals, in which selected articles were published together with count of articles published in every journal.

As we can see in the table above, in Table 2, our selected articles were published in various journals. Most journals contained only one article from our selected articles, but few journals were represented by four selected articles, e.g. Big Data, International Journal of Computer Applications or Procedia Computer Science.

2.6 Field/Discipline

Last part from Table 1 was a field or discipline related to published articles. We have chosen several main categories, other with small representation were merged into group 'others'.

Table 2 Journals, in which selected articles were published

Journal name	Number of publications
Big Data	4
International Journal of Computer Applications	4
Procedia Computer Science	4
Decision Support Systems	3
IEEE Intelligent Systems	3
Computers & Industrial Engineering	2
Expert Systems with Applications	2
International Journal of Emerging Trends & Technology in Computer Science	2
International Journal of Science and Research	2
Journal of Biomedical Informatics	2
Journal of Business Research	2
Journal of Fundamental and Applied Sciences	2
PloS one	2
Procedia CIRP	2
Academic Emergency Medicine	1
AMIA Annual Symposium Proceedings	1
Annals of Emergency Medicine	1
Applied Sciences	1
Asia Pacific Journal of Research	1
Biocomputing	1
Biological Forum—An International Journal	1
Children and Youth Services Review	1
Chronic Respiratory Disease	1
Clinical Therapeutics	1
eGEMs: The Journal of Electronic Health Data and Methods	1
Engineering	1
EPJ Data Science	1
European Heart Journal	1

(continued)

Table 2 (continued)

Journal name	Number of publications
European Journal of Operational Research	1
Future Generation Computer Systems	1
Future Internet	1
HEALTH AFFAIRS	1
Health Care Management Science	1
Health Management, Policy and Innovation	1
Human Pathology	1
IBM Journal of Research and Development	1
IEEE Internet of Things Journal	1
IFAC Proceedings Volumes	1
Indian Journal of Public Health Research & Development	1
Indian Journal of Science and Technology	1
International Journal of Advanced Research in Computer Science and Software Engineering	1
International Journal of Computer and Communication Technology	1
International Journal of Computer Science & Engineering Technology	1
International Journal of Computer Science and Network Security	1
International Journal of Computing and Business Research	1
International Journal of Data Analysis and Information Systems	1
International Journal of Database Theory and Application	1
International Journal of Digital Content Technology and its Applications	1

(continued)

Table 2 (continued)

Journal name	Number of publications
International Journal of Distributed Sensor Networks	1
International Journal of Engineering Trends and Technology	1
International Journal of Enhanced Research in Science Technology & Engineering	1
International Journal of Geo-Information	1
International Journal of Grid and Distributed Computing	1
International Journal of Information Management	1
International Journal of Information Technology	1
International Journal of Integrated Care	1
International Journal of Interactive Multimedia and Artificial Intelligence	1
International Journal of Logistics Management	1
International Journal of Medical Informatics	1
International Journal of Multimedia and Ubiquitous Engineering	1
International Journal of Research in Advent Technology	1
International Journal of Scientific & Technology Research	1
International Journal of Scientific Research in Computer Science, Engineering and Information Technology	1
International Journal on Information Technology	1
International Research Journal of Engineering and Technology	1
Journal of Infection and Public Health	1
Journal for Control, Measurement, Electronics, Computing and Communications	1
Journal of Big Data	1

(continued)

Table 2 (continued)

Journal name	Number of publications
Journal of Business Logistics	1
Journal of Computer Science & Information Technology	1
Journal of Economic Behavior & Organization	1
Journal of Economic Perspectives	1
Journal of Engineering and Applied Sciences	1
Journal of Mathematical Finance	1
Journal of Neurosurgery	1
Journal of Power Sources	1
Journal of the Eurographics Association	1
Kybernetes	1
Lecture Notes on Information Theory	1
MIS Quarterly	1
Natural Hazards and Earth System Sciences	1
Ocean Engineering	1
OMICS: A Journal of Integrative Biology	1
Online Journal of Public Health Informatics	1
Pattern Recognition Letters	1
Practical Assessment, Research & Evaluation	1
Procedia Engineering	1
Procedia Technology	1
Public Finance Review	1
Research & Practice in Assessment	1
The American Journal of Medicine	1
The American Journal of Surgery	1
The International Journal of Medical Robotics and Computer Assisted Surgery	1
The International Journal of the First Year in Higher Education	1

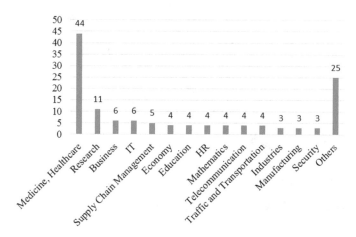

Fig. 3 Number of publications per field

As we can see in the figure above, Fig. 3, more than one third (rounded 37%) of publications were related to medicine and healthcare area. The second big group was researching area, where comparisons of various predictive models or techniques were performed. The third group was a business area with six articles, where predictive analytics was applied to business. The fourth group had also six articles published and it was IT area. We divided other articles into groups: Supply Chain Management, Economy, Education, HR, Mathematics, Telecommunication, Traffic and Transportation, Industries, Manufacturing, Security and rest of articles were included into group Others.

3 Conclusion

The aim of this paper was a detailed literature review and analysis. For this purpose, we used various databases, where we were searching for articles published in journals. Databases used were e.g. ScienceDirect, Google Scholar, IEEE Xplore or SAGE. After selecting final count of articles, 120, we created a selection (Table 1) with following structure: author(s), year of publication, title, topic or researched subject, challenge or problem, the methodology used, output, journal name, and field or discipline.

After a more detailed analysis of authors of published articles, we found few key players in Predictive Analytics field, e.g. Yongheng Wang with three publications, Basha, Hashimzade or Lee. During analyzing of years, when articles were published, we have come to several conclusions: (a) the articles began to be published at 2010, (b) except the year 2015, when the number of publication decreased, the number of publications was evenly increasing. We can see the timeline of Predictive Analytics on the figure below (Fig. 4).

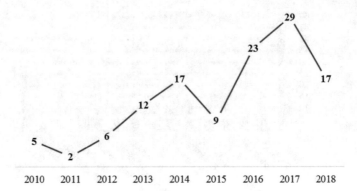

Fig. 4 Timeline of predictive analytics from 2010 to Date

In a deeper analysis of titles of publications, the most common words in titles were 'Predictive' and 'Analytics'. Words data, big, analysis, prediction, model or health-care were also common in titles of articles. The most common methodology used in articles was Analytical/Conceptual research. The second most frequent methodology was descriptive Research and the third was Literature Review. Our selected publications were published in 98 different journals. Journal Big Data published four of our articles, same as journals International Journal of Computer Applications and Procedia Computer Science. Journals Decision Support Systems and IEEE Intelligent Systems published for three articles, nine journals had for two publications and other had only one publication. The output from the analysis of field or discipline related to selected articles was that the most common field in our articles was Medicine and Healthcare (44 articles), the second big group was Research.

Our deep analysis and literature review can help another authors or researches to get an overview of Predictive Analytics and see its evolution. In the future, it could be interesting to see the future development of this topic.

References

1. Andrienko N, Andrienko G, Rinzivillo S (2015) Exploiting spatial abstraction in predictive analytics of vehicle traffic. ISPRS Int J Geo-Inf 4:591–606. https://doi.org/10.3390/ijgi4020591
2. Balkan S, Demirkan H (n.d) Information services for advanced marketing 10
3. Basha SM, Zhenning Y, Rajput DS, Caytiles RD, Iyengar NCS (2017a) Comparative study on performance analysis of time series predictive models. Int J Grid Distrib Comput 10:37–48. https://doi.org/10.14257/ijgdc.2017.10.8.04
4. Basha SM, Zhenning Y, Rajput DS, SN, INC, Caytiles RD (2017b) Domain specific predictive analytics: a case study with R. Int J Multimed Ubiquitous Eng 12:13–22
5. Benoit DF, Van den Poel D (2012) Improving customer retention in financial services using kinship network information. Expert Syst Appl 39:11435–11442. https://doi.org/10.1016/j.eswa.2012.04.016

6. Berecibar M, Devriendt F, Dubarry M, Villarreal I, Omar N, Verbeke W, Van Mierlo J (2016) Online state of health estimation on NMC cells based on predictive analytics. J Power Sources 320:239–250. https://doi.org/10.1016/j.jpowsour.2016.04.109

7. Bhalla A (2012) Enhancement in predictive model for insurance underwriting. Eng Technol 3:6

8. Bihani P, Patil ST (2014) A comparative study of data analysis techniques 3:7

9. Boukenze B, Mousannif H, Haqiq A (2016) Predictive analytics in healthcare system using data mining techniques. Academy & Industry Research Collaboration Center (AIRCC), pp 01–09. https://doi.org/10.5121/csit.2016.60501

10. Gold M, McClarren R, Gaughan C (2013) The lessons Oscar taught Us: data science and media & entertainment. Big Data 1:105–109. https://doi.org/10.1089/big.2013.0009

11. Gotz D, Stavropoulos H, Sun J, Wang F (2012) ICDA: a platform for intelligent care delivery analytics. AMIA Annu Symp Proc 2012:264–273

12. Gregus M, Kryvinska N (2015) Service orientation of enterprises—aspects, dimensions, technologies. Comenius University in Bratislava. ISBN: 9788022339780

13. Guidi G, Miniati R, Mazzola M, Iadanza E (2016) Case study: IBM watson analytics cloud platform as analytics-as-a-service system for heart failure early detection. Future Internet 8:32. https://doi.org/10.3390/fi8030032

14. Halim MIA, Hashim W, Ismail AF, Suliman SH, Yahya AS, Raj RMA (2018) Evaluating predictive analytics model performance accuracy for network selection mechanism. J Fundam Appl Sci 10:162–172

15. Hanumanthappa DM (2011) Predicting the future of car manufacturing industry using data mining. Techniques 01:3

16. Harvey A, Luckman M (2014) Beyond demographics: predicting student attrition within the bachelor of arts degree. Int J First Year High Educ 5. https://doi.org/10.5204/intjfyhe.v5i1.187

17. Hashimzade N, Myles G (2017) Risk-based audits in a behavioral model. Public Financ Rev 45:140–165. https://doi.org/10.1177/1091142115602062

18. Hashimzade N, Myles GD, Rablen MD (2016) Predictive analytics and the targeting of audits. J Econ Behav Org Tax Soc Norms Compliance 124:130–145. https://doi.org/10.1016/j.jebo.2015.11.009

19. Jacob MSG (n.d) Discovery of knowledge patterns in clinical data through data mining algorithms: multi-class categorization of breast tissue data. Int J Comput Appl 32:8

20. Kaczor S, Kryvinska N (2013) It is all about services—fundamentals, drivers, and business models. Soc Serv Sci J Serv Sci Res 5(2):125–154

21. Kapoor B, Sherif J (2012) Global human resources (HR) information systems. Kybernetes 41:229–238

22. Khan SS, Quadri SMK (n.d) Prediction of angiographic disease status using rule based data mining techniques. 5

23. Klindworth WA (n.d) Assessment of predictive modeling for identifying fraud within the medicare program. 29

24. Kraljević G, Gotovac S (2010) Modeling data mining applications for prediction of prepaid churn in telecommunication services. Automatika 51:275–283. https://doi.org/10.1080/00051144.2010.11828381

25. Kryvinska N, Olexova R, Dohmen P, Strauss C (2013) The S-D logic phenomenon—conceptualization and systematization by reviewing the literature of a decade (2004–2013). Soc Serv Sci J Serv Sci Res 5(1):35–94 Springer

26. Kryvinska N (2012) Building consistent formal specification for the service enterprise agility foundation. Soc Serv Sci J Serv Sci Res 4(2):235–269

27. Kryvinska N, Gregus M (2014) SOA and its business value in requirements, features, practices and methodologies. Comenius University in Bratislava. ISBN: 9788022337649

28. Kumar NMS, Eswari T, Sampath P, Lavanya S (2015) Predictive methodology for diabetic data analysis in big data. Procedia Comput Sci Big Data Cloud Comput Chall 50:203–208. https://doi.org/10.1016/j.procs.2015.04.069

29. Lee J, Kao H-A, Yang S (2014) Service innovation and smart analytics for industry 4.0 and big data environment. procedia CIRP, product services systems and value creation. In: Proceedings of the 6th CIRP conference on industrial product-service systems, vol 16, pp 3–8. https://doi.org/10.1016/j.procir.2014.02.001

30. Lee J, Lapira E, Yang S, Kao A (2013) Predictive manufacturing system—trends of next-generation production systems. In: IFAC proceedings, vol 46, pp 150–156. https://doi.org/10.3182/20130522-3-BR-4036.00107

31. Li X, Xie H, Song Y, Zhu S, Li Q, Wang FL (2015) Does summarization help stock prediction? a news impact analysis. IEEE Intell Syst 30:26–34. https://doi.org/10.1109/MIS.2015.1

32. Liu S, Shen Z, Mei J, Ji J (2013) Parkinson's disease predictive analytics through a pad game based on personal data 19:17

33. Mishra D (2010) Predictive data mining: promising future and applications, 2:9

34. Ng K, Ghoting A, Steinhubl SR, Stewart WF, Malin B, Sun J (2014) PARAMO: a PARAllel predictive MOdeling platform for healthcare analytic research using electronic health records. J Biomed Inform 48:160–170. https://doi.org/10.1016/j.jbi.2013.12.012

35. Renjith S (2015) An integrated framework to recommend personalized retention actions to control B2C E-commerce customer churn. Int J Eng Trends Technol 27:152–157. https://doi.org/10.14445/22315381/IJETT-V27P227

36. Shameer K, Johnson KW, Yahi A, Miotto R, Li LI, Ricks D, Jebakaran J, Kovatch P, Sengupta PP, Gelijns S (2017) Predictive modeling of hospital readmission rates using electronic medical record-wide machine learning: a case-study using mount sinai heart failure cohort. In: Pacific symposium on biocomputing 2017. World Scientific, pp 276–287

37. Shams I, Ajorlou S, Yang K (2015) A predictive analytics approach to reducing 30-day avoidable readmissions among patients with heart failure, acute myocardial infarction, pneumonia, or COPD. Health Care Manag Sci 18:19–34. https://doi.org/10.1007/s10729-014-9278-y

38. Sridhar P, Dharmaji N (n.d) A comparative study on how big data is scaling business intelligence and analytics, 2:10

39. Upendran D, Chatterjee S, Sindhumol S, Bijlani K (2016) Application of predictive analytics in intelligent course recommendation. Procedia Comput Sci 93:917–923. https://doi.org/10.1016/j.procs.2016.07.267

40. Wang Y (2013) A proactive complex event processing method for intelligent transportation systems. Lect Notes Inf Theory 1:109–113. https://doi.org/10.12720/lnit.1.3.109-113

41. Wang Y, Gao H, Chen G (2018) Predictive complex event processing based on evolving Bayesian networks. Pattern Recognit Lett Mach Learn Appl Artif Intell 105:207–216. https://doi.org/10.1016/j.patrec.2017.05.008

42. Zhang W, Li C, Ye Y, Li W, Ngai EWT (2015) Dynamic business network analysis for correlated stock price movement prediction. IEEE Intell Syst 30:26–33. https://doi.org/10.1109/MIS.2015.25

43. Zhu X, Kui F, Wang Y (2013) Predictive analytics by using bayesian model averaging for large-scale internet of things. Int J Distrib Sens Netw 9:723260. https://doi.org/10.1155/2013/723260

Information Processing from Unemployment Rates: Evidence from Spain, Switzerland and the European Union

Marina Fad'oš and Mária Bohdalová

Abstract The chapter describes the behavior of the unemployment rates and actions taken to control it in the country with the most turbulent unemployment rates as it is Spain, in the country with the lowest and steadiest unemployment rates as it is Switzerland, and in the European Union which represents the countries with average unemployment rates. Actions on the labor market, that could have impacted unemployment, were described. The hypothesis of hysteresis was validated for the univariate series of gender inequality and the unemployment rates using the LM test, and on panel data using ILT test. Employing Pesaran CD test, the cross-dependence of the series of distinct characteristics, in both, unemployment and the unemployment gender inequality series was confirmed. The OLS, GLS-RE and GLS-FE models were used to explain the influence of the distinct characteristics of the unemployment and unemployment gender inequality series in this chapter.

Keywords Unemployment · Gender · Inequality · Switzerland · Spain · European Union · EU

1 Introduction

Unemployment has been considered as an important issue, and it has been discussed by many economists, researchers, academics and policy makers since the time when it appeared for the first time during the industrial revolution. They were concerned about the impact of unemployment on the economy of the country. Investigations were focused on the identification of the source of the unemployment rates issue and how it should be efficiently tackled. Despite many theories, unemployment has been

M. Fad'oš · M. Bohdalová (✉)
Faculty of Management, Comenius University in Bratislava, Odbojárov 10,
82005 Bratislava, Slovakia
e-mail: maria.bohdalova@fm.uniba.sk

M. Fad'oš
e-mail: marina.fados@fm.uniba.sk

© Springer Nature Switzerland AG 2020
N. Kryvinska and M. Greguš (eds.), *Data-Centric Business and Applications*,
Lecture Notes on Data Engineering and Communications Technologies 30,
https://doi.org/10.1007/978-3-030-19069-9_9

introduced over the history that were focused on the diminishing of the unemployment, when implemented nowadays, they were proven to be ineffective. This was probably due to the unstoppable World evolution. The economic situation and development of the countries is different than it was a hundred years ago, when economists, that set the foundations of the economics, lived. Not even post-war teachings and theories had the same impact on the economy today. Therefore, it is still challenging for the researchers to discover how could be unemployment controlled during rapidly changing world, when a country's economy depends on the world economy. Country economy doesn't depend only on her own labor force, but the labor force freely moves from one territory to another: Labor market doesn't depend only on the country labor market policy but accepts the world labor policy or in this case, labor market policy of the European Union.

This chapter focuses on the analysis of the unemployment in the European Union. We compared unemployment of Spain and Switzerland. Unemployment in Spain was the most turbulent and high during the observed period and it was directly affected by the European Union Labor Market Policy. Unemployment rate in Switzerland was the lowest and steady and it was not directly affected by the European Union Labor Market Policy.

We started by describing the actions implemented in the Labor Market and by describing the significant global events that could have an influence on the unemployment of the Spain, Switzerland and the European Union. Labor market policies that lead to gender inequality, were also described. Further, we researched gender inequality in unemployment. Afterwards, the hypothesis of hysteresis and cross-dependence of the unemployment and the unemployment gender inequality by age, education, previous occupation of the unemployed and the duration of the unemployment was validated. Finally, unemployment and the unemployment gender inequality were estimated.

2 Literature Review

European Economic Community was established in 1957, and at the time, the unemployment rate in this community was below 2%. However, it has been increasing and by the year 1993 reached 10.9% [11]. High levels of the unemployment rates in the European Economic Community raised concerns on the EU level. As a result of the discussions between economists and the policy makers, The White Paper on Growth and Employment was introduced in 1993. This paper focused on the main issues that should be solved in order to decrease unemployment. Next goals that should be achieved were set. The legislative should be more flexible and business friendly, in order to facilitate the entrepreneurship in the country and to attract the investors, which could lead to an increase of the employment. The effective labor market should have been created. It was also established that the community should participate in the development of those countries were unsatisfied demand existed, since one of the goals was achieving open international economic environment [53].

However, unemployment rates were not decreased as they should, and in 1997, the European Employment Strategy was implemented. This strategy allowed the European Commission to be in charge of employment monitoring and coordination of each member country of the European Union. More specific actions, toward the unemployment issue solving, were described in the Lisbon Strategy for Growth. This Strategy was introduced in 2000. The main purpose of the part of this strategy that focused on employment was to increase the employment rate to 70%, to increase the employment rate of women to more than 60% and to increase employment of those 55–64 years old to 50%. The goals were planned to be accomplished by year 2010 [30]. However, the Economic Crisis in 2008 caused the rapid increase of the unemployment rates in the European Union, which put in danger the fulfilment of the goals of the Lisbon Strategy for Growth. This strategy was replaced by The Europe 2020 in 2010. The Europe 2020 strategy was focused on the unemployment issue more than previous strategies [11]. Decrease of the unemployment rates should be achieved by achieving 75% employment of 20–65 years old, by lowering school drop-out rates below 10%, by achieving 40% of 30–34 years old having tertiary education and by withdrawing from poverty and social exclusion 20 million people [56]. Listed objectives were supported by four employment guidelines, which should help accomplishing the goals by 2020. These guidelines were focused on increasing the employment rates and job quality, decreasing unemployment rates, increasing the educational attainment of the workforce, building skills, supporting lifelong learning, increasing quality of education, promoting tertiary education and reducing poverty [43].

Spain was the country with the highest and the most turbulent unemployment rates in the European Union. As a part of the preparations for entering the European Union, tax reforms and reforms concerning the flexibility of temporary employment were introduced in 1984 [18]. The reforms were effective, and the unemployment rates were decreased. However, this did not last for a long time, the ERM crisis in 1991 led to an increase of the unemployment rates once again. As a response to increased unemployment, reform concerning flexibility in the wage setting was introduced in 1994, that led to decrease of unemployment rates [18]. Unemployment rates were low in Spain by 2007. In this year, the unemployment rate in Spain was only by one percent higher than the average unemployment rate of all the European Union member countries. Economic Crisis in 2008 had a drastic impact on Spain. When occurred, crisis caused economic collapse in Spain. Unemployment rates achieved alarming levels. When compared with other countries of the European Union, Economic crisis caused in Spain the highest jump in rates of unemployment in between the two years. The consequences of the economic crisis in 2008 lasted longer in Spain than it was expected and by 2013, the unemployment rates of Spain reached the record highest rate in her history (26.1%). However, it has been decreasing since [47].

Switzerland was known as a wealthy country, in which unemployment did not exist. Therefore, it was for a long time labeled as "employment paradise". Unemployment appeared for the first time during the first oil shock. Even though, it has decreased by 1981, the recession in 1982 triggered it again and by 1984 increased to

1.1%. The recession in 1990 had the strongest impact on the unemployment rates in Switzerland, that caused it increasement to 4.1% by 1997 [21]. As a response to the recession in 1992, the country has risen the benefits in unemployment up to the 80% and prolonged the benefits compensation period prolonged to 400 days in 1993. Since the unemployment rates were increasing, the new unemployment law was introduced in 1995 and implemented in 1996 and 1997. This law modified compensation system during unemployment [17], and due to it, the unemployment rates in Switzerland decreased to 2.5% by 2001 [23]. Swiss law on unemployment insurance was revised in 2000 and implemented in 2001. The main objective of this law was to facilitate the employment process and to help unemployed finding employment faster [9]. Rising minimum wages for low skilled employees, due to union initiative, caused increase of unemployment rates in the period 1999–2001 [49]. Due to the Economic crisis in 2008, unemployment rates increased to 4.5%, and it had been increasing since [12].

2.1   Employment by Distinct Characteristics

Age

Youth has been at risk of being unemployed for a long time, and it appears this fact would not change. High levels of youth unemployment could be related to legislative restraints that are focused on the protection of the older employees. The law protects older employees more than younger employees. Therefore, in times of crisis, young employees are first to go, despite being more suitable for the position than their older colleagues. The ineffective replacement of the more suitable employees due to older protection law, leads to diminishment of the company's productivity company's growth [55].

The unemployment rates of the youth workforce could be decreased with the increase of their educational attainment. High levels of the unemployment rates of youth (20.5% in 2015) could be related to their attainment to part-time positions. High work fluctuation was related to the youth employment, which could be the reason they were unemployed for a shorter time than older workforce [42].

Integration of the unemployed youth in the labor market could be achieved by focusing on their training or a job assistance. Trainings were usually structured by increasing educational attainment of the youth workforce, by increasing their skills that match need for skills in the labor market, so they could become the perfect fit for the existing job positions. Besides trainings, young workforce could benefit from the job search assistance and a wage subsidy during the unemployment or they could participate in the public sector work programs [10].

When it comes to youth unemployment rates, Spain is the country that ranks as the second country with the highest youth unemployment rates. High levels of the youth unemployment rates could be related to high inflow from employment to unemployment than due to inflow from the inactivity to the workforce. Youth inflows to unemployment from employment and from inactivity to workforce was higher than the same inflows of the prime-age workers, even though they were increasing in this group as well but not so fast as it was increasing of youth workforce. The higher inflows to unemployment of youth workforce could be related to lower protection of youth in the labor market. Prime-age worker protection on the labor market was higher and kept their inflows to unemployment lower.

Inflows from inactivity to the labor market was high in Spain, and despite they easily find employment, they usually abandon them in a short period [20]. Unemployed Youth in Spain struggle finding employment for three to four months, while the unemployed prime-age workforce integrated to the employment within the first month of unemployment. Youth workforce preferred clerical jobs before the hard-physical job offers, that kept them unemployed longer. Youth women had less chances to find employment than men, however, having primary or secondary education had no effect on their employment chances [20]. Since the Youth has been protected less than the prime-age workforce, the policies should focus on granting the same rights for all workforce.

Young workforce in Spain, was usually employed on a part-time position and had temporary working contracts. After graduating, Youth in Spain struggled finding first employment for three to five years. When they failed to find fixed-term employment during that time period, they accepted the temporary contract job offers. Temporary jobs were usually created during the times of prosperity, when the demand for products was high and they needed additional workforce. Youth workers were employed in these temporary jobs, because they were more productive than the older workers, and they could comply with the demand on the market. However, after the period of prosperity ended, they were released from the employment. Released employees were not employable, due to their low educational attainment. This was due to higher job creations in periods of prosperity, that motivated 16 years old to abandon school and enter the labor market. After the period of prosperity, they were released, they were left without jobs and without education, that made them not employable. To tackle this issue and to decrease Youth unemployment, Estrategia de Emprendimiento y Empleo Joven was introduced and it was focused on supporting young people to start their own business [41].

The Youth unemployment rate was high in Switzerland as well. Therefore, to decrease the Youth unemployment, Swiss public services departments implemented reforms in 2003, that were focused on providing the access to suitable training to Youth workforce. To motivate young people finding employment faster, they reduced their compensation in unemployment. While prime-age employees received 70–80% of gross pay as benefits in unemployment, youth were not granted with such benefits. Benefits in unemployment received only 30% of unemployed younger than 25 years. However, the benefit period for unemployed prime-age workers was reduced in 2010, from 18 to 9 months and increased for those older than 55 years to 25 months [12].

Duration in unemployment
Women spent more time being unemployed than men [48]. Duration of time spent being unemployed was prolonged due to economic crisis [3] and kept youth workforce even longer unemployed that prime-age workforce. With the increase of time spent being unemployed, the probability of finding employment was decreasing [19].

With the increase of time spent being unemployed, unemployed people get a mislead-ing perception of wages on the labor market. With the increase of time, their wage expectations were high and did not match the wages offered on the labor market, that prolonged their duration in the unemployment [8]. Duration of unemployment could be diminished by the participation of the unemployed in training programs that were focused on helping them to find employment faster. Women participating in these programs were employed sooner than men who participated [2].

The European Union focused on the decreasing the long-term unemployment by implementing flexicurity policies. The aim of these policies was to build the skills of the long term unemployed and low-skilled unemployed to match the skills needed in the labor market. As a result, the unemployed would have the suitable skills to apply for the vacant position, and their probability of being employed would increase. The European Union was also focused on supporting the increasement of the employment by subsidizing employment to motivate companies to create new job positions. Another action which the European Union promoted was increasement of self-employment, which could also result in decreased unemployment rates [58].

The long-term unemployment was the most persistent one type of unemployment in Spain. The long-term unemployment of the young people was related to family support. Youth workers were not motivated to apply for jobs since their family cov-ered their basic needs. With the increased family income, young workforce was less motivated to apply for jobs. High levels of the long-term unemployment could be also related to low wage flexibility, high employment protection, generous unemployment benefits and dysfunctional system of collective agreements. Flexibility of the labor market as a measure for decreasing unemployment was introduced in 1980, as a part of the new amendment. The amendment was focused on replacing fixed-term contract with temporary contracts. In 1992, unemployment assistance policy was revised, and 80–70–60 scheme was replaced with 70–60 scheme (70% for the first six months, 60% for the rest of qualifying period). Assistance period was risen from 6 months to 12 months, which was the time period without job, during which the unemployed would receive assistance benefits. This resulted in increasing the unemployment rates in Spain. However, in 1994, unemployment assistance was increased again. After this period, unemployment protection got more restrictive, which remained until the cri-sis in 2008 [26]. Modest benefits and reduced level of protection was offered during the unemployment in this period. However, active employment policies were needed. Conditions for being eligible for unemployment benefits were made more rigorous, with the increase the of duration in unemployment. To be eligible for the unem-ployment benefits, unemployed were required to participate in actions organized by employment offices, that were designed to help them find employment faster. Rejection of the participation in these actions by unemployed person would result in sanction implementation, which in some cases excluded them from the unemploy-ment benefits receiving. The unemployed were motivated to accept the lower paid jobs, rather than being except from the benefits receiving. Benefit income depends on the number of children in the family and their independence. With the increase the number of children in a family, the benefits income was higher [1]. As a response to the long-term unemployment after the crisis, unemployment benefits were increased, subsidies were provided to companies to motivate them to hire long-term unemployed and to support job creation. Provided were also incentives for self-employment and employment in public services was increased [59].

The duration of the unemployment in Switzerland, was decreased by implementation of the active labor market policies (ALMP) in 1997. This policy was divided into four groups: Basic Training, Advanced Training, Employment programmes and Subsidized jobs. Basic Training and Employment programmes prolonged the duration of unemployment. During the first four to six months, advanced training and subsidized jobs also prolonged duration of unemployment, however medium and long-term advanced trainings and subsidized jobs led to decrease of the average unemployment duration [38].

Education
Increase of the educational attainment was related to decrease of the unemployment rates. Employees with higher education and highly skilled are being more productive and can achieve higher outcomes than the lower skilled employees, therefore, were more valuable for the employers than the lower skilled employees. The European Union focused on building the skills of employees that could match the very exigent labor market demand for skills. Since the labor market is rapidly changing, the employees could be competitive in the labor market only if the lifelong learning is promoted and implemented. Lifelong learning and building employees skills would result in decreased long-term unemployment [57]. In 2011, 26% of the workforce had higher education, and European Commission expects that by the year 2020, 35% of the jobs would demand highly skilled employees. There is also lack of the researchers in the European Union, therefore, focus on the increase of the researchers that would lead to more research-intensive economies, which could result in better forecasts of the future economic situation. Education, research and business should be related and combined. Combined cooperation and knowledge sharing would lead to faster development and growth of the economies. To increase educational attainment in the European Union, the European Commission also focuses on the promoting youth mobility, and experience and knowledge sharing across EU countries that could result in promising effective alliances [14].

Primary and secondary education had no effect on the employment rates in Spain during the period 1992–2000, however, tertiary education had a positive effect on the employment. Ladder effect phenomenon was present in Spain, that caused high unemployment among low-skilled workforce. Ladder effect is defined as the employment of the highly skilled employees in the positions suitable for the low-skilled employees. Ladder effect was caused by fresh graduate workforce entering the labor market and choosing as their first employment a low skilled job, in order to gain experience and on the-job-training. Employers preferred highly skilled fresh graduate before other groups of workforces on the labor market, due to their capability to increase the productivity of the company without much effort and with minimal trainings [22].

In distinction of the Spain, the unemployment rate was lower of those with secondary and tertiary education in Switzerland. However, the unemployment rates of those with primary educational attainment were significantly higher, which was due to Switzerland preference of high quality over quantity. The low unemployment rate in Switzerland was also related to the dual-track vocational and professional educational attainment and training system. Highly qualified workers were preferred by the industry sector, hereof people with poor education were excluded from the employment process. Switzerland is known as a country focused on quality achievement, which was yet another reason why unemployment of low-qualified workers was increasing [52].

Previous occupation of the unemployed

Employment probability depends on the previous occupational attainment of the unemployed. While some previous occupations could help the unemployed find the new employment sooner, others could diminish their chances. Re-employment was also related to the employer's perception of the previous occupation of the unemployed and motivate or demotivate him to employ the job candidate. Employers could also perceive some occupations being more suitable for women than for men and based on their perception they could employ or not employ the job candidate. On the other side, employees could also make wrong occupational decisions. They could accept employment in positions that are less suitable for them, just to stay employed. A decision like this, could have a negative impact on their future employment, and could make the employee less suitable for the job offering. When the employee decides to change occupational attainment, this could decrease its chances of being employed in the position he applied.

Gender inequality in unemployment has been related with the previous occupation as well. Women were usually employed in services positions, while men worked in industry or agriculture. Women participation on the labor market has increased over the last fifty years. When it comes to unemployment insurance benefits, women were less likely to receive them, since they were more likely being unemployed longer than men [44]. Despite social protection in unemployment was introduced to decrease unemployment, its implementation resulted in decrease of labor force participation. Women participated in the social transfer program more than men [60]. The probability of women entering the labor force increased by four percent in times when their husbands were involuntarily released from the employment. Worsening economic situation had only a negligible effect on the labor force participation of women [29]. The monetary shock could cause a decrease of employment across all sectors. The impact of the income shocks had a small effect on the unemployment in the short run. Significant impact on the unemployment had price shocks, that caused the consequences from it lasting longer [7]. The highest wage gender inequality was observed in the private sector, while in the public sector, wages were not related to gender [51]. The service-oriented problems in employment was solved in [28, 35–37, 45].

Applying for the job position from another occupation group, than their previous occupational group, caused unemployed being 2–3 months longer unemployed, and received 1.3–1.6% lower wages than unemployed that applied for the position similar to their last occupational attainment [5]. Young people were changing occupational attainment more often than the prime age workforce. Changing employments and occupational attainments could result in higher wage losses. Changing employment is however related with age, and with the increase of age the switching declines [24]. Re-employment depends on the situation on the labor market. Supply and demand on the labor market affects occupational employment and modifies the structure of the labor market by occupation [25]. The negative aggregate technological shock could cause people with higher skills, switching to lower skill jobs, and in this way increases their chances of staying employed [31].

Gender

Men have always been better off than women on the labor market. They can get re-employed faster than women. Probability of married men getting employed is even higher than the probability of not married men and married and not married women. Probability of married women getting employed is even lower than not married women [6]. When it comes to abandoning of the employment, women were those who abandoned employment more than men were. Even though, probability of women finding employment is lower than the probability of men, it increases with the increase of the educational attainment [34]. Highly educated women and men are equal when it comes to mobility from one job to another. This probability is high for both genders, when compared with a workforce with lower educational attainment [54]. Gender inequality in the labor market could be related to the wrong professional attainment and wrong educational attainment of women, who usually chose professions of which labor market had been already saturated. Wrong educational attainment decreases their chances of getting employed after graduating [33].

Gender equality is one of the goals of the European Union and it is secured by the Charter of Fundamental Rights. Some goals of the Strategy for equality between women and men are equal economic independence, equal pay for equal work, equality in decision-making, etc. [15].

The job finding opportunities were similar across most of the European countries, except in Spain. However, separation rates were higher for women than were for men. Discrimination probably wasn't the reason for higher women separation rates, but the accommodation between employment and household responsibilities was. To increase participation of women in the labor market, flexible working contracts should be offered to them [34].

Spain, as a member of the European Union, should follow the policies that were introduced by the European Commission. Therefore, policies focused on the gender equality in the European Union are also obligatory for Spain. Legal actions toward gender equality in Spain were introduced in the Equality Act in 2007 [40]. The unemployment rate in Spain was related to gender. Male employees were preferred before female employees, due to the women's lower commitment to the labor market than men. Gender inequality in Spain could be tackled by revising the family and institutional policy. Therefore, regulation of maternal leaving, flexible working hours, flexible wage determination and other rigid law that keeps women out of the labor market, should be revised in the way that provides women with the opportunity to participate in the labor market and being treated equally as men were treated [46].

When it comes to Switzerland, gender inequality had been always low, and almost inexistent. In 1981, the constitutional amendment was introduced that guaranteed the equal pay independent from gender for the same work. The amendment was constituted after was chosen by referendum. Shortcomings from the amendment were removed in 1996, by introducing the Gender Equality Act. In 1999, based on the referendum, the Agreement on the Free Movement of Persons between the European Union and Switzerland, was introduced. The agreement ended the unequal treatment of foreign employees in Switzerland [13].

3 Data and Methods

3.1 Data

We employed average unemployment rate and unemployment rate data disaggregated by gender, age, education, duration of unemployment and previous occupations of the unemployed, for the sample period 1996–2016, of Spain, Switzerland and the European Union, from the Eurostat database [16]. Unemployment rates of the Spain, Switzerland and the European Union, during the period 1996–2016, were depicted in Fig. 1.

Unemployment rate by gender, for each territory, was introduced in Fig. 2. Afterwards were described their implications and actions that were taken by each territory.

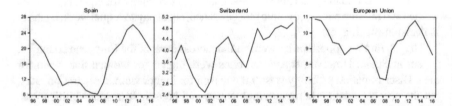

Fig. 1 Unemployment in Spain, Switzerland and the European Union. *Source* Own illustration based on data from OECD database [47]

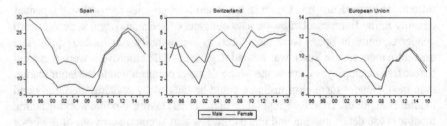

Fig. 2 Unemployment by gender of Spain, Switzerland and the European Union. *Source* Own illustration based on data from Eurostat database [16]

3.2 Methods

Stationarity of the univariate time series was tested using LM unit root test explained by Lee et al. [39] using following data generating process:

$$y_t = \delta' Z_t + e_t, e_t = \beta e_{t-1} + \epsilon_t, \tag{1}$$

where Z_t represents vector of the exogenous variables depending on used model. We consider two models, model without structural and the model with one structural break, where $D_t = 1$ if $t \geq T_B + 1$, zero otherwise and $DT_t^* = t - T_B$ if $t \geq T_B + 1$, zero otherwise. LM unit root test statistics $(\tilde{\tau})$ is estimated to validate the unit root null hypothesis $\phi = 0$ from the regression model

$$\Delta y_t = \delta' \Delta Z_t + \phi \tilde{S}_{t-1} + e_t, \tag{2}$$

where Δ is the first difference operator, $\tilde{S}_t$ is detrended form of y_t, $\tilde{\delta}$ is the coefficient of the exogenous variables in the regression. Structural break points T_B, were searched in a grid that is determined by $LM_\rho = Inf_\lambda \tilde{\rho}(\lambda)$ and $LM_\tau = Inf_\lambda \tilde{\tau}(\lambda)$ when the $\tilde{\tau}$ was minimized. Here $\tilde{\rho} = T \cdot \tilde{\phi}$ and $\lambda = T_B/T$ [32].

Stationarity in panel data was tested using ILT Unit Root Test explained by Im et al. [27]. This test is based on regression model

$$\Delta Y_{i,t} = \gamma_i' \Delta Z_{i,t} + \delta_i \hat{S}_{i,t-1} + \varepsilon_{i,t}, \tag{3}$$

where $\hat{S}_{i,t-1}$ is detrended variable of $Y_{i,t-1}$, $\varepsilon_{i,t}$ is error term.

We compute panel LM t-statistic $LM(\bar{t})$ based on t-statistics, denoted as t^* which validate the null hypothesis $\delta_i = 0$ for each unit of the panel data

$$\bar{t} = \frac{1}{N} \sum_{i=1}^{N} t_i^*, \quad LM(\bar{t}) = \frac{\sqrt{N}(\bar{t} - E(\bar{t}))}{\sqrt{V(\bar{t})}}. \tag{4}$$

$E(\bar{t})$ and $V(\bar{t})$ were tabulated in paper of Im et al. [27].

Half-life estimator is a measure the persistence of consequences from the impact of economic shocks on our series, we employ a half-life estimator denoted by Queneau and Sen [50] as

$$HL_\rho = \frac{\log(0.5)}{\log(\rho)}, \tag{5}$$

where ρ was obtained when we rewritten Eq. (4) as

$$y_t - y_{t-1} = \delta' \Delta Z_t + (\rho - 1)\tilde{S}_{t-1} + e_t, \tag{6}$$

from which $\rho = \phi + 1$.

Pesaran's Cross-Dependence test verify hypothesis that there is no cross-sectional dependence across panels data. The hypotheses for this test are:

H_0: *The panels data are not correlated (No cross dependence)*
H_1: *The panels data are correlated (Cross dependence).*

The null hypothesis was validated using *CD* statistic

$$CD = \sqrt{\frac{2T}{N(N-1)} \left(\sum_{i=1}^{N-1} \sum_{j=i+1}^{N} \tilde{\rho}_{ij} \right)}, \tag{7}$$

where $\tilde{\rho}_{ij}$ is an average of the pairwise correlation coefficients from the ADF regression residuals [4].

4 Unemployment Gender Inequality Results

When looking at the unemployment gender inequality (Fig. 3), Spain, Switzerland and the European Union have one common attribute, in all three, women were unemployed more than men. Unemployment by gender was calculated as the percentage of the workforce of each gender. The truth is, that percentage of unemployed women from the women workforce is higher than the percentage of unemployed men from the men workforce. But when looking at the numbers of the unemployed people, then there are more men than women. The labor market is changing, employment and labor force participation of men is decreasing while employment and labor force participation of women is increasing. Even when there are more and more women in the labor market, gender inequality was decreasing.

Fig. 3 Unemployment gender inequality of Spain, Switzerland and the European Union. *Source* Own illustration

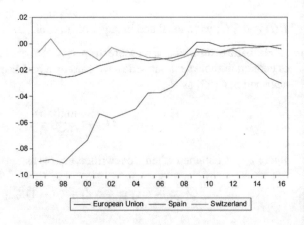

Policies toward gender inequality in Spain, Switzerland and the European Union, are well set and they are approaching to fulfil the set objectives. Higher success had policies of Spain, where gender inequality was the highest, and they achieved to decrease it. Policies of the European Union and Switzerland maintained the gender inequality low during the observed period. If we consider the intervals of gender inequality that we proposed, then in the European Union and Switzerland, during the observed time period, gender inequality didn't exist. While in Spain, decreased from high gender inequality to equality, and returned to low inequality with tendency do increase.

When economic disturbances were not included, unemployment gender inequality series of Spain, Switzerland and the European Union show that hysteresis is present, series are not-stationary (Table 1). Appears that policies implemented to control unemployment gender inequality doesn't have effects. Unemployment gender inequality would eventually reverse, since depends on time and can't be influenced. But what if the reverse process was triggered by something that happened in economy at some point? To investigate it, we include one structural break into analysis and measure its influence (Table 2).

When economic disturbances are taken account, it was clear that the structure of the unemployment gender inequality doesn't depend on time. Unemployment gender inequality was only impacted by economic disturbance that happened at one point in time and the consequences of it disappeared quickly, according to half-life estimators. An effort of the policy makers to control gender inequality was giving fruits. Gender inequality is decreasing, not even random economic shock had significant impact on it, but even is decreasing is still persistent and need to be controlled.

Table 1 LM test without break

Serie	ϕ	δ_1	k
Spain	−0.2077 (−1.0702)	0.0049 (0.9307)	4
Switzerland	−0.5388 (−2.2377)	−0.0013 (−1.0830)	0
European Union	−0.3658 (−1.6169)	0.0025 (2.0281)	0

Numbers in parenthesis under the estimated coefficients are t-statistics for the null hypothesis
Source Own calculations

Table 2 LM test with one break

Country	ϕ	δ_1	δ_2	δ_3	k	T_B	HL_ρ
Spain	−1.249*** (−4.433)	−0.0051 (−1.7874)	−0.0060 (−1.1092)	0.0025 (0.6200)	4	2009	0.4989
Switzerland	−3.8649*** (−7.4245)	0.0078 (6.3994)	0.0013 (0.7217)	−0.0060*** (−5.3719)	5	2005	0.6586
European Union	−2.3864*** (−5.7345)	0.0015 (2.1322)	−0.0056 (−2.6513)	0.0087 (4.7037)	5	2007	2.1216

***Significant at level 0.01; **Significant at level 0.05; *Significant at level 0.1
Numbers in parenthesis under the estimated coefficients are t-statistics for the null hypothesis
Source Own calculations

Cross-dependence results
Unemployment gender inequality depends on distinct characteristics only for Spain
and the European Union, this was not proven for Switzerland therefore we didn't
find a common behavior pattern between them. We conclude, that gender inequality
doesn't depend on the previous occupation of the unemployed, duration of unemployment,
age and education. But when we look at full panel, consisted from all series
from each category, there were all dependent on each other in all three territories
(Table 3).

The level of the unemployment rate depends on distinct characteristics, such
as age, education, duration of unemployment and the previous occupation of the
unemployed, this was proven for each territory. Also, when testing for dependence
between all characteristic series together, dependence between them was found for
all territories. Meaning that unemployment by each characteristic depend on each
other's.

4.1 Unemployment and Unemployment Gender Inequality by Distinct Characteristics

Unemployment rates of different characteristics were dependent on each other, therefore
we estimate them further (Table 4). We have also found that the level of the
unemployment gender inequality depends on the distinct characteristics, which we
therefore estimate (Table 5). For better estimation results, we compare two models.

Unemployment

Table 3 Pesaran CD test results

	Spain		Switzerland		European Union	
	u_t^{inq}	u_t	u_t^{inq}	u_t	u_t^{inq}	u_t
All	37.672 (0.00)	28.912 (0.00)	11.078 (0.00)	21.612 (0.00)	44.194 (0.00)	28.912 (0.00)
Previous occupations	8.238 (0.00)	8.919 (0.00)	1.148 (0.03)	3.597 (0.00)	11.222 (0.00)	8.008 (0. 00)
Duration in unemployment	12.065 (0.00)	9.301 (0.00)	1.227 (0.22)	7.243 (0.00)	14.799 (0.00)	5.537 (0.00)
Age	6.958 (0.00)	7.664 (0.00)	0.161 (0.16)	6.299 (0.00)	7.409 (0.00)	6.543 (0.00)
Education	7.257 (0.00)	7.026 (0.00)	1.328 (0.18)	4.641 (0.00)	7.067 (0.00)	4.339 (0.00)

Source Own calculations

Table 4 Unemployment estimation by distinct characteristics

		GLS			ML		
		Spain	Switzerland	EU	Spain	Switzerland	EU
Previous occupation of the unemployed	Managers, professionals, and technicians	−1.913	3.139 ***	0.617	9.311 ***	6.017 ***	6.019 ***
	Clerical, service, and sales workers	10.151 **	4.188 ***	5.035 ***	21.375 ***	7.067 ***	10.437 ***
	Skilled agricultural and trades workers	−0.073	0.205	1.528 **	11.150 ***	3.083 ***	6.931 ***
	Plant and machine operators, and assemblers	−6.499 *	−1.911 ***	−2.229 ***	4.725 *	0.968 **	3.173 ***
	Elementary occupations	10.597 ***	−1.606 **	1.785 **	21.820 ***	1.272	7.187 ***
	Armed forces	−11.205 ***	−2.879 ***	−5.366 ***	0.019	0	0.037
	Not elsewhere classified	40.158 ***	6.914 ***	26.414 ***	51.382 ***	9.793 ***	31.816 ***
Duration in unemployment	Less than 1 month	−3.435	−0.362	−1.306 *	7.788 ***	2.517 ***	4.096 ***
	From 1 to 2 months	4.991	1.758 ***	2.628 ***	16.215 ***	4.637 ***	8.030 ***
	From 3 to 5 months	5.151	1.885 ***	3.137 ***	16.375 ***	4.763 ***	8.539 ***
	From 6 to 11 months	7.517 *	1.702 ***	4.556 ***	18.741 ***	4.580 ***	9.959 ***
	From 12 to 17 months	−0.496	−0.468	1.275 *	10.727 ***	2.411 ***	6.677 ***
	More than 18 months	21.601 ***	2.385 ***	13.524 ***	32.825 ***	5.263 ***	18.927 ***

(continued)

Table 4 (continued)

		GLS						ML					
		Spain		Switzerland		EU		Spain		Switzerland		EU	
Age	From 15 to 24 years	23.231	***	4.457	***	14.074	***	34.455	***	7.336	***	19.476	***
	From 25 to 49 years	4.717		0.936		3.226	***	15.940	***	3.814	***	8.629	***
	From 50+	0.7		−0.029		1.240	*	11.924	***	2.85	***	6.643	***
Education	Primary education (levels 0–2)	10.586	***	4.038	***	8.648	***	21.809	***	6.917	***	14.05	***
	Secondary education (levels 3 and 4)	5.740		0.719		3.743	***	16.964	***	3.598	***	9.145	***
	Tertiary education (levels 5 and 6)							11.224	***	2.879	***	5.402	***
	Intercept	11.224	***	2.879	***	5.402	***						

*** Significant at level 0.01; **Significant at level 0.05; * Significant at level 0.1

Source Own calculations

Table 5 Unemployment gender inequality estimation by distinct characteristics

		GLS			ML		
		Spain	Switzerland	EU	Spain	Switzerland	EU
Previous occupation of the unemployed	Managers, professionals, and technicians	0.360	1.331 ***	0.325	−5.122 ***	0.137	−1.765 ***
	Clerical, service, and sales workers	−5.239 ***	−0.843 ***	−3.674 ***	−10.721 ***	−2.036 ***	−5.764 ***
	Skilled agricultural and trades workers	18.729 ***	4.209 ***	9.093 ***	13.247 ***	3.015 ***	7.003 ***
	Plant and machine operators, and assemblers	13.823 ***	3.936 ***	7.566 ***	8.341 ***	2.742 ***	5.476 ***
	Elementary occupations	2.343	−0.198	1.975 ***	−3.139 ***	−1.391 ***	−0.115
	Armed forces	11.816 ***	1.194 ***	8.133 ***	6.333 ***	0	6.043 ***
	Not elsewhere classified	−1.417	−0.409	−0.112	−6.899 ***	−1.603 ***	−2.202 ***
Duration in unemployment	Less than 1 month	4.109 ***	1.043 ***	1.089 **	−1.372	−0.151	−1.001 ***
	From 1 to 2 months	3.261 **	0.765 **	1.240 **	−2.221 **	−0.429 *	−0.849 **
	From 3 to 5 months	2.537 *	1.056 ***	1.430 ***	−2.945 ***	−0.138	−0.659 *
	From 6 to 11 months	1.657	0.799 **	0.926 *	−3.825 ***	−0.395 *	−1.165 ***
	From 12 to 17 months	1.083	0.715 *	0.920 *	−4.399 ***	−0.479 **	−1.169 ***

(continued)

Table 5 (continued)

		GLS			ML		
		Spain	Switzerland	EU	Spain	Switzerland	EU
	More than 18 months	−0.796 **	−0.314	0.841 *	−6.278 ***	−1.508 ***	−1.249 ***
Age	From 15 to 24 years	3.389 **	1.352 **	2.045 ***	−2.093 *	0.158	−0.045
	From 25 to 49 years	0.855 *	0.217	0.468	−4.628 ***	−0.977 ***	−1.623 ***
	From 50+	2.429 *	0.997 ***	2.068 ***	−3.053 ***	−0.197	−0.023
Education	Primary education (levels 0–2)	0.992	1.065 ***	1.082 **	−4.490 ***	−0.129	−1.008 ***
	Secondary education (levels 3 and 4)	0.052	0.943 ***	0.525	−5.430 ***	−0.251	−1.565 ***
	Tertiary education (levels 5 and 6)				−5.482 ***	−1.194 ***	−2.090 ***
	Intercept	−5.482 ***	−1.194 ***	−2.090 ***			

*** Significant at level 0.01; ** Significant at level 0.05; * Significant at level 0.1

Source Own calculations

People who were never employed before, had the worst chances of getting employed than people who were employed before. People who previously worked as clerical, service and sales workers were in the second worst position in the labor market. The more time people spent unemployed the less chances they have, to get employed again. The reverse situation was with age of unemployed. Young people had considerable problems when entering the labor market. Since majority of the young people was never employed before, which also decreases their chances of getting employed. Higher educational level also increases chances of the unemployed in the labor market.

Unemployment gender inequality

Unlike unemployment, unemployment gender inequality was the lowest among young people. Opposite effect had also the educational attainment of the unemployed, the lowest gender inequality was among those with the lowest educational attainment. Surprisingly, the more time people spent in unemployment, women were those who got employed before men that caused that unemployment gender inequality was decreasing with time. It is curious that unemployment gender inequality is almost non-existent among those who were not employed before. Highest gender inequality was among those who previously worked as skilled agricultural and trade workers. We also observe that gender inequality depends on the perception of the working position. There are more men unemployed who previously worked in positions that are perceived as women's professions, therefore, there are more men who were previously employed as clerical, service and sales workers and more women that were previously employed as skilled agricultural and trade workers. There are no women unemployed that were previously employed in the armed forces.

5   Conclusions

Chapter summarize teachings and theories of the important economists, since the beginning of the unemployment researching until this very day. Labor market policies and actions toward fighting unemployment and unemployment inequality were described next. Policies toward unemployment gender inequality were proven to be efficient and stood up steadily even against the economic crisis. This was proven with analysis of the hysteresis, where it was found that the unemployment gender inequality for all three territories was a stationary process when one break included. Consequences of this economic disturbance disappeared quickly, and the unemployment gender inequality series recovered their behavior from before disturbances appeared.

Women are still in disadvantage compared to men, even gender inequality is persistent, and it is decreasing with time. During the very high unemployment rates, unemployment gender tends to be lower, which is the consequence of increase of men unemployment rates. Even it was proven that level of the unemployment gender inequality doesn't depend on the level of the unemployment rates, or on annual changes. Unemployment gender inequality doesn't depend on distinct characteristics such as previous occupation of the unemployed, duration in unemployment, age or education, but when all put together and analyzed, correlation between them was found. On the other hand, unemployment rates depend on distinct characteristics, in both cases.

Unemployment rate decreases when increase education level and age but it increases with time spent in unemployment. While, unemployment gender inequality increases when increase education level and time spent in unemployment. Both, unemployment and unemployment gender inequality highly depend on the previous occupation of the unemployed. The worst position in the labor market had people with no previous working experience, or those who worked in elementary occupations or at clerical, service and sales positions, while the least problems had people previously employed in agricultural sector or at managerial positions.

Unemployment also depend on the employers' perception of the working position. There are more men unemployed who previously were employed at clerical, service or sales positions, which are perceived as positions suitable for women, and there are more women unemployed who previously worked at agricultural or trade positions, which are perceived as positions suitable for men. The lowest probability of getting employed had youth women with previous elementary occupation or non-occupation, who have been unemployed for more than 18 months and have primary education. The highest probability of getting employed have men, from 25 to 49 years old, who previously worked as a plant and machine operators or in armed forces, with secondary education.

Acknowledgements First author is very grateful to prof. Cecilio Tamarit for accepting her to do part of her research at the University of Valencia. Many thanks to the University Valencia as well, for providing her access to the University resources and facilities.

References

1. Arcanjo M (2012) Unemployment insurance reform—1991–2006: a new balance between rights and obligations in France, Germany, Portugal and Spain. Soc Policy Adm 46(1):1–20
2. Arellano FA (2010) Do training programmes get the unemployed back to work? a look at the Spanish experience. Revista de Economia Aplicada 18(53):39–65
3. Bachmann R, Sinning M (2016) Decomposing the ins and outs of cyclical unemployment. Oxf Bull Econ Stat 78(6):853–876
4. Bakas D, Papapetrou E (2014) Unemployment by gender: evidence from EU countries. Int Adv Econ Res 20(1):103–111
5. Bauer A (2016) Reallocation patterns across occupations in Germany. Econ Lett 148(2016):111–114
6. Baussola M, Mussida C, Jenkins J, Penfold M (2015) Determinants of the gender unemployment gap in Italy and the United Kingdom: a comparative investigation. Int Labour Rev 154(4):537–562

7. Berument MH, Dogan N, Tansel A (2009) macroeconomic policy and unemployment by economic activity: evidence from Turkey. Emerg Mark Financ Trade 45(3):21–34
8. Boheim R, Horvath GT, Winter-Ebmer R (2011) Great expectations: past wages and unemployment durations. Labour Econ 18(6):778–785
9. Bonvin JM, Moachon E (2007) The impact of contractualism in social policies: The case of active labour market policies in Switzerland. Int J Sociol Social Policy 27(9/10):401–412
10. Caliendo M, Schmidl R (2016) Youth unemployment and active labour market policies in Europe. IZA J Labor Policy 5(1):1–30
11. Campbell M (2011) Social Europe guide: employment policy, 1st edn. Publications Office of the European Union, Luxembourg
12. Chabanet D, Giugni M (2013) Patterns of change in youth unemployment regimes: France and Switzerland compared. Int J Soc Welf 22(3):310–318
13. Erne R, Imboden N (2015) Equal pay by gender and by nationality: a comparative analysis of Switzerland's unequal equal pay policy regimes across time. In: Auth D, Hergenhan J, Holland-Cunz B (eds) Gender and family in European economic policy. Palgrave Macmillan, Cham, London, pp 81–107
14. European Commission (2011) Communication from the commission to the European parliament, the council, the European economic and social committee and the committee of the regions. Supporting growth and jobs—an agenda for the modernisation of Europe's higher education systems. http://eur-lex.europa.eu/LexUriServ/LexUriServ.do?uri=COM:2011:0567: FIN:EN:PDF Accessed 6. 1. 2017
15. European Commission (2017) Strategy for equality between women and men 2010–2015. Publication Office of the European Union, Luxembourg
16. Eurostat (2017) Eurostat [database]. http://ec.europa.eu/eurostat/data/database
17. Falter JM, Fluckiger Y, Silber J (1998) From employment to exclusion a statistical analysis of unemployment in the Geneva area in 1995–1996. Int J Manpow 19(6):424–437
18. Faria JR, León-Ledesma MA (2008) A simple nonlinear dynamic model for unemployment: explaining the Spanish case. Discret Dyn Nat Soc 2008
19. Flek V, Hála M, Mysíková M (2015) Duration dependence and exits from youth unemployment in Spain and the Czech Republic. Ekonomska Istrazivanja 28(1):1063–1078
20. Flek V, Mysikova M (2015) Youth labour flows and unemployment in great recession: comparing Spain and the Czech Republic. Rev Econ Perspect 15(2):179–195
21. Fluckiger Y (1998) The labour market in Switzerland: the end of a special case? Int J Manpow 19(6):369–395
22. Fournier MC, Mercier CS (2009) Economics of employment and unemployment. Nova Science Publishers Inc, New York
23. Gerfin M, Lechner M (2002) A microeconometric evaluation of the active labour market policy in Switzerland. Econ J 112(482):854–893
24. Gervais M, Jaimovich N, Siu H, Yedid-Levi Y (2016) What should I be when I grow up? occupations and unemployment over the life cycle. J Monet Econ 83(2016):54–70
25. Hagglund AE, Bachmann A (2017) Fast lane or down the drain? does the occupation held prior to unemployment shape the transition back to work? Res Soc Stratif Mobil 49(2017):32–46
26. Howell DR (2005) Fighting unemployment: the limits of free market orthodoxy. Oxford University Press, New York
27. Im KS, Lee J, Tieslau M (2005) Panel LM unit root tests with level shifts. Oxf Bull Econ Stat 67(3):393–419
28. Kaczor S, Kryvinska N (2013) It is all about services—fundamentals, drivers, and business models. Soc Serv Sci J Serv Sci Res 5(2):125–154
29. Karaoglan D, Okten C (2015) Labor-force participation of married women in Turkey: a study of the added-worker effect and the discouraged-worker effect. Emerg Mark Financ Trade 51(1):274–290

30. Kesner-Skreb M (2010) Employment policy in the European Union. Financ Theory Pract 34(3):315–317
31. Khalifa S (2012) Job Competition, Crowding Out, and Unemployment Fluctuations. Macroeconomic Dynamic 16(1):1–34
32. Khraief N, Shahbaz M, Heshmati A, Azam M (2015) Are unemployment rates in OECD countries stationary? evidence from univariate and panel unit root tests. http://ftp.iza.org/dp9571.pdf
33. Klugman J, Kolb H, Morton M (2014) Persistent gender inequality in the world of work. Fletcher Forum World Aff 38(2):133–152
34. Koutentakis F (2015) Gender unemployment dynamics: evidence from ten advanced economies: gender unemployment dynamics. Labour 29(1):15–31
35. Kryvinska N (2012) Building consistent formal specification for the service enterprise agility foundation. Soc Serv Sci J Serv Sci Res 4(2):235–269
36. Kryvinska N, Gregus M (2014) SOA and its business value in requirements, features, practices and methodologies. Comenius University, Bratislava in Bratislava
37. Kryvinska N, Gregus M (2015) Service orientation of enterprises—aspects, dimensions, technologies. Comenius University, Bratislava in Bratislava
38. Lalive R, Van Ours JC, Zweimuller J (2008) The impact of active labour market programmes on the duration of unemployment in Switzerland. Econ J 118(525):235–257
39. Lee J, Strazicich MC, Meng M (2012) Two-step LM unit root tests with trend-breaks. J Stat Econom Methods 1(2):81–107
40. López Díaz E, Santos del Cerro J (2015) Gender discrimination in the Spanish labour market and regulation of the public policies. J Public Programs Policy Eval 5:63–82
41. Maguire S et al (2013) Youth unemployment. Intereconomics 48(4):196–235
42. Maloney CB (2011) Understanding the economy: unemployment among young workers. Unemployment: a closer look. Nova Science Publishers Inc, New York, pp 137–150
43. Marlier E, Natali D, Van Dam R (2010) Europe 2020: towards a more social EU?. P.I.E. Peter Lang, Brussels
44. Michaelides M, Mueser PR (2013) The role of industry and occupation in recent US unemployment differentials by gender. Race Ethn East Econ J 39(3):358–386
45. Molnár E, Molnár R, Kryvinska N, Gregus M (2014) Web intelligence in practice. Soc Serv Sci J Serv Sci Res 6(1):149–172
46. Mussida C, Fabrizi E (2014) Unemployment outflows: the relevance of gender and marital status in Italy and Spain. Int J Manpow 35(5):594–612
47. OECD (2016) Unemployment rate (indicator). https://data.oecd.org/unemp/unemployment-rate.htm. Accessed 15 Aug 2016
48. Pašic P, Kavkler A, Boršič D (2011) Gender disparities in the duration of unemployment spells in slovenia. South East Eur J Econ Bus 6(1):99–110
49. Puhani PA (2003) Relative demand shocks and relative wage rigidities. KYKLOS 56(4):541–562
50. Queneau H, Sen A (2009) Further evidence on the dynamics of unemployment by gender. Econ Bull 29(4):3162–3176
51. Seshan GK (2013) Public-private-sector employment decisions and wage differentials in peninsular Malaysia. Emerg Mark Financ Trade 49(sup5):163–179
52. Strahm R (2013) Switzerland: a dual track to low unemployment. http://www.elmmagazine.eu/articles/what-are-they-doing-right-3-cases. Accessed 3 Jan 2017
53. Talani LS (2014) European political economy: issues and theories, 2nd edn. Ashgate Publishing Limited, Farnham
54. Theodossiou I, Zangelidis A (2009) Should I stay or should I go? the effect of gender, education and unemployment on labour market transitions. Labour Econ 16(5):566–577
55. Turner D (2012) Euro zone jobless rate hits new high. Institutional Investor
56. Van Stolk C et al (2011) Life after Lisbon: Europe's challenges to promote labour force participation and reduce income inequality. RAND Corporation, Santa Monica

57. Weiss G (2000) Researching the European Union: data and ethnography. European Union discourses on un/employment. John Benjamins Publishing Company, Amsterdam, pp 51–72
58. Whitley BL (2010) European response to the financial crisis. Nova Science Publishers Inc, New York
59. Wolfl A, Mora-Sanguinetti JS (2011) Reforming the labour market in Spain. In: OECD Economics Department Working Paper, no 845, pp 5–33
60. Yildirim J, Dal S (2016) Social transfers and labor force participation relation in Turkey: a bivariate probit analysis. Emerg Mark Financ Trade 52(7):1515–1527

Innovative Activities Development of Industrial Enterprises in Ukraine

Iryna Moyseyenko, Mariya Fleychuk and Mariya Demchyshyn

Abstract This paper deals with the main problems and perspectives of the innovation sector of the Ukrainian economy development. Usage of econometric tools is proven. The main obstacles to the introduction of innovations in the activities of industrial enterprises of Ukraine are institutional factors. Thus, described analytical toolkit is the basis of the method for the strategic monitoring of entrepreneurship innovation development.

Keywords Innovation policy · Innovation activity · Innovation development · Statistical analysis · Econometric analysis

1 Introduction

The variability of innovative development can raise the issue of ensuring its manageability through the creation of targeted, interrelated changes with the allocation of key characteristics and factors of this type of development, to which the authors of research [1] include:

- innovative goals, innovative results of production activity (goods, services, efficiency, competitiveness, etc.) in a tactical and strategic plan;
- innovative means of achieving goals (factor-innovation in the form of new technology and technology, new organization and motivation of labor force and production);

I. Moyseyenko (✉)
Lviv State University of Internal Affairs, Lviv, Ukraine
e-mail: iruna_m2015@ukr.net

M. Fleychuk (✉) · M. Demchyshyn
Lviv University of Trade and Economics, Lviv, Ukraine
e-mail: fleychukm@gmail.com

M. Demchyshyn
e-mail: mariya.demchyshyn@gmail.com

© Springer Nature Switzerland AG 2020
N. Kryvinska and M. Greguš (eds.), *Data-Centric Business and Applications*,
Lecture Notes on Data Engineering and Communications Technologies 30,
https://doi.org/10.1007/978-3-030-19069-9_10

- possibilities of reproduction of innovation orientation on a balanced basis for the management system to possess all organizational, managerial, resource and motivational conditions for it;
- specific methods of marketing research;
- a sharp increase in the volumes of information and its continuous accumulation;
- increase in the depth of forecasting and its multivariate nature.

Analyzing the results of studies [1, 2], we can conclude that the effectiveness of innovation development is achieved in the terms of mandatory increase in the depth of forecasting and planning of each option. At the same time, in our opinion, it is necessary to take into account and determine the driving force of such development.

Ukraine's economy has its own traits, which impose additional restrictions on both the range of instruments used and the ultimate effectiveness of the policies being implemented. An oligarchic component means—when the choice of policy instruments in Ukraine is made in favor of a limited group of stakeholders.

The most typical example of this is the discreteness of policies and instruments that are focused on current, not future goals. In Ukraine, the predominance of the political component is more pragmatic when adopting strategies for innovation development, fiscal and economic decisions.

From the point of view of the methodology of making strategic decisions, due to the fact that traditional management technologies based on the construction of formal models and the exchange of formal knowledge, such instruments are not present today—there is a need to use economic and mathematical tools for controlling existence and development of socio-economic systems. The quality of management decision-making requires new tools for adequate forecasting and control that allow to coordinate interests and influence on decision-making.

The question of the current state and development of innovation activity is the subject of research by many leading foreign and domestic scientists. An analysis of recent scientific publications has shown that some important aspects of the problem under study remain insufficiently considered and require further study: features and factors of innovation activity, distribution of innovation costs, etc.

At the end of the twentieth century P. Krugman, based on the modeling of the influence of innovation development and technology transfer on international trade established [3]:

- in the presence of the same level of labor productivity, higher incomes will be observed in a country with a higher level of innovation development;
- innovative development on equal terms (political, economic, institutional) contributes to the flow of capital into technologically growing industries and activities.

Thus, if technology transfers from developed countries to less developed, then due to the difference in wage levels this will contribute to the growth of the production of innovative products in less developed economies and thus reduce inequalities in their development.

Relevance of the outlined problem and its practical significance implies research of tendencies and directions of innovative development of industrial enterprises, diagnostics of the external environment and identification of problems and limitations of their economic growth [4]. For the introduction of technologies V and VI technological processes it is advisable to establish a special procedure for stimulating the innovation activity of enterprises, in particular through tax policy and lending.

Solving the tasks of the transition to an innovative way of development requires the development of methodological tools and the implementation of system analytical studies and predictive assessments, especially in terms of identifying the parameters of innovation activity and stimulating the introduction of innovations.

The purpose of the paper is to investigate the state of innovation activity of industrial enterprises of Ukraine in the long-term perspective, to identify the main tendency of development and to determine the directions of stimulation of innovation entrepreneurship using econometric methods and models.

2 Task

– to carry out a statistical analysis of innovation activity of industrial enterprises of Ukraine;
– determine the list of growth parameters and the average rate of growth of innovation activity;
– identify important factors of innovation activity of industrial enterprises and estimate the level of influence of growth parameters on macroeconomic indicators.

3 Features of Industrial Enterprises Innovative Activity Formation in Ukraine

Traditionally, the government proposes the same list of recommendations: to increase financing, to improve existing laws and programs, to create a new government body and to accelerate the creation of innovative infrastructure—various parks, technopolices and venture funds.

1. Enterprises are turning to science and new technologies and introduce innovations to maintain competitiveness and increase the efficiency of their activities. In the big business of Ukraine, there is practically no competition between producers, but there is competition for proximity to power and public resources. In this case, it is more appropriate to invest in corruption schemes and in selective enforcement schemes, which is also an element of corruption. The opaque business environment created in the country is quite familiar and even comfortable for Ukrainian business leaders, which allows them to function successfully without introducing innovations [5].
2. This is the main reason for the non-innovation of the big Ukrainian business. Other factors of innovation activity of industrial enterprises include the following:

the overall technological backwardness of the Ukrainian industry in relation to the level of production of advanced countries; modernization of the production base at the expense of foreign purchases of technological second-hand; low business interest in domestic science because of the lack of sufficient experience in the creation of samples of modern technology and technology "turnkey", with the necessary guarantees of their quality and a high level of service [6].

3. The ineffectiveness of transformational changes in economic policy, innovation activity stimulating are determined by the effect of such factors. First, because they represent and protect the interests of the innovative passive big business [7]. Secondly, most resources are spent to support the functioning of the country's socio-economic complex, not its development, which requires entirely other information and analytical tools using. Thirdly, in the legislature and in the executive branch there is a conviction that the basis of innovative economies is a well-funded science, and the problems of innovation development can be solved by adopting correctly written laws. Fourth, domestic legislators and civil servants are not responsible for long-term prospects for the country's subordination, but only for its state of 1–3 years in advance, while innovative economies are created over a decade. Fifthly, the concentration of almost all resources of the country in several financial-industrial groups, despite the fact that determining element of such development is the small and medium innovative business.

Thus, the listed factors of innovation activity of industrial enterprises of Ukraine formation do not allow to solve complex problems of long-term innovation development [2, 7–10].

Overcoming innovation-simulation activity and attempts to expand the economic system, innovation transformation will be useless without the presence of relevant institutions and institutional environment.

Innovative economies of the most successful countries are created and operated, relying on the existence fundamental conditions: entrepreneurs; highly competitive environment; advanced science; financial institutions and financial resources; innovation infrastructure; institutional environment for innovation development.

Thus, for the successful deployment of innovative transformations in the economy, the subjects of innovative behavior (entrepreneurs) should be necessarily represented, as well as there is a competitive business environment that stimulates such behavior, that is, the necessary first two conditions from the list below are necessary.

Obstacles to the innovations introduction are:

– an alternative way of making a profit because the benefits of economic power to the rental income maximization path have become a cost for innovation development;
– significant transaction costs of innovation activity, due to the lack of an organizational scheme for managing the process of promoting innovation and innovation management in modern business [11]. In economic practice, the needs of development and innovation of production renewal, and short-term financial interests, are not dominated. The introduction of innovations through those network structures, in which the modern business is executed, simply plummet into a large number of organizational problems and their associated costs;

– inconsistencies in the timing of economic planning and the short-term interests of entrepreneurs through the domination of arbitrariness and the possibility that income, property, enterprise position and personal freedom will be lost, that is, low level of economic security;
– profit or individual income can only be guaranteed by incorporating into the system of government, but not by investing in innovation or personal investment in knowledge;
– lack of appropriate educational infrastructure for innovation activities (training systems for science, qualified engineers, technologists, designers, highly skilled workers, etc.);
– tools of innovation policy, which can be implemented through the definition of priorities, goals, tasks, tools and the influence of institutions in the direction of ensuring the innovation development of all spheres.

The basic principles of the formation of the policy of innovation development are:

1. A program-targeted approach that identifies priorities and strategies for innovation activity, from which regional and sectoral development strategies and programs must be agreed upon.
2. Balance of development, aimed at optimal use of internal and external benefits and ensuring conditions for the transition to innovation development.
3. Institutional system of consistency of organizational, economic, legal, infrastructural and managerial regulatory influences in the direction of promoting innovative development.
4. Financial security requires a balanced formation of sources of financing for innovation development with the involvement of public and private investment.
5. Information and consultation provision providing for the full access to information about innovative objects and subjects and a wide range of consulting services for innovation activities.
6. Social and environmental orientation. The toolkit for regional innovation policy should be based on the priority of improving living standards and environmental safety of the environment.

Given the priority of the institutional system in the implementation of innovation policy, the production should be based on the target orientation, vertical and horizontal integration of regulatory influences, innovation infrastructure that covers the horizontal links of the innovation process subjects. The system of institutional regulation of innovation development should include its own subjects of governance (authorities, innovation infrastructure, social and public institutions) and their own institutional regulation of innovation development.

4 Overview of the Models of Description of Innovative Processes

Formation of an innovative type economy is extremely important, especially in the context of globalization, taking into account the role of innovation as a determining factor of competitiveness, scientific and technological and social progress, and ensuring a positive dynamics of economic development in general.

Historical review of the evolution of innovative processes from simple linear to more complicated nonlinear models was carried out by R. Roswell [4].

The first stage (50–60-ies of the twentieth century)—approaches to the analysis of the innovation process as a linear research, scientific and technical, production and marketing activity, which later became known as the "technological push" concept, the main ideas of which were limited to assertion, that the output for the innovation process is the scientific and technical preconditions. The theoretical foundations of this concept were laid down in the writings of M. Tugan-Baranovsky, J. Schumpeter, K. Freeman, N. Rosenberg, R. Nelson and others [4, 7, 8].

The Austrian economist J. Schumpeter in 1912, in the book "The Theory of Economic Development", proposed a concept of innovation, on the basis of which he took the idea of "new combinations". Among the combinations that generally form the structure of the innovation process, the scientist called the following: the launch of a new product or a known product of a new quality; the introduction of a new, still unknown, method in production in a specific field; penetration to a new market—known or unknown; obtaining new sources of raw materials or semi-finished products; organizational restructuring, in particular, the creation of a monopoly or its elimination. In further works by J. Schumpeter, the term "new combination" is replaced by the term "innovation", which became a scientific category [7].

The second stage is connected with the emergence in the late 1960s of the concept of "demand challenges", whose supporters, in particular, G. Less, E. von Hippel and J. Shmuckler noted that the determining factor of innovation development is market demand, that is, the market defines new inquiries, and in the linear model, application development leads to new opportunities [4].

The third stage—the beginning of the 70s—the mid 80-ies of the twentieth century—the synthesis of previous approaches led to the emergence of a model of interaction of technological opportunities and market needs, which required the consideration of the relationship between elements of the innovation process.

The fourth stage—from the mid 80s—is associated with a model that describes the integration of enterprises with suppliers, buyers, conducting fundamental research.

The fifth stage—the 90s and the end of the twentieth century—with a model of strategic integration, according to which the company organizes a permanent innovation process, reacting to changes in the external environment. At the same time, in the latter model, a special role is played by close interaction between market participants, feedback between producer and consumer, transfer of technologies and scientific and technical cooperation [9, 12–15].

However, the sixth stage (the beginning of the XXI century—and till today) can be singled out, which is associated with the acceleration of the rate of scientific and technological progress on the basis of the development of the IT sphere and due to the transition to the so-called sixth technological way. The sixth technological way is based on systems of artificial intelligence, global information networks and will be characterized by the comprehensive role of science and technology in the social development. Taking into account these factors is a guarantee of the development of any state of the world and Ukraine in particular [16, 17].

5 Parameters of Diagnostics of Industrial Enterprises Innovation Activity Factors in Ukraine

At the present stage of economic activity, the study of the innovative activity of enterprises indicators is of great importance, as the definition of modern trends, problems of the current stage of development, and the estimation of promising indicators of this type of activity are the key to the competitiveness of enterprises in the future. The selection and application of effective forms of innovation implementation can increase enterprises innovative activity and improve economic performance.

At present, the innovation activity of enterprises is measured by several indicators: the number of innovative enterprises, the volume of sold innovation products, the amount of innovation costs, the number of new technological ways introduced and innovative types of products [8, 9].

The development of the economy involves the intensification of the industrial enterprises innovative activity to form and maintain competitive advantages and ensure sustainable development. One of the areas of state regulation of innovation activity activation in Ukraine should be the definition of this concept in the legislation and the development of indicators that would allow to analyze innovation activity and identify the most effective directions of innovation development. The most effective will be the application of a complex combination of statistical, resource-cost and effective approach to the analysis of the state of innovation activity [16].

The current state of innovation activity of Ukrainian enterprises can be characterized first by the indicators such as the share of enterprises engaged in innovations and implemented them and the share of realized innovation products in the total volume of industrial products (Fig. 2). The relatively low level of the share of enterprises engaged in innovations, as well as the share of industrial enterprises in the total number of industrial enterprises that introduced innovations, indicates negative trends in the innovation sector.

Taking into account the peculiarities of domestic statistical accounting, in which it is necessary to have only a minimal level of novelty, to include any changes to innovations, essentially two thirds of innovations are products that are new only for the enterprise where it was developed. At the same time, it can exist in the world market for many years. The national domestic practice considers organizations to be innovative only in the case if its had the costs of innovation in the reporting period, despite its size, the stage of innovation process and the level of its completeness [13, 16–18] (Fig. 1).

The analysis of the structure of expenditures on innovative activity by enterprises of Ukraine by sources of financing in 2000–2016 showed that the main source of Ukrainian enterprises innovation financing were their own financial resources (see Fig. 2).

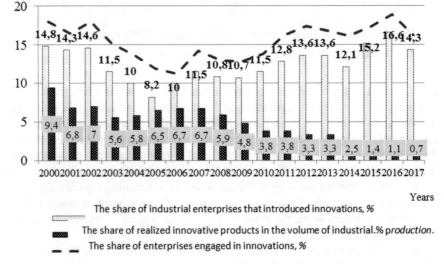

Fig. 1 Dynamics of enterprises innovative activity indicators of Ukraine in 2000–2017. *Source* Built by authors using [9]

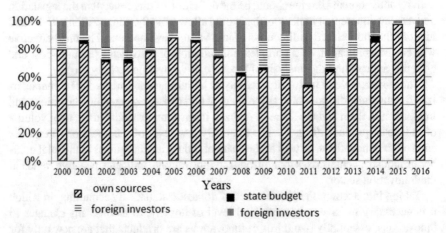

Fig. 2 Structure of expenses for innovative activity by industrial enterprises of Ukraine by sources of financing in 2000–2016. *Source* Built by authors using [9]

Table 1 Average rates of growth/decline of enterprises innovative activity indicators of Ukraine for 2000–2017 years. *Source* Built by authors using [9]

Indicator	Annual growth rate
The share of enterprises engaged in innovations, %	1.003
Total cost, *mln. UAH*	1.175
Including directions	
Research and development	1.149
Acquisition of other external knowledge	0.992
The purchase of machinery equipment and software	1.200
Other expenses	1.103
Share of enterprises that introduced innovations, %	1.007
Introduced new technological processes, *processes*	1.059
Incl. low-waste, resource-saving	1.035
Introduced production of innovative types of products, *points*	0.921
Including new types of technology	1.046
The share of realized innovative products in the volume of industrial, %	0.973

The amount of funding from public funds and foreign investment is very low. The dominance of the Ukrainian enterprises innovation activities financing at the expense of its own funds, especially in recent years (2014–2016), reflects an extremely negative tendency towards indifferent attitude of the state towards financing innovative activity and reluctance of foreign investors due to high risks of loss of their funds.

Statistical analysis of the dynamics of innovation activity of Ukrainian enterprises by the average growth or decline (Table 1) showed that during 2000–2017 indicators were characterized by uneven rates of development.

The average annual growth was: the share of enterprises engaged in innovations—by 0.3%, the total amount of expenses—by 17.5%, including in directions—for research and development—by 14.9%, for the purchase of machinery equipment and software—by 20%, other expenses—by 10.3%; expenses for innovation activity at the expense of own sources—by 18.8%, at the expense of the state budget—by 21.7%, other sources—by 10%. The share of enterprises that introduced innovations grew annually on average by 0.7%, introduction of new technological processes—by 5.9%, including low-waste, resource-saving—by 3.5%. The decline in the indicators of innovation activity was only observed for such indicators as: the cost of acquiring other external knowledge—by 0.8%, financing by foreign investors—by 10.3%, the number of implemented innovations—7.9%, specific the weight of the realized innovative production in the volume of industrial—by 2.7% (Fig. 3).

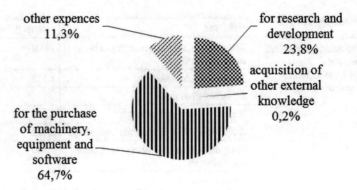

other expences
11,3%

for research and
development
23,8%

acquisition of
other external
knowledge
0,2%

for the purchase
of machinery,
equipment and
software
64,7%

Fig. 3 Structure of Ukrainian enterprises expenses innovation activity in 2017, in directions. *Source* Built by authors for [9]

Table 2 Distribution of enterprises and organizations of Ukraine by types of innovations (% of the total number of enterprises [9])

Types of innovation	Years		
	2010–2012	2012–2014	2014–2016
Total number of enterprises and organizations	100.0	100.0	100.0
Innovatively active	20.4	14.6	18.4
Were engaged only in product innovations	1.5	1.6	1.6
Were engaged only in process innovations	3.2	3.6	5.8
Were engaged in product and process innovations	4.2	3.6	4.5
Engaged only in marketing and organizational innovation	10.4	5.1	5.1
Did not do any of the innovations	79.6	85.4	81.6

The analysis of Ukrainian enterprises by types of innovations allowed to establish that during 2010–2012 the share of enterprises engaged in marketing and organizational innovations amounted to 10.4%, for the period 2012–2014 and 2014–2016—5.1% (Table 2).

In Table 3 we shows the distribution of enterprises that consider that the mentioned factors significantly influenced their decision to develop innovative projects or constrained the implementation of innovation activities during 2012–2014. The list of these reasons illustrates the lack of interest of domestic enterprises in enhancing innovation activity, as well as the impossibility of innovation development through financial status.

Innovative businesses face a number of challenges. First, in the conditions of dynamic development of manufacturing technologies, the question arises about the choice of the necessary technology.

Secondly, the question arises about the alternative: to buy technology in the market, or to implement their own developments.

Table 3 Distribution of non-innovative enterprises for reasons which hampered the implementation of innovations during 2012–2014, % [9]

There are no compelling reasons to innovate	82.2
Including Low demand for innovations in the market	6.0
Because of previous innovations	3.9
Because of the very low competition of enterprises in the market	3.0
Lack of good ideas or opportunities for innovation	7.4
Possible introduction of innovations is hampered by weighty factors	17.8
Including Lack of funds within the enterprise	11.4
Lack of loans or direct investments	6.1
Lack of skilled workers within the enterprise	1.7
Difficulties in obtaining state aid or subsidies for innovation	5.8
Lack of cooperation partners	1.9
Uncertain demand for innovative ideas	2.1
Too much competition on the market	4.3

Thirdly, since the technology itself can generate profits, enterprises must solve the following issue: supply their products to the market, or use them only for internal needs. Within the enterprise itself there are also a number of obstacles for the implementation of innovations: lack of funds, loans or direct investments in the enterprise; lack of skilled workers; difficulties in obtaining state aid or subsidies for innovation; lack of cooperation partners.

Low innovative activity is due to the following: lack of financial resources; lack of motivation of R&D staff; low level of stimulation; migration of researchers, including because of low wages; a long-term process of introducing innovations into production and a significant payback period, low competition of enterprises in the market and lack of motivation for the introduction of innovations.

As a result of the research, a number of factors hindering innovation activity were identified, among which the main are: positive dynamics of the number of enterprises that introduced innovations, which are confirmed by statistical data; enterprises of the Ukrainian industry are focused on the introduction of technological innovations, and the share of marketing and organizational innovations is insignificant; insufficient amount of innovative activity financing hamper innovative development of enterprises; The main source of financing for the development and implementation of innovations at domestic enterprises was and remains its own resources, and the amount of financing from public funds and foreign investments is extremely low [19].

Solving problems of innovative development of domestic industrial enterprises requires: development and implementation of effective programs of support and stimulation of innovation development at the state level; provision of state guarantees for projects that involve the introduction of modern technological processes, including low-waste, resource-saving and non-waste; creation of a favorable investment climate for the introduction of innovations of different types.

6   Modeling Factors of Industrial Enterprises Innovation Activity

In order to provide a more detail and reliable analysis of the dynamics of innovation activity of Ukrainian enterprises, we conducted an econometric analysis based on linear, power and parabolic trend models that were built according to statistical data for 2001–2017. Adequate trend models can be used for forecasting only in relation to enterprises, which were engaged in innovations and enterprises that introduced innovations. Data of these models are presented in the Table 4, and the value of their influence on the effectiveness of economic development in the form of GDP is presented in Fig. 4.

According to the equation of the pair linear regression $= 210{,}803x - 1{,}910{,}438$ it can be argued that for an increase in the share of Ukrainian enterprises engaged in innovations, by 1%, one can expect an increase in GDP on average by 210,803 million of UAH. The value of the determination coefficient $R2 = 0.767$ indicates that the increase of the share of Ukrainian enterprises engaged in innovations by 76.7% leads to an increase in GDP.

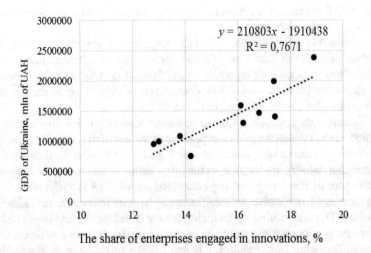

Fig. 4 Dependence of GDP of Ukraine on the share of enterprises engaged in innovations. *Source* Built by authors using [9]

Table 4 Econometric analysis based on linear, power and parabolic trend models

The share of realized innovative products in the volume of industrial, %

Type of model	Trend equation t-Student's criterion	R^2	F-Fisher criterion	DW-criterion Darbin-Watson	Forecast for year		
					2018	2019	2020
Linear	$\bar{y}=8.09-0.31t$ 12.63^* -5.03^*	0.628	25.3*	1.50*	2.42	2.11	1.79
Power	$\bar{y}=10.88t^{-0.4137}$ 10.40^* -3.81^*	0.492	14.5**	1.05	3.29	3.22	3.15
Parabolic	$\bar{y}=8.86-0.56t+0.01t^2$ 8.49^* -2.09^{**} 0.94	0.650	13.0**	1.50*	3.19	3.13	3.10

Costs for the purchase of machines, equipment and software, million of UAH

Linear	$\bar{y}=533.24+630.61t$ 0.31 3.80^*	0.490	14.4*	2.21*	11884.30	12514.92	13145.53
Power	$\bar{y}=1048.04t^{0.785}$ 28.47^* 6.8^*	0.755	46.2*	2.09*	10128.09	10567.10	11001.16
Parabolic	$\bar{y}=379.88+679.04t-$ $2.690.01t^2$ 0.13 0.93 -0.07	0.490	6.7**	2.21*	11730.94	12310.44	12884.55

Costs for research and development, million of UAH

Linear	$\bar{y}=-35.81+129.56t$ -0.33 12.19^*	0.908	148.6*	1.11*	2296.33	2425.90	2555.46
Power	$\bar{y}=141.46t^{0.938}$ 39.03^* 15.65^*	0.942	245.1*	0.97	2130.67	2241.55	2352.08
Parabolic	$\bar{y}=172.0+63.94t+$ $3.65t^2$ 1.02 1.48 1.57	0.922	82.7*	1.32*	2504.15	2702.98	2909.11

Source Calculated by authors using [9]

Note * statistical probability with probability
$p = 0.99$; ** with probability $p = 0.95$ (tabular values $F(1;15; 0.99) = 8.68$; $F(1;15; 0.95) = 4.54$; $t(15; 0.99) = 2.602$; $t(15; 0.95) = 1.753$; $d_1(0.05) = 1.08$; $d_2(0.05) = 1.36$; $d_1(0.01) = 0.81$; $d_2(0.01) = 1.07$)

Table 5 Regression analysis of the influence of indicators of innovation activity on macroeconomic indicators of Ukraine

№	Indicator	Equation	Coefficient R	R^2
1	The share of enterprises engaged in innovations, %	$\bar{y} = 210{,}803x - 1{,}910{,}438$	0.876	0.767
2	Total cost	$\bar{y} = 76.689x + 477{,}867$	0.708	0.502
3	Research and development	$\bar{y} = 893.82x + 117{,}133$	0.942	0.888
4	Domestic R&D	$\bar{y} = 999.05x + 247{,}579$	0.942	0.886
5	External R&D	$\bar{y} = 2161.6x + 783{,}445$	0.478	0.229
6	Acquiring other external knowledge	$\bar{y} = -2164.5x + 1{,}910{,}438$	−0.596	0.355
7	Purchase of equipment and software machines	$\bar{y} = 83.029x + 681{,}617$	0.753	0.566
8	Other expenses	$\bar{y} = -490.31x + 1{,}910{,}438$	−0.737	0.544
9	Introduced new technological processes, *processes*	$\bar{y} = 373.48x + 653{,}216$	0.487	0.237
10	The share of soled innovative products in the volume of industrial, %	$\bar{y} = -100{,}208 + 2{,}000{,}000$	−0.338	0.114

Source Calculated by authors using [9]

Regression analysis of the influence of the parameters of innovation activity on the macro parameters of economic development is reflected in Table 5.

Within analyzing the factors of industrial enterprises innovation activity formation of Ukraine, it was revealed that the highest level of dependence of its level on sources of financing showed the model of own funds (Table 6), which can be recommended for forecasting.

Creating an attractive investment climate and the introduction of mechanisms and incentives that will increase the interest of enterprises in innovations implementing include:

(1) improvement of tax legislation of innovation activity promoting;
(2) the system of state subsidization of scientific developments according to the determined priority directions;

Table 6 Econometric analysis based on linear, power and parabolic trend models

Costs of own sources, mln of UAH

Type of model	Trend equation t-Student's criterion	R^2	F-Fisher criterion	DW-criterion Darbin-Watson	Forecast for year		
					2018	2019	2020
Linear	$\bar{y} = -518.32 + 777.71t$ $-0.32\ 4.93^*$	0.618	24.3**	0.82	13480.5	14258.2	15035.9
Power	$\bar{y} = 1005.51t^{0.8234}$ $32.13^*\ 8.10^*$	0.814	65.6**	0.73	10864.1	11358.7	11848.7
Parabolic	$\bar{y} = 2976.6 - 325.96t + 61.31t^2$ $1.22\ -0.52\ 1.82$	0.691	15.7*	0.88	16975.4	18918.1	20983.4

Source Calculated by authors using [9]

(3) raising the level of financial support of innovation activity and foreign investors attraction;
(4) development of infrastructure for financial, informational, consulting, marketing and other types of innovation support;
(5) the formation of a training system for the innovative processes and projects implementation [12, 20].

7 Research Results

The study solves an important scientific and practical task to further identify key trends and factors of influence on the industrial enterprises innovative development. The main results of the study are as follows.

Realization of innovative development of the country's economy depends on a considerable number of interests and factors, determined first of all by the use of appropriate macrofinancial indicators, which are necessary to provide an adequate comprehensive assessment of the conditions for making managerial decisions, conducting a preliminary analysis of their consequences, and choosing the best scenarios for achieving the set strategic goals. The choice of appropriate indicators ensures the efficiency of state regulation, management of potential risks of macroeconomic development. One of the main prerequisites for the implementation of the economic and fiscal policy of innovation development is clearly defined intermediate and strategic policy objectives, measured by means of separate indicators or their systems, limited by time frame, realistic and justified.

The need for indicators arises at the moment of assessing the current state and identifying economic trends. Since management is a continuous cyclical process, in fact, during one cycle, indicators are used twice: in the process of assessing the situation for the formulation of a plan of measures, forecasting and evaluating the results of the implementation of these measures.

Analyzing the results of studies [1, 19, 21], one can conclude that the effectiveness of innovation development is achieved provided that the prediction depth forecasting and planning increasing for each option are mandatory.

This led to the choice as an analytical tool for this study of trend models and regression analysis. Trend models of selected growth parameters show the dynamics of their growth, and models of regression analysis—the level of their influence on macroeconomic parameters.

The statistical analysis of the dynamics of innovative entrepreneurship indicators in Ukraine and their structure is made possible to distinguish the main directions of innovation activity in Ukraine. The results obtained on the basis of econometric analysis of the dynamics of these indicators reflected growth trends. An adequate level of statistical probability of the obtained trend equations for individual indicators allows us to recommend them for practical application of determining the directions of development of enterprises innovation activity.

Thus, the described analytical tools are the basis of the method of entrepreneurship innovation development strategic monitoring, which involves the implementation of the following stages: the choice of innovation activity indicators to determine their level of influence on macroeconomic indicators; selection of trend models and determination of innovation activity predictive data.

8 Conclusions

As a result of the study, a number of factors hindering the implementation of innovation policy in industrial enterprises were identified, among them the main ones are: lack of state support, outdated material and technical base, lack of financial resources, lack of information provision and lack of incentives for innovation. The indicators of innovation activity having the highest growth dynamics were determined and models of the trend of these indicators were constructed for prediction of the direction of change.

The proposed analytical tools for making strategic decisions based on statistical and econometric analysis allows us to determine the main trends.

References

1. Vyshnevskyy VP, Dement'ev VV (2010) Pochemu Ukrayna ne ynnovatsyonnaya derzhava: ynstytutsyonal'nyy analyz. Zhurnal ynstytutsyonal'nykh yssledovanyy 2(Tom 2):81–17
2. Pavlenko IA (2007) Innovatsiyne pidpryyemnytstvo u transformatsiyniy ekonomitsi Ukrayiny : monohrafiya / Pavlenko IA. – K.: KNEU, 248 s
3. Krugman P (1979) A model of innovation, technology transfer, and the world distribution of income. J Polit Econ 87(2)

4. Баранов Д.Н. Эволюция развития моделей ынновационного процесса[Електроний ресурс]. – Режим доступу. https://www.muiv.ru/vestnik/pdf/eu/eu_2015_3_14_015-020.pdf

5. Kaczor S, Kryvinska N (2013) It is all about services—fundamentals, drivers, and business models. Soc Serv Sci J Serv Sci Res Springer 5(2):125–154

6. Kryvinska N, Gregus M (2014) SOA and its business value in requirements, features, practices and methodologies. Comenius University in Bratislava. ISBN: 9788022337649

7. Kasych AO (2013) Dosvid formuvannya natsional'nykh innovatsiynykh system v krayinakh, shcho rozvyvayut'sya / Kasych AO // Aktual'ni problemy ekonomiky 5(143):46–49

8. Miroshnychenko OYU (2013) Innovatsiyna aktyvnist' promyslovykh pidpryyemstv Ukrayiny: stan i tendentsiyi / Miroshnychenko OYU / Visnyk Kyivs'koho natsional'noho universytetu im. Tarasa Shevchenka. Seriya: Ekonomika. Vyp. 10(151):73–78. http://dx.doi.org/10.17721/1728-2667.2013/151-10/16

9. Ofitsiynyy sayt Derzhavnoyi sluzhby statystyky Ukrayiny. Rozdil Nauka, tekhnolohiyi ta innovatsiyi. [Elektronnyy resurs]. – Rezhym dostupu. http://www.ukrstat.gov.ua

10. Hrechan AP (2006) Osnovy vyznachennya innovatsiynoho rozvytku ekonomiky / Hrechan AP // Ekonomika ta derzhava 8:12–14

11. Molnár E, Molnár R, Kryvinska N, Greguš M (2014) Web Intelligence in practice. Soc Serv Sci J Serv Sci Res Springer 6(1):149–172

12. Poberezhna NM. Analiz faktoriv vplyvu na innovatsiynyy rozvytok promyslovykh pidpryyemstv: praktychnyy aspekt. repository.kpi.kharkov.ua/…/2016_Poberezhna_Analiz_faktoriv_v

13. Rohoza MYE, Verhal KYU (2011) Stratehichnyy innovatsiynyy rozvytok pidpryyemstv: modeli ta mekhanizmy: monohrafiya. RVV PUET, Poltava, 136 s

14. Kryvinska N (2012) Building consistent formal specification for the service enterprise agility foundation. Soc Serv Sci J Serv Sci Res Springer 4(2):235–269

15. Gregus M, Kryvinska N (2015) Service orientation of enterprises—aspects, dimensions, technologies. Comenius University in Bratislava. ISBN: 9788022339780

16. Reustov AYU (2011) Analyz resursnoy, rezul'tatnoy y statystycheskoy komponent ynnovatsyonnoy aktyvnosty orhanyzatsyy // Ynnovatsyy. Ynvestytsyy. № 33. [Elektronyy resurs]. – Rezhym dostupu. www.uecs.ru/component/content/article/65017

17. Yastrems'ka OM (2016) Innovatsiyna diyal'nist' promyslovykh pidpryyemstv: rezul'taty otsinyuvannya / Yastrems'ka OM, Dourtmes PO // Biznes Inform 4:161–168

18. YeleikoVI (2017) Innovatsiina aktyvnist pidpryiemstv Ukrainy: analiz dynamiky / Ieleiko VI, Demchyshyn MIa, Kozhan DV // Visnyk Lvivskoho torhovelno-ekonomichnoho universytetu. Ekonomichni nauky. Vypusk 53. LTEU, Lviv, S. 42–48

19. D'yachkova YUM (2017) Innovatsiyna aktyvnist' promyslovykh pidpryyemstv Ukrayiny v konteksti yikh staloho rozvytku / D'yachkova YUM, Tokareva AO, D'yachkov AM // VISNYK Donbas'koyi derzhavnoyi mashynobudivnoyi akademiyi 1(40):107–115

20. Hryn'ko TV (2011) Sutnist' innovatsiynoyi aktyvnosti pidpryyemstva ta yiyi otsinka / Hryn'ko TV, Yermakova HV // BIZNESINFORM 11:60–64

21. Honcharova NP (2002) Innovatsyonnyy typ razvytyya kak faktor sbalansyrovannosty ékonomyky // Stratehiya ekonomichnoho rozvytku Ukrayiny 1(8):125–132

Determinants of Employment in Information and Communication Technologies and Its Structure

Ján Huňady, Peter Pisár and Peter Balco

Abstract Economic sector of information and communication technologies (ICTs) is one of the most innovative and has several positive consequences on productivity and economic growth. Our main aim was to identify potential determinants affecting the employment in ICT in short-run and long-run. We also examine its structure according to subsectors, gender, age and education in EU countries. We find significant differences between countries in the structure of ICT employment and also indentify several trends. Based on the results there is in general a decreasing trend of women employed as ICT specialists in the EU. Significant drop was recorded especially during economic crisis in 2009–2010. On the other hand the share of ICT specialist with higher education is growing in recent years. Furthermore, we applied correlation, panel Granger causality tests and panel cointegration regression in order to find potential determinants of employment ICT. Our results suggest that there is a positive effect of ICT skills and education on employment in ICT. This is especially true in the case of employment in ICT services. Furthermore, economic situation as well as share of people using internet in the country seems to be both significant determinants of employment in ICT in the long-run.

J. Huňady · P. Pisár
Faculty of Economics, Matej Bel University in Banska Bystrica, Tajovskeho 10, 975 90 Banska Bystrica, Slovakia
e-mail: jan.hunady@umb.sk

P. Pisár
e-mail: peter.pisar@umb.sk

P. Balco (✉)
Faculty of Management UK, Odbojárov 10, P.O. Box 95, Bratislava, Slovakia
e-mail: peter.balco@fm.uniba.sk

© Springer Nature Switzerland AG 2020   277
N. Kryvinska and M. Greguš (eds.), *Data-Centric Business and Applications*,
Lecture Notes on Data Engineering and Communications Technologies 30,
https://doi.org/10.1007/978-3-030-19069-9_11

1 Introduction

The paper focuses its attention on Information and communication technologies (ICT) sector for serval reasons. First of all, ICTs in general have several positive effects on economy. ICTSs have been also recognized as one of the key factors determining innovation, economic growth [1, 2].

There are three main impacts of ICT on economy [3]:

- ICT industries contribute directly to productivity and growth through their own technological progress,
- ICT use improves the productivity of other factors of production,
- there are some spill-over effects'on the rest of the sectors in the economy as ICT.

Perhaps the most important effect of ICT is on productivity and economic growth. The most ICT-intensive industries appeared to experience significantly larger productivity gains than other industries [4].

Furthermore, ICT could affect economic growth both directly or indirectly. As mentioned direct effect of ITC sector on product is evident but there are several other effect.

There are three channels through which ICT can affect economic growth [2]:

- fostering technology diffusion and innovation,
- enhancing the quality of decision-making by firms and households,
- increasing demand and reducing production costs, which together raises the output level.

Innovation drives economic competitiveness and sustained long-term economic growth. Hence, the positive effect of ICT [5]. As stated by Hall et al. [6] research and development (R&D) together with information and communication technology (ICT) investment have been both identified as main sources of relative innovation underperformance in Europe vis-à-vis the USA. In knowledge-based economy, ICT plays important role in many types of innovation which could increase productivity [7, 8]. According to Lall [9] ICT sector is the source of more than 50% of innovation worldwide and has an even greater role in developing countries. Thus, based on mentioned previous findings we can say that ICT sector is one of the key innovation leaders among all sectors of the economy. Hence, higher share of ICT sector on economy could mean higher innovation potential of the country.

In our paper we also examine factors potential determining the share of ICT sector. There are several factors, which could be seen as those having effect on ICT sector and its share. Ngwenyama and Morawczynski [10] argue that the deregulation is not enough for ICT sector expansion. They found that that existing there are several existing socio-economic determinants such as economic development, human capital, geography, and civil infrastructure. Especially these factors have to be taken into account when setting policy frameworks for ICT sector. Moreover, Rai and Kurnia [11] based on the example of Buthan economy found that especially foreign direct investment (FDI), policy, infrastructure and human resources affect the growth of ICT sector.

The chapter aims to examine the share of ICT sector in selected European countries and identify its potential determinants of this share. This research is into some extent based on our previous publication Pisar et al. [12] presented and published at conference INCoS 2018: Advances in Intelligent Networking and Collaborative Systems. This is significantly extended (by more than 50%) and improved version of mentioned previous publication. Among other things we are focus more on analysis of ICT employment structure and improved the methodology using panel cointegration test and panel cointegration regressions.

2 Data and Methodology

In this section we describe the methodology and shortened step-by-step procedure of our analysis. We firstly examine the structure of ICT employment and identify the differences between EU member states. We are focused on differences between two subsectors of ICT namely ICT services and ICT manufacturing [13]. Further we compared gender, education an adage structure of ICT specialist between countries and describe certain trends during seceded period. After this comparison and discussion, we identify its potential determinants and consequences of ICT employment share. We used panel data with cross-sectional and period dimensions. The dataset consists of 28 EU member states plus Norway and Switzerland during the period 2005–2015. Together we get maximum of 330 observations. However, the missing observations pose a problem in the case of several variables. Some of the missing observations were replaced by the nearest available data from the same country. Consecutive missing observations were excluded from the sample during the analysis. In some graphs we used cross-sectional or period dimension separately. All variables used in the analysis are described in Table 1.

Table 2 shows basic descriptive statistics for variables used in the analysis.

We compared cross-sectional data for European countries and based on the scatter plots and Pearson correlations coefficients we examine potential relationships between selected pairs of variables with the focus on the employment in ICT. Further we also used panel Granger causality test in order to identify the significance and direction of potential effect between selected variables. However, this informs us only about the causality in Granger sense in the short-run. It is likely that several determinants of ICT employment are more potent in the long-run. This is especially true with respect to educational outcomes, infrastructure and economic development. There could be some long-run relationship between these factors and employment in ICT. Thus we decided to test this kind of potential relationship using panel cointegration regression approach. We used overall employment in ICT as well as employment in ICT services as dependent variable.

We firstly tested the weak stationary and the order of integration for all variables, which we want to use in the cointegration model. After we managed to satisfactorily demonstrate the same level of integration by unit root tests, we tested for the existence of cointegration by panel cointegration tests. Cointegration between the dependent

Table 1 Short description of variables used in the analysis

Variable	Short description	Source
ICT employment	Total employment in ICT sector (% of active population)	Eurostat
ICT manufact. employment	Employment in ICT manufacturing (% of active population)	Eurostat
ICT services employment	Employment in ICT services (% of active population)	Eurostat
GDP per capita	GDP per capita in PPP	World bank
Internet users	Individuals using internet on population (%)	World bank
ICT education	Persons with ICT education on active population (%)	Eurostat
ICT infrastructure	Proxy variable for ICT infrastructure—secure Internet servers (per 1 million people)	World bank database
FDI inflows	Foreign direct investment, net inflow (% of GDP)	World bank
Openness	Import + Export (as % of GDP)	World bank
Political stability	Index of political stability—world bank worldwide governance indicators	World bank
Regulations quality	Index of regulatory quality—world bank worldwide governance indicators	World bank
R&D expenditure	Intramural R&D expenditure (GERD) as % of GDP	Eurostat
Total tax rate	Total tax rate (% of company profit)	World bank

Source Authors

Table 2 Basic descriptive statistics for selected variables used in the analysis

Variable	Obs.	Mean	Std. Dev.	Min.	Max.
ICT employment	290	2.75	0.85	1.28	4.82
ICT manufact. employment	275	0.49	0.40	0.03	1.89
ICT services employment	314	2.32	0.75	1.01	4.33
ICT infrastructure	330	637.15	673.5	5.44	3101
GDP per capita	330	102.41	44.39	35.00	270.0
Internet users	330	67.97	17.58	19.97	97.3
ICT education	321	1.03	0.45	0.21	2.65
FDI inflows	330	12.54	43.62	−58.32	451.7
Openness	330	118.1	64.66	45.61	410.2
Political stability	330	0.80	0.41	−0.47	1.59
R&D expenditure	330	1.55	0.89	0.37	3.75
Total tax rate	330	42.61	12.93	18.40	76.70

Source Authors

and independent variables has been tested for using panel cointegration tests developed by Pedroni [14], which are widely used in the empirical literature. They are testing the null hypothesis of no cointegration between selected variables. Obviously, the panel cointegration tests allow us to identify the presence of cointegration, but cannot estimate any long-run coefficients. For this purpose we use panel cointegrated regression models. The long run parameters are estimated by the fully modified OLS (FMOLS) and the dynamic OLS (DOLS) panel cointegration estimators. Both types of estimators have been used in their two forms referred to as a pooled estimator and group-mean estimator. While pooled estimators are based on the "within dimension" of the panel, the group-mean estimators are based on the "between dimension of the panel". The pooled FMOLS estimator is proposed in [15]. The pooled DOLS estimator is introduced by Kao and Chiang [16] and the concept of the group-mean estimator is extended from FMOLS to DOLS by Pedroni [17]. Both estimators are robust with respect to the potential problems of serial-correlation and endogeneity, which are potential problems with common OLS panel data estimators. The FMOLS estimator solves this by nonparametric corrections, while the DOLS estimator uses parametric correction, in effect adding leads and lags of differenced regressors into the regression.

3 Results

The main focus of our analysis is on differences in ICT employment size and structure in the EU. Furthermore, our aim is to identify some potential determinants of ICT employment share.

Firstly we compared total employment in ICT and its structure in EU member states. We divided total ICT employment into two main sub-sectors employment in ICT services and employment in ICT manufacturing. The sum of employment in both sub-sectors represents the total employment in ICT in every country expect Cyprus and Luxembourg due to unavailability of data for employment in ICT manufacturing. As it can be seen in the Fig. 1, there are rather significant differences between countries. The highest overall share of employees in ICT is in Sweden followed by Ireland and Luxembourg. Ireland and Luxemburg have both very high employments in ICT services. Together with Malta, Sweden and Denmark are these five countries leaders in this statistic. On the other hand ICT sector is developed significantly less in Greece, Portugal and Romania. It is also evident that employment in ICT services is much higher compared to manufacturing in every EU country.

Furthermore, we also compared the development of employment in both subsectors. We calculated as the average of all EU countries during the selected period selected period from 2005 to 2015. As shown in Fig. 2 the employment in ICT services has significantly rising trend, the employment in ICT manufacturing is gradually decreasing over this period in the EU. Thus available data demonstrate increasing importance of ICT services over manufacturing in the EU.

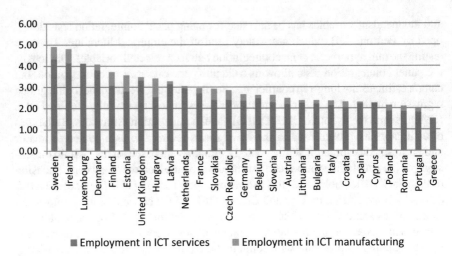

■ Employment in ICT services   ■ Employment in ICT manufacturing

Fig. 1 Total employment in ICT sector and ICT services (% of active population) in European countries in 2015. *Source* Authors based on data from Eurostat database. *Note* The share of employment in ICT manufacturing is not available for Luxembourg and Cyprus. Missing data for Denmark, Ireland and Portugal in year 2015 have been substituted by older values

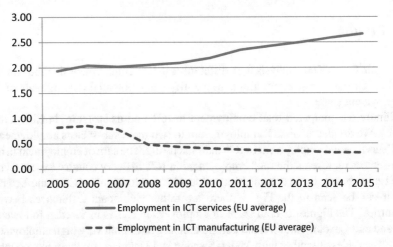

———— Employment in ICT services (EU average)

– – – Employment in ICT manufacturing (EU average)

Fig. 2 The development of total employment in ICT sector and ICT services (% of active population) as EU countries average. *Source* Authors based on data from Eurostat database

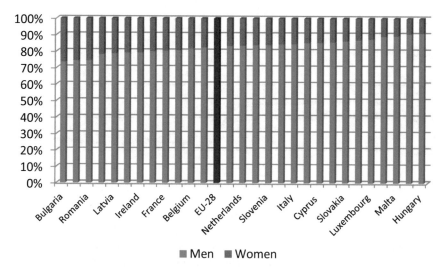

Fig. 3 ICT specialists by gender in EU member states (year 2017). *Source* Authors based on data from Eurostat database

Following part of our analysis is focused more in detail on the gender, educational and age structure of ICT specialist in EU member states. Gender classification can is compared in Fig. 3. As it could be expected, men have significantly larger proportion on total IC specialists in every country.

On one hand, countries such as Bulgaria, Lithuania and Romania reached almost 30% representations of women. On the other hand, the proportion of women in ICT in Hungary and Czech Republic is only about 10% level. The average share of women on total number ICT specialists is approximately 20%.

However, the development of this share over recent years is also interesting for us. Based on data shown in Fig. 4 we can make some conclusions. There is more or less evident downward trend in the employment of female ICT specialists during selected period. Interestingly, there is the very strong drop is evident especially between years 2010 and 2011. This drop is in general especially evident in new EU member states (enlargement in 2004 and after) rather than in 15 original EU states. This could imply the fact that women in ICT were significantly more affected by the financial and economic crisis during 2009–2010. This is especially true for new member states. Hence, it is likely that women are often employed in more vulnerable positions in ICT with respect to effects of economic cycle. In general, we can say that the average share of women employed in ICT appears to be in the EU considerably lower in recent years than ten years ago.

Figure 5 further illustrate the situation in gender structure and its development. There are substantial differences in average annual rate of change in person employed as ICT specialist between countries as well as between men and women. Gender aspect seems to play rather important role here. Employment of male ICT specialist

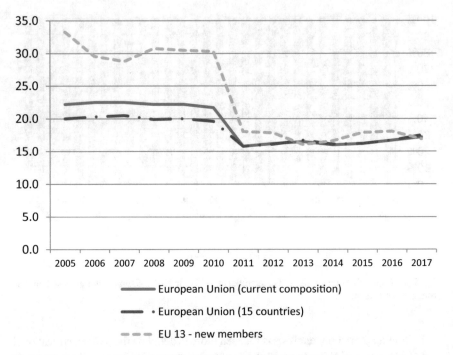

Fig. 4 Share of women on total ICT specialists in the EU (year 2017). *Source* Authors based on data from Eurostat database

is in average growing over selected period in each EU member state. However, the situation is more complicated in the case of women. There are many countries such as Slovakia, Czech Republic and Hungary with substantial drop in employment of women as ICT specialists. However, the situation is different in France, Belgium and Netherlands where the employment of women as ICT specialist is rising even more then in the case of men. The strongest overall growth can be seen in Ireland where employment is increasing almost equally for both sexes.

Next, we focus our attention on educational structure of ICT specialist. This kind of structure in all EU countries can be seen in Fig. 6.

Despite the fact, that this type of work can require a high school education degree, it is not always the case. Interestingly, only less than 40% of ICT specialist in Italy has higher education. On contrary, there is more than 80% share of ICT specialists with higher education degree in countries such as Lithuania, Ireland, Cyprus and Spain. In average the proportion of higher educated ICT specialist is slightly higher than 60% in the year 2017. This share has been increasing trend at least during last 12 years as we can see in Fig. 7. The growth has been higher in new member states, where the share of ICT specialist with higher education rises from less than 50% to more than 65%. However, it is important to mention that the share of employees with higher education seems to have a growing tendency in many other sectors in economy.

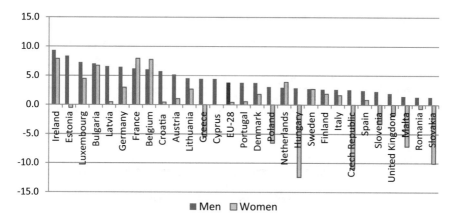

Fig. 5 Average annual rate of change for the number of persons employed as ICT specialists by sex, 2007–2017

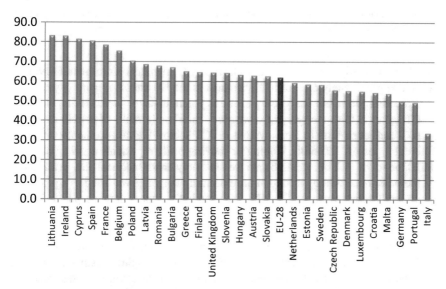

Fig. 6 The share of ICT specialist with tertiary education (%) in 2017. *Source* Authors based on data from Eurostat database

As well as in the case of education there also appears to be differences in the age structure of ICT specialists between EU countries. This is graphically illustrated in Fig. 8. We distinguish only between two age groups of ICT specialists. First group include those which are between 15 and 35 years old. This group represents young employees. Second group are those between 35–74 years. As we can see younger ICT specialist make up a larger group in Malta, Latvia, Lithuania and Estonia. However, in the most countries specialists older than 35 years represents significant majority.

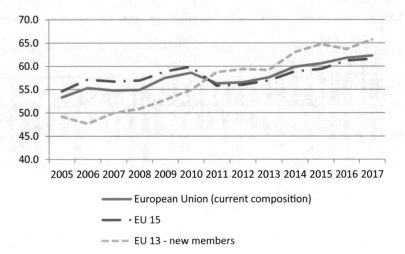

Fig. 7 Share of ICT specialist with tertiary education (%). *Source* Authors based on data from Eurostat database

This is particularly true in Italy and all three EU members from Scandinavia. In average approximately 65% of ICT specialist in the EU are older than 35 years.

In the next part of our analysis we examine potential factors determining the share of ICT sector in the country. Firstly, we are focused on IT skills of population in the country and its relation to Employment in ICT sector. We take to the account two different types of ICT skills. Coping or moving a file represents a rather basic IT skill. Due to the lack of available data we used only cross-sectional data in this case. On the other hand writing a code in programming language was used as a proxy for significantly more advanced IT skills. We used both of them in two separate Scatter plots on X axis together with the employment in ICT sector on Y axis (see Figs. 9 and 10).

As it can be seen basic IT skills are the most developed in population of Denmark followed by Latvia and Norway. There seems to be slightly positive correlation between basic IT skills and employment in ICT sector in the country. Similar relationship is also evident in the case of more advanced IT skills as it can be seen in Fig. 9. However, in this case the slope of the fitted line appears to be mostly affected by two top performing countries which are Finland and Sweden. Slovakia is approximately at the average of all countries in the case of basic IT skills, but performs bellow the average in advanced IT skills. Thus, it skills of population could be potentially important factor determining the share of ICT sector in the country. On the other hand, it is also possible that higher employment in ICT causes better IT skills in the population.

We also take to the account other potential determinants besides IT skills. One of them is ICT education, which is of course closely related to IT skills. We can assume that higher share of population with ICT education is related to higher share of ICT

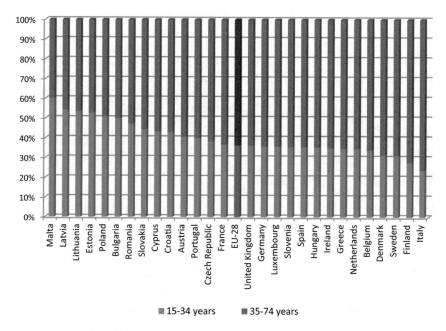

Fig. 8 The structure of person employed as ICT specialists according to age groups. *Source* Authors based on data from Eurostat database

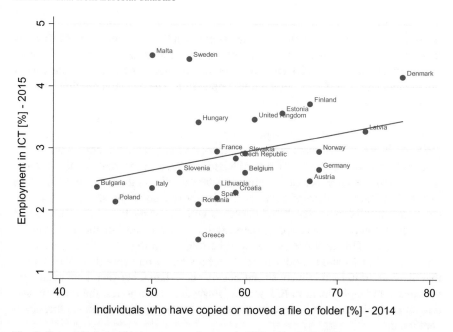

Fig. 9 Potential relationship between employment in ICT and basic IT skills. *Source* Authors based on data from Eurostat database

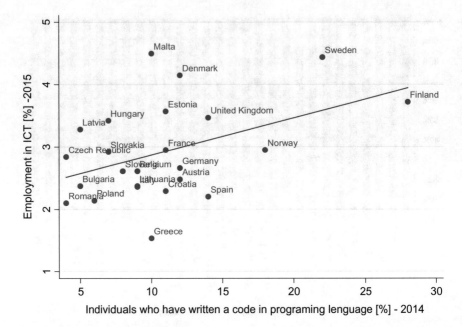

Fig. 10 Potential relationship between employment in ICT and advanced IT skills. *Source* Authors based on data from Eurostat database

sector in the country. On one hand, more people with ICT education may lead to new domestic firms in this sector as well as attract more foreign ICT firms. On the other hand, higher demand for ICT employees on labour market may cause in the longer run the increase in ICT education within population. The potential relation between these two variables is shown in Fig. 11.

This time we are able to use panel data. This means that each dot on the graphs represents the observation for single European country and single year from 2005 to 2015. As it can be seen there seems to be a positive correlation in both cases.

Finally, we selected another two indicators related to external economic environment, namely GDP per capita and regulatory quality in the country. We believe that both of these indicators could be positively related to the share of ICT sector. Economically more developed countries with better quality of institutions and regulations could represent appropriate external environment for the development of ICT sector. This assumption seems to be true according to our empirical data. Figures 12 and 13 show motioned relationships based on panel data. We can see that there seem to be relatively strong positive correlation between both economic variables and employment in ICT sector. However in order to test the intensity and statistical significance of these correlations we further applied Pearson correlation coefficients.

The results of Pearson correlation coefficients together with their statistical significance are shown in Table 3.

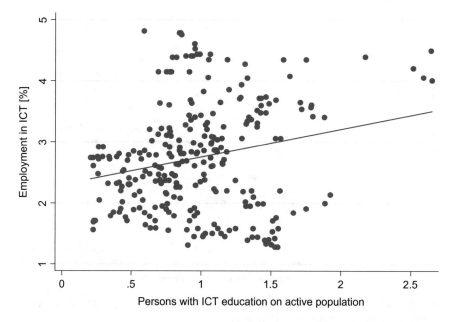

Fig. 11 Potential relationship between the share of population with ICT education (left), R&D expenditure (right) and employment in ICT. *Source* Authors based on data from Eurostat database

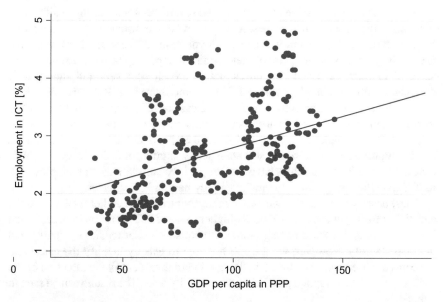

Fig. 12 Potential relationship between GDP per capita and employment in ICT sector. *Source* Authors based on data from Eurostat database

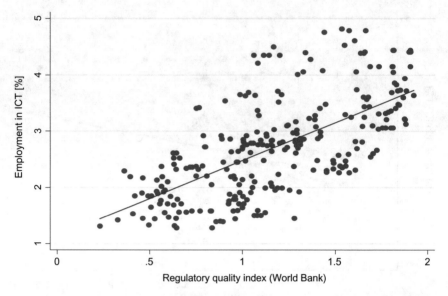

Fig. 13 Potential relationship between GDP per capita and employment in ICT sector. *Source* Authors based on data from Eurostat database

As we can see in this table there seem to be several relatively strong positive correlations between selected variables. As expected there is strong positive correlation between total employment in ICT and employment in ICT services. However, what is more important from our point of view is that GDP per capita, R&D expenditures, share of internet users and ICT infrastructure are all positively and significantly correlated with both employment in ICT and employment in ICT services. These relationships appear to be even stronger in the case of ICT service rather than in the total ICT employment. The positive correlations between political stability, regulatory quality and trade openness on one hand and employment in ICT on another hand are all slightly less intensive but still statistically significant at 5% level of significance. All mentioned variables have weaker positive correlations with employment in ICT manufacturing, rather than with employment in ICT services.

Thus, based on the results, we can say that countries with higher GDP per capita and higher R&D per capita have mostly higher share of ICT sector on total economy. The same is true for those having more internet users, better regulatory quality and political stability. However, we are still not able to exactly identify the cause and the consequence based on this methodology. In order to identify the direction of the causality we further applied panel Granger causality tests. The results can be seen in Table 4.

Despite our main aim is to identify the determinants of ICT employment, based on Granger causality we are able to make some conclusions also about potential consequences. With respect to potential consequences of ICT sector we can say that employment in ICT services appears to have significant effect on GDP per capita

Table 3 Results of Pearson correlation coefficients for selected variables (panel data)

	Employment in sectors		
	ICT total	ICT manufacturing	ICT services
FDI inflows	0.26***	0.14**	0.19***
GDP per capita	0.44***	0.01	0.70***
ICT education	0.23***	−0.05	0.24***
ICT employment	1.00***	0.56***	0.87***
ICT infrastructure	0.51***	−0.12**	0.64***
ICT manufacturing employment	0.56***	1.00***	0.15***
ICT services employment	0.87***	0.15**	1.00***
Internet users	0.58***	0.03	0.74***
Openness	0.37***	0.28***	0.42***
Political stability	0.55***	0.42***	0.47***
Regulations quality	0.56**	0.25***	0.55***
R&D expenditure	0.67**	0.22***	0.72***
Total tax rate	0.04	0.23***	−0.22**

Source Authors calculations

Note */**/*** means significance at the 10%/5%/1% levels, %

Table 4 Results of panel Granger causality tests

	F-Statistic (2 lags)	F-Statistic (3 lags)
ΔICT employment does not Granger Cause ΔGDP per capita	1.73	1.07
ΔGDP per capita does not Granger Cause ΔICT employment	0.99	1.95
ΔICT services employment does not Granger Cause ΔGDP per capita	2.77*	7.56***

(continued)

Table 4 (continued)

	F-Statistic (2 lags)	F-Statistic (3 lags)
ΔGDP per capita does not Granger Cause ΔICT services employment	0.72	0.64
ΔICT employment does not Granger Cause ΔICT education	0.58	0.41
ΔICT education does not Granger Cause ΔICT employment	2.48*	3.21**
ΔICT manufact. employment does not Granger Cause ΔICT education	0.01	0.03
ΔICT education does not Granger Cause ΔICT manufact. employment	1.07	0.72
ΔICT employment does not Granger Cause ΔICT infrastructure	6.22***	0.24
ΔICT infrastructure does not Granger Cause ΔICT employment	0.91	0.21
ΔICT employment does not Granger Cause ΔInternet users	1.03	1.11
ΔInternet users does not Granger Cause ΔICT employment	1.14	0.03
ΔICT manufact. employment does not Granger Cause ΔInternet users	0.76	4.93**
ΔInternet users does not Granger Cause ICT manufact. employment	8.95***	2.08

(continued)

Table 4 (continued)

	F-Statistic (2 lags)	F-Statistic (3 lags)
ΔICT employment does not Granger Cause ΔOpenness	19.40***	13.01***
ΔOpenness does not Granger Cause ΔICT employment	3.41**	0.89
ΔICT manufact. employment does not Granger cause ΔOpenness	31.43***	26.36***
ΔOpenness does not Granger cause ΔICT manufact. employment	0.55	0.11

Source Authors calculations

Note */**/*** means significance at the 10%/5%/1% levels, %. Numbers of lag was selected based on the results of Akaike and Schwarz criterion

and as far as we know from our previous analysis this effect spouse to be positive. However, in the case of total ICT employment we fail to prove statistically significant effect on GDP by using the Granger causality test. On the other hand, we found positive and significant effect of ICT employment on ICT infrastructure measured by number of servers per inhabitant. This is of course in line with our expectation, because higher share of ICT sector could be related to better ICT infrastructure. Moreover, there is positive and significant effect of both total ICT employment and employment in ICT manufacturing on trend openness of the economy. This effect is highly significant on both cases, but it is even more intensive for employment in ICT manufacturing. Thus we can say that countries higher share of ICT sector in the economy could positively affect the GDP per capita as well as increase the trade openness of the economy. First may be due to higher productivity and innovation in this sector and second could be the consequences of potentially higher import needs and more export orientation of firms in this sector. There is also some slightly weaker evidence for potential bi-directional effect of internet users and employment in ICT manufacturing.

With respect to potential determinants of ICT sector employment there is also some evidence that higher openness of the economy could have also some backward positive effect on the ICT employment in the country. We can also conclude, that share of people with ICT education may be the determining factor. Hence, education in ICT field of study seems to be important for increasing the employment in this sector, which could be due to either attracting foreign ICT companies or increasing the number of home-grown enterprises.

In the final part of our analysis we are focused on potential determinants of employment in ICT in the long-run. In order to identify potential long-run relationships we applied panel cointegration tests and panel cointegration regression as already described in the methodological section. Firstly we used total ICT employment as dependent variable and try to examine long-run effect of selected variables. Secondly, we changed this variable for employment in ICT services mostly in order to check for robustness. We assume that economic situation proxied by GDP per capita, ICT infrastructure proxied by share of internet users and ICT education proxied by persons with ICT education in population could all have some effect on ICT employment and ICT employment in services in the long-run. Results of panel cointegration tests are shown in the Table 5.

We used both tests using within dimension as well as between dimension approaches. Despite somewhat mixed results, tests mostly suggest that selected variables are cointegrated in both cases. Hence, we can further perform two cointegration

Table 5 Results of Pedroni panel cointegration test

Cointegration: ICT_employment, GDP, InternetUsers, ICT_education

		Statistic	Weighted Stat.
Pedroni test (Engle-Granger based) tests Automatic lag length selection based on AIC Null hypothesis: no cointegration	Panel v-Statistic (within dimension)	−0.57	−1.45
	Panel rho-Statistic (within dimension)	2.34	2.39
	Panel PP-Statistic (within dimension)	−2.73***	−3.84***
	Panel ADF-Statistic (within dimension)	−3.14***	−3.98***
	Group rho-Statistic (between dimension)	4.59	
	Group PP-Statistic (between dimension)	−6.38***	
	Group ADF-Stat. (between dimension)	−5.37***	

Cointegration: ICT services employment, GDP, InternetUsers, ICT_education

		Statistic	Weighted Stat.
Pedroni test (Engle-Granger based) tests Automatic lag length selection based on AIC Null hypothesis: no cointegration	Panel v-Statistic	−0.32	−1.29
	Panel rho-Statistic	1.90	1.59
	Panel PP-Statistic	−3.71***	−6.54***
	Panel ADF-Statistic	−4.93***	−7.38***
	Group rho-Statistic	3.85	
	Group PP-Statistic	−10.30***	
	Group ADF-Statistic	−8.71***	

Source Authors calculations

Note */**/*** means significance at the 10%/5%/1% levels, %

regression analyses using the same variables. Based on panel dynamic OLS (DOLS) as well as fully modified OLS (FMOLS) approach we further test potential long-run causalities between selected variables. First group of regressions shown in Table 6 represents long-run effects of three independent variables on total ICT employment. In all three models we included constant and select optimal number of leads and lags included into regression according to the value of Akanke criterion, which is common procedure.

Variables capturing GDP per capita and share of internet users appear to have both positive a statistically significant effect at 1% level in the long-run. ICT education is statistically significant only in group-mean DOLS model. Hence, we find some empirical evidence for the effect of economic situation and ICT infrastructure on employment in ICT in the long-run. Problems with potential endogeneity were eliminated by using DOLS and FMOLS estimators as mentioned in the methodology section. The significance of potential effect of ICT education could be questionable based on the results.

Further, we substitute the overall employment in ICT by employment in ICT services only. We again used same specifications of regression models as in previous case. Results are similar as in previous case as it can be seen in Fig. 7. Hence, there are several differences. This time ICT education appears to be statistically significant and positive in all three models. Thus, more people with ICT education could have positive effect on employment in ICT services in the long-run. Thus, the education in ICT is more important factor with respect to employment in ICT services than in the case of general ICT sector capturing also ICT manufacturing. This could be interpreted also by the fact that employment in ICT services requires in general

Table 6 Long-run effects on total ICT employment based on the results of DOLS and FMOLS regressions

Dependent variable: Total ICT Employment (% of total employment)			
	DOLS—group-mean estimator (constant included, automatic leads and lags based on AIC)	DOLS—pooled estimator (constant, automatic leads and lags based on AIC)	FMOLS—pooled estimator (constant included, automatic leads and lags based on AIC)
GDP	0.014*** (5.31)	0.016*** (4.14)	0.014*** (7.44)
Internet users	0.002*** (1.13)	0.007*** (2.93)	0.004*** (3.62)
ICT education	0.190*** (3.04)	0.107 (0.984)	0.02 (0.43)
Log-run variance	0.031	0.031	
Total panel observations	250	250	250

Source Authors calculations

Note */**/*** means significance at the 10%/5%/1% levels, %

Table 7 Long-run effects on employment in ICT services based on the results of DOLS and FMOLS regressions

Dependent variable: ICT services employment (% total employment)			
	DOLS—group-mean estimator (constant included, automatic leads and lags based on AIC)	DOLS—pooled estimator (constant, automatic leads and lags based on AIC)	FMOLS—pooled estimator (constant included, automatic leads and lags based on AIC)
GDP	0.003 (0.50)	0.013*** (4.94)	0.013*** (4.89)
Internet users	0.018*** (5.45)	0.02*** (11.43)	0.002*** (8.76)
ICT education	0.513*** (4.59)	0.41*** (4.40)	0.327*** (3.67)
Log-run variance	0.004	0.02	
Total panel observations	277	277	277

Source Authors calculations

Note */**/*** means significance at the 10%/5%/1% levels, %

higher level of education. GDP per capita is not statistically significant in group-mean DOLS regression. Despite this fact all other results are very similar to previous ones (Table 7).

Based on our results we can say that we found empirical evidence that macroeconomic situation proxied by GDP per capita as well as number of internet users are potential determinants of employment in ICT in the long-run. Similar effect could be also present in the case of share of people with education in ICT. However, this effect is more significant with respect to employment in ICT services rather than general employment in ICT sector.

4 Conclusions and Policy Implications

Companies in ICT sector are often seen as the most innovative ones [18–20]. The performance of this sector therefore has significant effect one of the economies in general. For example it can accelerate the productivity and economic growth by several ways. Higher share of ICT sector could therefore be beneficial for a country with respect to increasing its innovation potential. Hence, we in detail examine the share and structure of this sector in EU countries. The relative size of ICT sector was in our case proxied by its employment.

We examine and compared the employment in ICT services and ICT manufacturing in European countries. We identified the countries with highest overall share of ICT sector as well as those with higher proportion of employment in ICT services.

Moreover, we also compared the structure of persons employed as ICT specialists according to their gender, education and age. The average share of female ICT specialists is approximately 20% in the EU. However, there are several countries such as Bulgaria, Lithuania and Romania with much higher proportion of women. We can also say that the share of women employed as ICT specialists are decreasing in the EU during the period 2005–2017. On the other hand, the share of ICT specialists with higher education is steadily growing during the same period. This is especially true for new EU member states which joined EU after the year 2004. The negative effect of financial and economic crisis is particularly evident in the case of women employed in ICT.

Furthermore, we are focused on examining the potential determinants of the relative size of ICT sector in the country. Based on the correlation analysis, we found some positive correlation between ICT skills and employment in ICT sector in the country. Based on the panel data analysis we further found that the employment in ICT sector is significantly positively correlated also with several other country characteristics such as its GDP per capita, trade openness, R&D expenditure, number of internet users, number of servers per inhabitant, quality of regulation and political stability. Policies dealing with support of ICT sector should take to the account these findings. However, based on correlation analysis we were of course not able to identify casual relationships and its direction. Hence, we also applied panel Granger causality test in order to find potential causalities in the short-run. However, we also expected that some effect of potential determinants could be more present in the long-run. Hence, we also used panel cointegration regression to test assumed potential causalities in the long-run. We found relatively strong evidence for the positive long-run effect of GDP per capita and the share of internet users on total employment in ICT. Furthermore, we also find positive effect of education in ICT on employment in ICT services.

Based on our results we can make also several policy implications. First of all, education in ICT appears to be important determinant of employment in ICT and this is particularly true for ICT services. In order to support the level of employment in ICT sector as the sector with high added value, policies must first focus on ICT education. Financial or non-financial support aimed to increase the share of students in ICT could be one of the first and important steps. This is relevant especially for higher education with respect to trend of rising share of employees with higher education in ICT. On the other hand, mostly due to decrease during economic crisis, the share of women in ICT sector is in the most EU countries significantly lower than ten years ago. Perhaps, public policies at national and the EU level should try to deal with this issue as well. We have also found that ICT infrastructure, ICT skills and the share of people using internet could represent another important determinants for further development of ICT sectors. Hence, policies focused on development of ICT infrastructure and improving ICT skills in the population could be also seen as important support for increasing the employment in ICT sector.

Acknowledgements This work has been supported by the Scientific Grant Agency of Slovak Republic under project *VEGA No.* 1/1009/16 "Innovation potential of the regions of Slovakia, its measurement and innovation policy at the regional level".

References

1. Bauer JM, Shim W (2012) Regulation and innovation in telecommunications. Quello Center Working Paper 01-2012. Presented at scientific seminar entitled communications & media markets: emerging trends & policy issues, pp 1–26
2. Pohjola M (2003) The adoption and diffusion of ICT across countries: patterns and determinants. The new economy handbook, pp 77–100
3. Doucek P (2010) Human resources in ICT–ICT effects on GDP. In: IDIMT-2010: information technology–human values, innovation and economy, pp 97–105
4. Stiroh KJ (2002) Information technology and the US productivity revival: what do the industry data say? Am Econ Rev 92(5):1559–1576
5. Cardona M, Kretschmer T, Strobel T (2013) ICT and productivity: conclusions from the empirical literature. Inf Econ Policy 25(3):109–125
6. Hall BH, Lotti F, Mairesse J (2013) Evidence on the impact of R&D and ICT investments on innovation and productivity in Italian firms. Econ Innov New Technol 22(3):300–328
7. Brynjolfsson E, Saunders A (2010) Wired for innovation. How information technology is reshaping the economy, Cambridge, Massachusetts-London
8. Brynjolfsson E (2011) Innovation and the E-economy. Unpublished paper. MIT, Sloan School of Management
9. Lall S (1996) Learning from the Asian Tigers: studies in technology and industrial policy. Springer
10. Ngwenyama O, Morawczynski O (2009) Factors affecting ICT expansion in emerging economies: an analysis of ICT infrastructure expansion in five Latin American countries. Inf Technol Dev 15(4):237–258
11. Rai D, Kurnia S (2017) Factors affecting the growth of the ICT industry: the case of Bhutan. In: International conference on social implications of computers in developing countries. Springer, Cham, pp 728–739
12. Pisar P, Hunady J, Balco P (2018) Employment in information and communication technologies in European countries and its potential determinants and consequences. In: Xhafa F, Barolli L, Greguš M (eds) Advances in intelligent networking and collaborative systems. INCoS 2018. Lecture notes on data engineering and communications technologies, vol 23. Springer, Cham
13. Kaczor S, Kryvinska N (2013) It is all about services—fundamentals, drivers, and business models. Soc Serv Sci J Serv Sci Res Springer 5(2):125–154
14. Pedroni P (2004) Panel cointegration: asymptotic and finite sample properties of pooled time series tests with an application to the PPP hypothesis. Econ Theory 20(3):597–625
15. Phillips PCB, Moon HR (1999) Linear regression limit theory for nonstationary panel data. Econometrica 67:67-1057-111
16. Kao C (1999) Spurious regression and residual based tests for cointegration in panel data. J Econ 90(1):1–44
17. Pedroni P (2001) Purchasing power parity tests in cointegrated panels. Rev Econ Stat 83:727–731
18. Gregus M, Kryvinska N (2015) Service orientation of enterprises—aspects, dimensions, technologies. Comenius University in Bratislava. ISBN: 9788022339780
19. Kryvinska N (2012) Building consistent formal specification for the service enterprise agility foundation. Soc Serv Sci J Serv Sci Res Springer 4(2):235–269
20. Kryvinska N, Gregus M (2014) SOA and its business value in requirements, features, practices and methodologies. Comenius University in Bratislava. ISBN: 9788022337649

Simulation Model of Planning Financial and Economic Indicators of an Enterprise on the Basis of Business Model Formalization

Stepan Vorobec, Vasyl Kozyk, Olena Zahoretska and Viktoria Masuk

Abstract The methodical and applied aspects of the simulation model implementation to the system of planned financial and economic indicators of the enterprise are considered. It is performed with a usage of a formal description of its main business processes. By means of both classical economic-statistical methods and the newest methods of analysis, in particular neural network technologies, the dynamics of key factors of the simulation model are studied, which are crucial for the formation of metrics of such planning indicators. Possibility of implementing invariant scenarios for the development of financial and economic situation with a strict identification of the system of financial and economic indicators and the dynamics of their metrics on a certain horizon of planning is demonstrated.

Keywords Business process · Value added chain · Financial and economic indicators · Neural network technologies · Network profile · Modeling notation · Efficiency booster · Simulation model · Transact · Model variables

1 Introduction

The aim of the study is to analyze the methodical and applied aspects of the system metrics for planned financial and economic indicators of the enterprise activities based on a formal description of its key business processes, with their subsequent transformation into an imitation model, while investigating the dynamics of key factors of influence on them using both classical economic and statistical methods of neural network technologies.

S. Vorobec · V. Kozyk · O. Zahoretska (✉) · V. Masuk
Lviv Polytechnic National University, Bandera Street, 12, Lviv, Ukraine
e-mail: zagoreckao@gmail.com

S. Vorobec
e-mail: stepan.y.vorobets@lpnu.ua

V. Kozyk
e-mail: Vasyl.V.Kozyk@lpnu.ua

© Springer Nature Switzerland AG 2020
N. Kryvinska and M. Greguš (eds.), *Data-Centric Business and Applications*,
Lecture Notes on Data Engineering and Communications Technologies 30,
https://doi.org/10.1007/978-3-030-19069-9_12

Setting goals. The study of methodical and applied aspects of the system metrics planned financial and economic indicators of the enterprise determines the following objectives:

- consideration of methodological peculiarities of constructing a process model of the creation of value added chain in an enterprise;
- the substantiation of the methodology for studying the dynamics of key factors of influence on the financial and economic indicators of the enterprise's activity, with their subsequent use in the simulation model;
- applied aspects of the implementation of the simulation model and the peculiarities of its use in controlling systems, in particular, the planning of the enterprise.

2   Presenting of the Main Material

The construction of effective control systems in a context of fierce competition for most Ukrainian enterprises is possible only on the basis of strict formalization of their business model, in which the system of key business processes is clearly identified on the principles of dominance and hierarchical subordination, a system of indicators of efficiency of their implementation is defined. All this can be realized on the basis of a process approach to the enterprise management system [1]. Achievement of desired target indicators of the enterprise activity is possible only in case of building an optimal integral chain of the whole system—basic, managerial, investment or auxiliary business processes. The significance of the process approach to enterprise management is growing substantially as business complexity grows, its scale [2, 3]. The implementation of a process-oriented approach to enterprise management will help to streamline its business processes, make them transparent and manageable [4, 5]. An important link in such a management system is given to the system of control, the basis of which is the system of planning target indicators of the enterprise. Many factors that determine the performance of enterprises to date can be considered as random variables. In such conditions, the construction of an effective system of planning target indicators of the enterprise requires the use of effective methods of forecasting the above factors. In view of all this, we propose a methodology for calculating planned financial and economic indicators of activity, the logic of which is presented in Fig. 1.

We share the opinion of most scholars in the field of economics and management that achievement of the desired target indicators of enterprise activity is possible only in case of building an optimal integral chain of the whole system—basic, managerial, investment or auxiliary business processes. In [4, 6, 7] gives the following definition of the business process—"the process is a stable and focused set of interrelated activities, which according to a certain technology transforms inputs into outputs that are of value to the consumer." This definition of the business process, as noted in [8–10], is a reflection of the ISO 9000: 2000 standard and is a rather generalized notion.

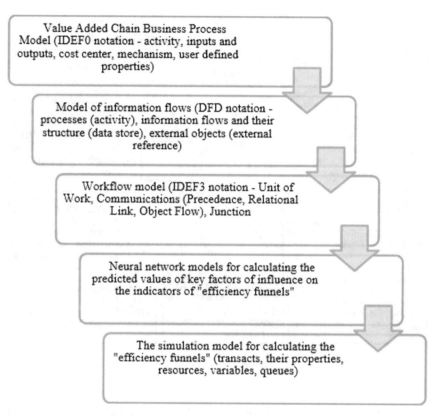

Fig. 1 The logic of the implementation of the system of planning indicators of "efficiency" of the enterprise. *Note* Developed by authors

Formalized description of the key value chain processes for the company under study is performed using the Case tool, namely All Fusion Process Modeler, which supports three modeling notations—IDEF0, IDEF3 and DFD. The process model implemented in the system All Fusion Process Modeler using the notation IDEF0, in which the subject domain is presented as a set of interrelated works or functions. Means of IDEF0—notations in the functional model determine the objects or information that is the raw material for the process to get the results of the process that it serves as the control factors for it and what resources are used in this. A complete formalized representation of the business process involves identifying the clients of the process, the objectives of its implementation, the results of the process, its suppliers, process resources, executors, process owners, indicators, which assesses the effectiveness of its implementation, as well as the content and structure [11–14]. Such a business model represented by a system of integrated business processes is the basis for elaboration and substantiation of basic decisions on management of production, economic, financial or investment activity of the enterprise.

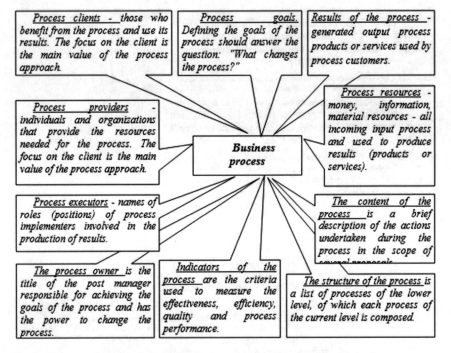

Fig. 2 Structure of the description of the business process. *Note* Developed by authors

As a result, we obtain a scheme of processes that can be called a map of processes. The process map gives a general look at the main processes of the enterprise and allows you to see what the activities of the enterprise constitute, which structure of the chain of processes, which are aimed at customer service, etc. (Fig. 2).

All further work on creating a process management system is aimed at deepening, detailing and improving this vision.

To date, the market of software products is represented by a sufficiently large number of CASE-systems that allow a formal description of business processes of the enterprise, through the use of relevant notations (simulation constructs). As a software implementation tool, the All Fusion Process Modeler, which supports three methodologies—IDEF0, IDEF3 and DFD, each of which solves its specific tasks, has been used. The fragment of the leader of the hierarchical three-level business process model implemented in the IDEF0 notation is presented in Fig. 3. At the same time at the level of the context diagram strictly defined subject modeling, goals and point of view on the model. As a subject, a system is defined that combines the processes of production and financial and economic activity of the enterprise [3].

Such a formalized representation of the business model at the IDEF0 level—the modeling notation makes it possible to assess their effectiveness using the Activity Based Costing (ABC) tools and the evaluation of User Defined Properties (UDP) system performance targets. Functional evaluation—ABC—is a method of studying

Fig. 3 Window of the model leader, representing the hierarchical structure of processes

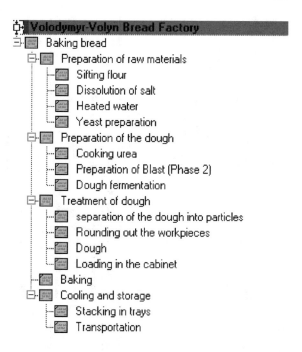

the cost of performing a function (action). The source for functional evaluation is the cost of resources (materials, personnel, etc.). In the framework of cost analysis, the following tasks are solved: the definition of the real cost of production of a product, the cost of marketing costs, the identification of the most expensive works, etc. The partial results of the implemented functional-cost domain analysis on the basis of the developed business model by means of All Fusion Process Modeler are presented in the Table 1.

Considering the subsystem of planning in the system controlling the enterprise, it should be emphasized that it is to a significant extent a sequence of processes of work with information flows. If we consider directly the subsystem of management processes, then to a certain extent, they represent work with information flows.

As a complement to the IDEF0 notation for modeling information flows that accompany the core processes of the value-added chain, the DFD notation is used. At the level of this notation, the following constructs of modeling are used: processes of information processing (activity), data warehousing (data store), external references to objects that are outside the model (external references) and which model their connection with information processes. But the very logic of the interaction of information flows is described using the notation of IDEF3 (workflow diagramming).

The means of this notation simulate the relationships between processes of information processing and objects that are part of these processes. IDEF3 complements the notation IDEF0 and is an intermediate link when constructing an imitation model for calculating predictive values of key financial and economic indicators, on the basis of which can be further used for simulation analysis.

Table 1 Partial results of functional and cost analysis based on the formalized business model

Activity						
Name	Number	Cost	Duration	Frequency	Parent	Children
Baking	A4	390.00	1.30	1.00	Baking bread	None
Baking bread	A0	6,670.00	15.25	1.00	None	Preparation of raw materials Preparation of the dough Treatment of dough Baking Cooling and storage
Cooling and storage	A5	380.00	1.85	1.00	Baking bread	Stacking in trays Transportation
Dissolution of salt	A12	65.00	1.20	1.00	Preparation of raw materials	None
Dough	A33	340.00	1.30	1.00	Treatment of dough	None
Dough fermentation	A23	120.00	1.20	1.00	Preparation of the dough	None
Heated water	A13	40.00	0.50	1.00	Preparation of raw materials	None
Loading in the cabinet	A34	250.00	0.00	1.00	Treatment of dough	None
Preparation of Blast (Phase 2)	A22	430.00	1.50	1.00	Preparation of the dough	None
Preparation of raw materials	A1	4,155.00	5.20	1.00	Baking bread	Sifting flour Dissolution of salt Heated water Yeast preparation
Preparation of the dough	A2	880.00	4.40	1.00	Baking bread	Cooking urea Preparation of Blast (Phase 2) Dough fermentation

(continued)

Table 1 (continued)

Activity						
Name	Number	Cost	Duration	Frequency	Parent	Children
Rounding out the workpieces	A32	135.00	0.00	1.00	Treatment of dough	None
separation of the dough into particles	A31	140.00	1.20	1.00	Treatment of dough	None
Sifting flour	A11	3,900.00	2.00	1.00	Preparation of raw materials	None
Stacking in trays	A51	300.00	1.15	1.00	Cooling and storage	None
Transportation	A52	80.00	0.70	1.00	Cooling and storage	None
Treatment of dough	A3	865.00	2.50	1.00	Baking bread	separation of the dough into particles Rounding out the workpieces Dough Loading in the cabinet
Yeast preparation	A14	150.00	1.50	1.00	Preparation of raw materials	None
Cooking urea	A21	330.00	1.70	1.00	Preparation of the dough	None

The functional value chain of value creation created in the IDEF0, DFD, and IDEF3 annotations with its defined target values, set in the form of its User Defined Properties (UDP) parameters, is transformed into an imitation model to represent the simulation system in the environment—Arena Rockwell Automation. The key advantage of simulation models is that the logic of process implementation takes into account the time factor. Such models allow to "lose" the processes in time and the results of their implementation, to get the meaning of their characteristics, with the assumption that this occurs in reality.

The essence of imitation modeling is the computer realization of the mathematical model of the investigated system for its use in order to simulate the behavior of the real system [15]. Among the numerous simulation methods one should note exclusively imitative modeling as the most important direction in the study of the dynamics of complex socio-economic systems on the basis of their prototyping [11]. In a simulation model, process and data changes are associated with events. Since the enterprise itself is considered as a complex socio-economic system, the construction of its simulation model allows obtaining practically meaningful results of its financial and economic activity at the appropriate time horizon. Realizing this using other simulation methods is virtually impossible. A significant advantage of simulation techniques is also the ability to display feedback loops in models and to take into account the interactions between individual parameters of the system. And this is extremely important when constructing models of such socio-economic systems as an enterprise. Simulation simulations allow us to investigate such behavior of complex systems, using the possibilities of computer simulation [16, 17].

Reproduction of the main processes of the chain of creation of added value on a computer (simulation) allows you to investigate the state of the system and its individual elements at certain moments of model time. The model time is the time that simulates the time of the real system. The development of a system-dynamic model of financial and economic activity of the enterprise under investigation involves the development of the following sequence of stages. At the same time, some of them have already been implemented at the presentation level of the process modem in the notices IDEF0, IDEF3:

1. Identification of causal relationships in the system being modeled; by the results of the analysis of the data of the system of managerial accounting (the definition of the system of financial and economic indicators, the relationships between them and the factors that affect them);
2. Cognitive simulation—development of a causal relationship card;
3. Formalized representation of the mathematical model of relations of financial and economic indicators in the form of a dynamic system of simultaneous equations;
4. Analytical representation of the dynamics of key factors that affect the system of controlled financial and economic indicators, using statistical and statistical methods with the means of Statistical for Windows;
5. Realization of the mathematical model on the platform of simulation, supporting the methods of system dynamics (in our case Arena);

6. Conduct numerical experiments. Scaling the model for its use on different horizons of management;
7. Verification of the model on retrospective data of managerial accounting (check it on adequacy);
8. Finding of optimal managerial decisions with the help of system dynamical simulation model.

Partially separate elements of such logic have already been implemented at the level of representation of the process model in the notes IDEF0, DFD, IDEF3. Implementation of the simulation model enables to study the behavior of complex systems at the same time taking into account its key causal relationships. Using the system-dynamic approach in simulation modeling provides the following advantages over other methods that are widely represented in the scientific literature [18]:

- modeling of various scenarios—calculation of various variants of promising indicators of enterprise activity when changing input data (in the first place—prices for raw materials, market prices for products of the enterprise, actual volumes of procurement of products by the end consumers, the risk of untimely payment for the delivered products, etc.);
- identification of the most critical factors of influence on the production and financial and economic activity of the investigated enterprise, followed by their ranking on the degree of threats of their onset. To a large extent, for the subject area, this refers to the equipment, its failure, non-compliance with the contractual discipline by the main suppliers, etc.);
- interactivity of the simulation model through the visualization of the course of events throughout the simulation time interval and the simultaneous visibility of the data entry into the model and obtaining results;
- the possibility of graphical interpretation of the model of the domain, which makes a simple reflection in the form of a graph of causal relationships between the objects of the model.
- universality of application of systems dynamics technologies.

In developing the simulation model of the enterprise under study, the task was to introduce into it a system of financial and economic indicators that directly from its "Efficiency funnel" (Fig. 4). It is this financial-economic model of the enterprise most systematically reflects the efficiency of the functioning of a holistic chain of value-added processes [1]. The system of profits that form the "efficiency funnel" below is presented in an analytical form, as a system of simultaneous equations, and which are directly represented in the simulation model.

The system of simultaneous Eqs. (1–8) for all types of profits that form a "funnel of efficiency"

$$marginal\ profit^i(t) = gross\ income^i(t) - variable\ costs^i(t) \qquad (1)$$

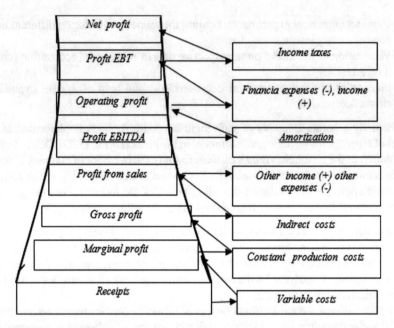

Fig. 4 Key financial and economic indicators of the enterprise that form the "efficiency funnel" of the enterprise. *Note* Developed by authors

where, Marginal profit on the i-th commodity nomenclature at time t (t) is the difference between the gross income of the same commodity position (= and *variable costs* (*t*) in the same time.

In this case, it is calculated as the product of the physical quantity of the corresponding element of variable costs (j ∈ J) at its price in the period t. J defines a plurality of elements of variable costs for the production of the i-th heading. At the same time, the price of the corresponding element of variable costs in the planned period is pre-projected on the basis of retrospective management accounting data using forecasting methods of neural network technologies.

The use of this method is due to the necessity of taking into account in the model simultaneous effects on the price of elements of variable costs of seasonal phenomena and autocorrelation simultaneously with the consistently determined tendency of their constant growth in the conditions of permanently high inflation, which is taking place today in Ukraine. Formulas (2)–(7) reflect the well-known logic of calculations for each of the remaining 6—and the profits of the "efficiency funnel". Without going into the details of the calculation of the corresponding costs, we note that for each of them in the model is determined by their expected value in each of the periods (t) for the defined horizon of the run of the simulation model.

$$Gross\,Profit(t) = Marginal\,Profit(t) - Direct\,Production\,Costs(t) \quad (2)$$

where, GrossProfit(t) is the gross profit of the enterprise in the t-th period of time, respectively Direct production Costs(t)—direct production costs. In the system of managerial accounting for the investigated enterprise strictly identifies the list of direct production costs and the algorithm of their distribution to the costs of a particular type of its commodity products. According to the results of the study of retrospective data on certain types of direct production costs using economic and statistical methods, their predicted values for the corresponding planning horizon are determined.

$$Profit\ From\ Sales(t) = Gross\ Profit(t) - Indirect\ Production\ Costs(t)$$
$$(3)$$

where Revenue(t) is the profit from sales in the t-th period, respectively, NeprVitr(t)—indirect production costs. The logic of their reflection in the model is exactly the same as it was done for direct production costs.

$$\underline{Profit\ EBITDA}(t) = Profit\ From\ Sales(t) + Other\ Income(t)$$
$$+ Other\ Expenses(t) \qquad (4)$$

where *Profit EBITDA(t)* is the EBITDA earnings of the enterprise in the t-th period of time, respectively, *Other Income(t)*—other company's income, *Other Expenses(t)*—other expenses. In the presented model, these two parameters are represented using the TRIA method (Min, Mode, Max), Min, Mode, Max values obtained based on retrospective management accounting data.

$$Operating\ Profit(t) = \underline{Profit\ EBITDA}(t) - Depreciation(t) \qquad (5)$$

where *Operating Profit(t)* is the operating profit for the t-th period, respectively *Depreciation(t)*—depreciation charges for the same period. The algorithm of calculation of which is strictly defined in the accounting policies of the enterprise.

$$Profit\ EBT(t) = Operating\ Profit(t) + FinProfit(t) - FinExpence(t)$$
$$(6)$$

where Profit EBT(t)—profit EBIT of the enterprise in the t period, respectively FinProfit(t)—income from the financial activity of the enterprise, FinExpence(t)—the cost of its financial activities. In the presented model, a simplified approach is used to determine the values of these parameters, which is similar to how it was done to calculate EBITDA profits. It is obvious that for their calculation, it would be advisable to use specialized techniques that are not the subject of our consideration.

$$Net\ Profit(t) = Profit\underline{EBT}(t) - Profit\ Tax(t) \qquad (7)$$

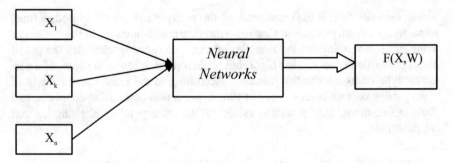

Fig. 5 Model of artificial neural network [19]. *Note* Developed by authors

where NetProfit(t) is the net profit of the enterprise in the t period, respectively, Profit Tax(t) is a profit tax, which is strictly determined at the level of the interest rate.

It should be noted that for all parameters of the model, for which the prediction of their predictive values is foreseen, the forecast period coincides with the calibration interval of the simulation model.

Using the functionality of the system of imitative modeling Arena implemented a model of calculation of profits system of the investigated enterprise, which form its "efficiency funnel". The logic of implementing such a model is as follows. The enterprise produces bakery products, the set of commodity nomenclatures of which is more than 100 species. The pool of end-users has more than 400 units, in the vast majority it is a system of regional stores in which the finished products are delivered. According to the results of marketing research conducted using ABC analysis, XYZ-analysis, BCG analysis in terms of both product and product categories, the range of products of the enterprise significantly reduced, which should be introduced into the developed simulation model. An important moment in implementing the simulation model is the task of selling physical sales volumes in the context of commodity nomenclature. The analysis of the retrospective data, implemented in Statistica for Windows, indicates their nonlinear dynamics, which is difficult enough to represent analytically using classical economic-statistical models. The most adequate reflection of such a dynamics was obtained using models based on neural networks. The work of the neural network consists in transforming the input vector into the output vector, and this transformation is determined by the weights of the neural network. The neural network (NN) is represented as a mathematical tool, using which you can practically simulate any continuous function—f(x; W) (Fig. 5).

But the main advantage of NN is that they allow to simulate nonlinear dependencies, which in the general case have the form

$$F(\dot{x};W)=\varphi(\sum(wk *\gamma(\dot{x};W))) \tag{8}$$

The estimation of the efficiency and profile of the neural network model of dynamics of daily physical volumes of sales of one of the commodity nomenclatures of the enterprise are presented in Fig. 6. The effectiveness of the neural network (type one-

Profile : MLP s6 1:6-1-1:1 , Index = 1
Train Perf. = 0,967017 , Select Perf. = 0,950651 , Test Perf. = 0,986960

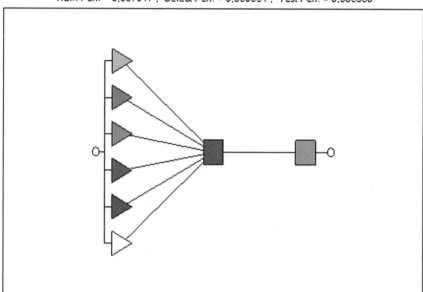

Fig. 6 Profile of the neural network and its performance on different sub-sections (educational, test and control). *Note* Developed by authors

level perceptron) is determined by its performance (for teaching, control and test choices). Productivity is defined as the correlation coefficient between the actual and theoretical data of the dynamic series. Sufficiently high values (close to 1) indicate a well-defined network architecture and initially defined parameters for its initialization (Fig. 6).

On the basis of the selected neural network, the predicted values of daily physical sales volumes on this product line are determined for the next scheduled month (Bulka fragrant) (Fig. 7).

A similar approach is also used to assess a number of other parameters that are involved in the calculation of the system of financial and economic indicators, and are also considered as stochastic variables. Accordingly, the generation of their values in each particular case is realized on the basis of pre-determined by the data of economic-statistical analysis for them empirical functions. This mainly refers to prices for basic and secondary raw materials, fuel and electricity, the percentage of returning unrealized products, volumes of production shortage and other types of losses. According to the above formulas for calculating each of the seven types of profits of the "efficiency funnel" and for formalizing the determined functions, changing the values of the factors through which they are determined, according to the results of the model run, we obtain a full range of financial and economic indicators of the activity of the investigated enterprise. In fact, any financial and economic

Fig. 7 Daily predictive values of physical volumes of fragrant loaf

indicator is reflected in the model, if it has logic for calculation. In addition to the characteristics of financial and economic activity, the model provides an opportunity to obtain predicted values of physical volumes of necessary raw materials, other materials, volumes of consumed fuel, electricity, etc. A fragment of the implemented model by means of the system of simulation modeling Arena in the form of its graphical representation is shown in Fig. 8.

The simulation model in the Arena system is represented by a directed graph whose nodes are different types of modules. In the presented model, a separate module defines the primary rule of generating transacts (objects) of the model, in the future an algorithm for generating its individual parameters or the formula for calculating the controlled financial and economic indicator. The main modules of the model represent the sequence of the main processes of the "value added chain" with the simultaneous identification of the resources necessary for their implementation, with a detailed definition of their properties, which characterize them in terms of productivity and cost of use. In addition to the modules that have a visual representation in the model, modules are used in the system that are not graphically interpreted and define a set of variables, on the one hand, both the definition of the results of the functioning of

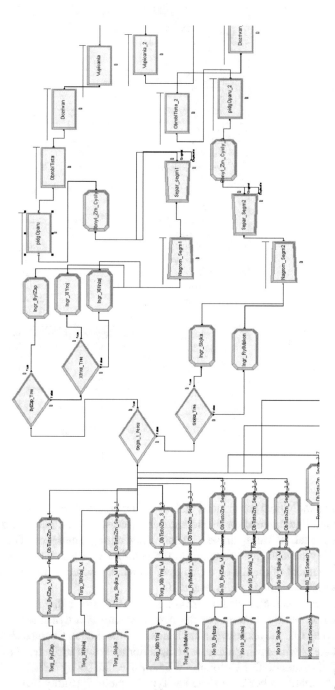

Fig. 8 Graphic representation of the simulation model of the calculation of key financial and economic indicators of the enterprise in the context of the formation of its "efficiency funnel"

Fig. 9 Variable models and their properties

	Name	Rows	Columns	Data Type	Clear Option	Initial
					Variable - Basic Process	
1	boro_total			Real	System	0 ro
2	Doxid_BylkaZap			Real	System	0 ro
3	VutrZm_BylkaZap			Real	System	0 ro
4	price_boro			Real	System	1 ro
5	price_drig			Real	System	1 ro
6	price_olia			Real	System	1 ro
7	price_sil			Real	System	1 ro
8	sil_total			Real	System	0 ro
9	drig_total			Real	System	0 ro
10	olia_total			Real	System	0 ro
11	PrubMarg_BylkaZap			Real	System	0 ro
12	den_no			Real	System	0 ro
13	Doxid_BatonNarizn			Real	System	0 ro
14	VutrZm_BatonNarizn			Real	System	0 ro
15	PrubMarg_BatonNariz			Real	System	0 ro
16	Doxid_BatonStol			Real	System	0 ro
17	VutrZm_BatonStol			Real	System	0 ro
18	PrubMarg_BatonStol			Real	System	0 ro
19	price_pat			Real	System	1 ro
20	price_boro_j			Real	System	1 ro
21	patoka_total			Real	System	0 ro

Fig. 10 System resources represented in the model and their properties

	Name	Type	Capacity	Busy / Hour	Idle / Hour	Per
		Resource - Basic Process				
1	TistoZmich1	Fixed Capacity	3	0.0	0.0	0.0
2	TistoZmich2	Fixed Capacity	2	0.0	0.0	0.0
3	Tistopodrib	Fixed Capacity	1	0.0	0.0	0.0
4	TistoOkrygl	Fixed Capacity	1	0.0	0.0	0.0
5	Tistozakat	Fixed Capacity	1	0.0	0.0	0.0
6	VidsChafa	Fixed Capacity	2	0.0	0.0	0.0
7	Resource 1	Fixed Capacity	1	0.0	0.0	0.0
8	PichGazova	Fixed Capacity	1	0.0	0.0	0.0
9	PichEletro	Fixed Capacity	1	0.0	0.0	0.0

the model and its individual parameters, and, on the other, the plurality of resources necessary for the implementation of the another process. So Variable is a global variable of the model. Its value can be read and varied during simulation. The value of a variable may depend, for example, on the possibility of transacting parts of the model. A fragment of the list of variables of a model, resources, and properties defined for them is presented in Figs. 9 and 10.

The modules are interconnected by arrows. It is on the arrows that moves transacties. Transaction (entity, in terms of the Arena system) is a dynamic object of an imitation model that moves between static nodes of the model. Characteristics of TRANSACTs are specified using attributes. In addition to its own attributes, TRANSACT has a set of system-defined attributes whose values are set by default, such as, "Total Time", "Serial Number", "Transaction Type", " picture" (Entity Picture). It is possible to change some of them programmatically [16, 17].

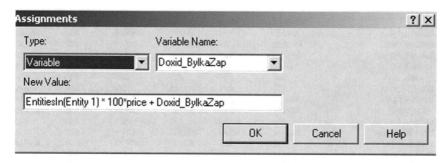

Fig. 11 Doxid_Bylka Zap module Assign type, which calculates the value of gross income for a separate product line

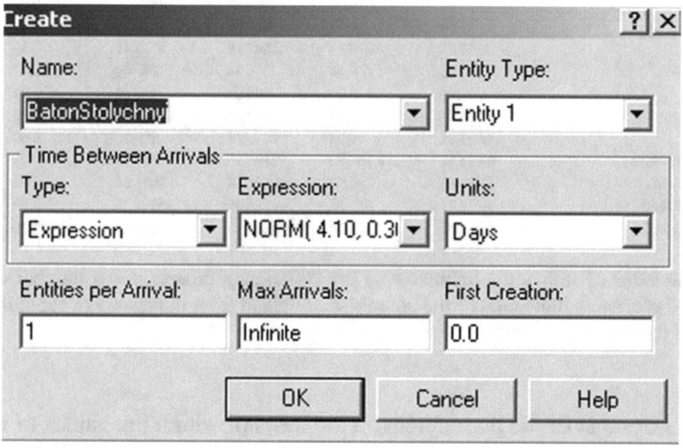

Fig. 12 Determine the parameters of the module for generating one of the transactional models

Assign type modules provide assignment of values to a variable model for TRANSACT attributes that pass through the module. In our model, they rely on a certain algorithm to value the value of profits that form the "efficiency" of the enterprise and a number of other financial and economic indicators. An example of determining this type of module is shown in Fig. 11.

When constructing the model of the process of processes "chain of creation of added value" in order to calculate the profits of the enterprise, which form its "efficiency funnel" in the system simulation Arena used the following logic of its implementation. At the first stage, formal algorithms for transactional generation (volume of orders for individual products from specific buyers) are given. The dialog for determining the parameters for generating one of the transactional models is shown in Fig. 12.

On the basis of such orders, the values of the variables that form the variable cost pool are calculated. And the results of their calculations are determined by the projected values of marginal revenue, marginal profits, the need for raw materials and

Variable	Average	Half Width	Minimum Average	Maximum Average	Minimum Value	Maximum Value
BatonNar_PrReal	9502.54	1 699,85	8373.17	11481.09	0.00	28110.60
BatonNar_ValPru	13122.55	2 347,41	11562.95	15854.83	0.00	38819.40
BatonStol_PrReal	2140.60	0,00	2140.60	2140.60	0.00	2140.60
BatonStol_ValPru	2251.00	0,00	2251.00	2251.00	0.00	2251.00
boro_total	0.00	0,00	0.00	0.00	0.00	0.00
BylkaZap_PrReal	0.00	0,00	0.00	0.00	0.00	0.00
BylkaZap_ReaPr	1631.81	160,04	1480.47	1826.40	0.00	5148.90
BylZap_ValPru	5568.01	546,08	5051.60	6231.98	0.00	17568.90
den_no	0.00	0,00	0.00	0.00	0.00	0.00
Doxid_BatonNarizn	27871.49	4 985,76	24558.98	33674.68	0.00	82450.00
Doxid_BatonStol	4500.00	0,00	4500.00	4500.00	0.00	4500.00
Doxid_BylkaZap	19966.22	1 958,20	18114.45	22347.16	0.00	63000.00
Doxid_XlibYroj	4400.00	0,00	4400.00	4400.00	0.00	4400.00
Doxid_XlibZapachn	17825.36	2 366,03	15635.81	20853.18	0.00	42550.00
Doxid_ZXlibKniaj	2550.00	0,00	2550.00	2550.00	0.00	2550.00
drig_total	83.2130	5,19	79.6430	89.1478	0.00	143.34
Kil_BatonNar	3279.00	586,56	2889.29	3961.73	0.00	9700.00
Kil_BatonStol	500.00	0,00	500.00	500.00	0.00	500.00
Kil_BylkaZap	2852.32	279,74	2587.78	3192.45	0.00	9000.00
Kil_XlibKniaj	300.00	0,00	300.00	300.00	0.00	300.00
Kil_XlibYroj	400.00	0,00	400.00	400.00	0.00	400.00
Kil_XlibZapachn	1550.03	205,74	1359.64	1813.32	0.00	3700.00
olia_total	95.6702	13,34	88.1279	111.71	0.00	265.18
PrubMarg_BatonNarizn	15614.59	2 793,20	13758.81	18865.74	0.00	46191.40
PrubMarg_BatonStol	2631.00	0,00	2631.00	2631.00	0.00	2631.00
PrubMarg_BylkaZap	8277.71	811,84	7509.99	9264.81	0.00	26118.90
PrubMarg_XlibKniaj	2550.00	0,00	2550.00	2550.00	0.00	2550.00

Fig. 13 The value of individual indicators of the "efficiency funnel" and a number of auxiliary financial and economic indicators of the enterprise, obtained from the results of running one of the scenarios of the simulation model

materials. For most of the parameters on the basis of which the values of the profits of the "efficiency funnels" are determined, analytical formulas for their calculation are determined on the basis of the analysis of retrospective data obtained from the management accounting system of the investigated enterprise.

The results of the simulation model are presented in a special report, which consists of separate sections. In our study, the most important results are given in the rubric Variable. It is here that the values of all the variables of the model are reflected, and first of all those that form the "efficiency funnel" of the enterprise. The report snippet for one of the scenarios of the model run is shown in Fig. 13.

So the Prub Marg_Baton Narizn, Prub Marg_Baton Stol, Prub Marg_Bylka zap, Prub Marg_Xlibkniaj variables reflect the margin revenue. Other variables represent the value of gross revenues, sales revenue for each of the commodity items that are reflected in the simulation model. At the same time for each variable its average, minimum or maximum value is received at a certain time interval of imitation.

3   Conclusion

The success of an enterprise in a highly competitive environment, which today operates most of Ukrainian enterprises, and first of all, those that produce food, depends on how highly effective the management system is. Such a system can be built on the basis of process modeling. In this case, the system of processes must be strictly formalized [20–27]. Formalization of the business model of the enterprise is proposed to be realized using the All Fusion Process Modeler software system, which belongs to the class of business analytics systems and implements the international standards for business process modeling—IDEF0, IDEF3 and DFD. It is on the basis of the formalized business model of the enterprise that the target management system is being built, an important component of which is the planning system as a key element of the control system. In turn, the planning system is proposed to be implemented on the basis of the simulation model of the main business processes of the enterprise, which form a "chain of value added." This simulation model allows you to "play" different scenarios of the enterprise and, accordingly, evaluate them through a system of performance indicators that form the "efficiency" of the enterprise. By analyzing different scenarios of activity, enterprise management can determine well-founded development goals, while taking into account the potential risks of their achievement. Technological aspects of implementing such an approach using modern business intelligence systems, including All Fusion Process Modeler, Statistica for Windows, Arena system software simulation.

References

1. Drury K (1997) Introduction to management and production accounting. Audit, UNITI, Moscow, 560 pp
2. Robson M (2003) Reengineering of business processes: practical guidance. In: Ullach F (ed). UNITY-DANA, Moscow, 222 pp
3. Kryvinska N (2012) Building consistent formal specification for the service enterprise agility foundation. Soc Serv Sci J Serv Sci Res Springer 4(2):235–269
4. Scheer AV (2000) Business processes. Basic concepts. Theory. Methods (Per from the English). The Logic of Business, Moscow, 182 pp
5. Edward Deming U (2014) Out of the crisis. New people management, systems and processes paradigm. Alpina Publishers, Moscow, 344 pp
6. Kaczor S, Kryvinska N (2013) It is all about services—fundamentals, drivers, and business models. Soc Serv Sci J Serv Sci Res Springer 5(2):125–154
7. Björn A (2003) Business processes. Tools for perfection (Per. from English Arinichev SV; scientific Ed. Adler YuP. Standards and quality). RIA, Moscow, 272 pp
8. Fayol A (1992) Management is a science and art. Republic, Moscow, 352 pp
9. Ponomarenko VS, Minukhin SV, Znakhur SV (2013) Theory and practice of modeling business processes: monograph. Vyd. KHNEU, Kharkiv, 244 pp
10. Kryvinska N, Gregus M (2014) SOA and its business value in requirements, features, practices and methodologies. Comenius University in Bratislava. ISBN: 9788022337649
11. Shannon R (1978) Simulation of systems—art and science. Mir, Moscow, 417 pp
12. Tools for describing business processes [Electronic resource]. Access mode: http://process.siteedit.ru/page30

13. Description of the business process [Electronic resource]. Access mode: http://process.siteedit. ru/page54
14. Gregus M, Kryvinska N (2015) Service orientation of enterprises—aspects, dimensions, tech-nologies. Comenius University in Bratislava. ISBN: 9788022339780
15. Low A, Kelton V (2004) Simulation modeling. Peter, St. Petersburg, 848 pp
16. http://www.arenasimulation.com/—official site of the system Rockwell Arena
17. Altiok T, Melamed B (2007) Simulation modeling and analysis with ARENA. Academic Press, 456 pp
18. Molnár E, Molnár R, Kryvinska N, Greguš M (2014) Web Intelligence in practice. Soc Serv Sci J Serv Sci Res Springer 6(1):149–172
19. Haykin S (2008) Neural networks. Full course. M. Williams, 1104 pp
20. Planning, directing and controlling [Electronic resource]. Access mode: http://www. principlesofaccounting.com/
21. Rotter M (2005) Learn to see business processes. The practice of building maps of value creation streams (Per. from english). Alpina Business Bucks, CBSD, Center for Business Skills Development, Moscow, 144 pp
22. Description of the standard IDEF0—automation of company management [Electronic resource]. Access mode: www.insapov.ru/idef0-standard-description.html
23. Mikhailovich KS, Mikhailovich KV. Modern methodologies for describing business processes are simply complex. IDEF3 methodology [Electronic resource]. Access mode: http://www. betec.ru/index.php?Id=6&sid=30
24. Barsky AB (2004) Neural networks: recognition, management, acceptance decisions. Finances and Statistics, Moscow, 176 pp
25. Kaplan Robert S, Norton David P (2003) Balanced scorecard. From strategy to action. ZAO Olimp-Business, Moscow
26. Kochnev A. How to build a company's target management system. http://company.iteam.ru/ action/seminar/venture_management/webinar_record.htm
27. Modeling of business processes with BPwin 4.0 M, "DIALOGMIFI" * 2002

How Global Is Your Business? A Business Globalization Index to Quantify a Business' Globalization Degree

Lennart C. Naeth-Siessegger

Abstract Globalization as a buzz word is present in many publications, being scientific, journalistic or even literarily. This work defines the scope of globalization with regards to its meaning for businesses and within this scope, a measure is proposed to determine a firms' degree of globalization. Starting with a review of scientific work, a multi-layered, modular approach is proposed built on three core aspects of globalization. By using a vector-based form, this index is able to provide both: a general measure for a company's degree of globalization and a detailed evaluation of a company's different states with regards to economical, spatial and regulatory aspects of globalization. The index developed is validated using a sample of four multi-national-companies (MNC). Results show, that firms, which are deemed global, only reach half of the overall possible degree of globalization.

Keywords Globalization · Index · Company analysis · Measure of globalization

1 Introduction

1.1 Relevance

How can the degree of globalization of a business be measured? In this regard, the subject is what globalization is and how this can be measured. Globalization occurs as a topic broadly known within the media during the past. This is accompanied by various definitions and understandings of the characteristics of globalization. In this relation there seems to be a blurred line between internationalization and globalization. Sometimes they are even used as synonyms. As a first approach international may be defined as

L. C. Naeth-Siessegger (✉)
Faculty of Business, Economics and Statistics, University of Vienna, Oskar - Morgenstern - Platz 1, 1090 Vienna, Austria
e-mail: l.c.n.s@me.com

© Springer Nature Switzerland AG 2020
N. Kryvinska and M. Greguš (eds.), *Data-Centric Business and Applications*,
Lecture Notes on Data Engineering and Communications Technologies 30,
https://doi.org/10.1007/978-3-030-19069-9_13

existing, occurring, or carried on between nations [1]

and globalization as

the process by which businesses or other organizations develop international influence or
start operating on an international scale [2]

Moreover, globalization is a work in progress and is characterized by its influences on economical, political, social and cultural processes of societies [3]. Also, legal aspects influence globalization, but this may describe a link between geographical and economical aspects. Furthermore, a differentiated understanding of geographical spread is part of globalization [3]. Consequently, globalization is a construct of higher order in contrast to internationalization, in which internationalization plays a pivotal role [3].

As part of globalization the electronic business and subsequently various topics around information are present in businesses throughout every sector of an economy. As showed with the work *Agile Information Business*, the sourcing, combination, processing, analyzing and subsequent use of information within a business is vital, and the degree and professionality are the degree with which a company succeeds [4].

1.2   Goals and Objectives

First, this chapter shall develop an understanding of the concept of globalization by using scientific publications on this topic.

Secondly, the relation of businesses to the concept of globalization will be assessed. Firms starting activities internationally have the opportunity to open up new markets for their products. Additionally, the company may explore and utilize foreign knowledge, resources of the foreign countries and much else. These aspects of modern societies enable a company to prosper and grow.

A globalized company can use these advantages as well. But an internationalized company may not use all of them or only on a limited scale. In contrast, a globalized company deals with all these aspects in a world-spanning perspective. This may activate potentials and may give a global company a competitive advantage.

How to assess and quantify the degree of globalization of a company? The difference between all its potential and the actual state of a company's globalization may initialize a growth to the full level of globalization. As a result, a company should have an inherent interest for decision making to know its position in the process of globalization. The objective of this work is to develop and substantiate a solution for such a quantification as well as its operationalization and the operationalized measurement shall be tested on a sample of multi-national-companies (MNC).

In conclusion, globalization should be measured and afterwards the implications for the company, especially for the management, have to be derived. As a result, this work proposes a globalization index for the measurement of a company's degree of globalization in a first step. Secondly, the work derives implications for the management of a MNC.

Criteria for this globalization index are, that it will consider the aspects of globalization. Such an index should have an architecture that is modular, expandable and adaptable by incorporating new measures. As the concept of globalization might change over time, more or different measurement options will be developed. This index architecture ensures adaptability to them. Finally, the result of the developed index should be an easy-to-understand degree of globalization for businesses.

The remainder of this chapter will be organized as follows: Initially Sect. 2 develops the scope of a global business by using the preceding sub-sections deriving the scope of a business and the concept of globalization. Section 3 refers to related work, while in Sect. 4 the Business Globalization Index (BGI) will be developed.

Section 5 will illustrate the index developed using a sample of MNC. This sample will be analyzed, and on this base Sect. 6 will develop management implications, both in general and for the sample analyzed in Sect. 5. Section 7 provides a critical review and discussion. Also, areas for further work will be identified in Sect. 8. Finally, Sect. 9 will conclude the findings of this work.

2 Global Business

Global Business is a term often used and understood in many different ways and sometimes even misunderstood as a synonym for international business. Google Scholar lists nearly one million hits for a definition of globalization [5]. As not all these hits are entirely about the construct of globalization on its own and some may be listed multiple times, the mere number shows the diversity of the topic and its interpretations.

A business is a planned and organized entity where resources of production are combined in the purpose of producing and selling the businesses products and services [6].

When dealing with globalization, one aspect is always stated: the spatial aspect of globalization. This aspect is closely linked to the concept of internationalization. In this context often International Corporations MNCs are stated. These firms have a common basic characteristic: they operate in several countries and not just in their home country [7].

Globalization has its origins in the meaning of world spanning and as it is used in many languages, the word seems to be important [8–10]. But this does not capture all of its aspects. Furthermore, the characteristics of world spanning may be divided into two segments:

First it can be defined as "Territorial" "World Spanning".

Secondly it may be seen as spanning the different marketplaces of the world.

This brings up the question how to define a measurable construct on this basis. Sometimes markets are not bound to nation-states or supranational organisations. In contrast "territorial world spanning" can be measured. Nations still form the fundamental structure of the world [3]. National borders confine these nations. Therefore, they will be used as foundation for the evaluation in contrast to the blurred definition of markets.

In [11] Scholte lists four misleading definitions for globalization: globalization as internationalization, as liberalization, as universalization and as westernization.

Internationalization has its origins in "inter-national". It can be translated as occurring between countries, but comprises less variables than Globalization [11].

The concept of Liberalization is to remove market restrictions in any way. Also called "Laissez-Faire Policy", Liberalism aims to have a market-based regulation with no further restrictions for the economy. This concept arose before the word globalization was coined. It is debated by scholars if this approach supports or hinders the ideas of globalization [11].

Universalization indicates a worldwide standardization of culture, politics, jurisdictions and economies. A form of universalization is westernization [11]. Concepts such as glocalisation, where global strategies are adapted to local requirements, show that this equalization may be wrong [12].

Scholte identifies that the result for the current western lead characteristic of globalization is historically explained and reducing it to these historical reasons is not sufficient [11].

2.1   Six Aspects of Globalization

Starting with the article by Scholte, he identifies the importance of six aspects when defining the concept of globalization: Economy, space (geographics), legal, culture, social life and politics. He emphasizes in his work all listed aspects of globalization. But he does not comment on the interdependencies between the aspects [11]. Peter Dicken also emphasizes these six aspects, but focuses on the economical aspect [3].

The reasons for the influence of these aspects on the concept of globalization are manifold. Furthermore, some justifications apply to more than one aspect or the justification necessitates another aspects' justification.

2.1.1   Economy

Economy is important for providing the human basic needs, such as food, not provided if the economy would not exist. Further economic sourcing does require the human resource for its existence. Accordingly, humans need economy and vice versa.

A tendency of economies is to concentrate production [3]. This tendency is accompanied with a dual trend: The modern world also known as *developed countries* or *advanced economies* [13] produces goods and services which require highly skilled labour and are intense in the use of technology. In contrast developing or low-cost countries tend to produce products that require low-skilled labour. This happens according to national resources of production. Further firms' products are more and more depending on intermediate products, which are supplied by other firms [14]. On this economic base webs of production or tangled webs emerge [3]. Webs of production are concentrated deposits of firms on a limited geographical area. These tangled webs may support the evolvement of economies of scope and in a second step the evolvement of economies of scale. In the *developed countries* this evolvement is supported by historical developments in these countries. For example, the clock and watch making industry in Switzerland is a case for such historical influence [14].

2.1.2 Geographics

As stated above nations are the foundation of the evaluation. The argument, that states do not matter anymore, seems untrue [3]. States are regulators and a major instance in global relations. After the financial crisis in 2007/2008 nation states revealed themselves as highest instance in the world with their national help- and regulators-programmes [3].

The financial crisis showed two trends: First, the modern world is highly interconnected. Secondly, national concerns still are the cornerstone of worldwide geographies [15].

Additionally, most natural resources are bound to national geographies. In case of national resources, such as coal, the resource must first be sourced before it can be transported. Another crucial resource is the human capital. People live in particular places. There may be more and more possibilities in transportation, but the vast majority of people live near to their work [14].

2.1.3 Legal

On the one hand, as previously explained, globalization is influenced by people and vice versa. Governments of nations, as part of politics, make laws. According to that laws are influenced by globalization.

On the other hand, laws differ from nation to nation. Using these differences may provide advantages for firms on various levels. Often taxation differs and the legal basis for patents may differ as well. But, of course in order to obtain these advantages for a MNC, efforts need to be made otherwise legal aspects will hinder success.

2.1.4 Culture

The culture in one country can be seen as a set of common values and norms [12]. There may be more than one culture in one country. Subcultures can differentiate the cultural deposit of a nation even more. A result of these common values and norms is that the attitudes towards work, money and risks may differ in each cultural background [16].

These attitudes are reflected in requirements for firms. A company, who starts operating beyond the home culture, may be faced with different cultures. A manufacturing company stemming from a western background would have to rethink their corporate actions when confronted with other norms [17].

There also is a trend to relocalize in order to meet customer needs [18]. This so-called glocalisation does explore the local advantages of the subsidiary. Moreover, this enables firms to produce goods just in time and the company minimizes with this type of production among others storage and transportation—costs. This trend then counteracts to the potential of Economies of Scale [18].

2.1.5   Social Life

Besides the cultural aspect influencing the social life of people, there are numerous other influences. Social life changes with every generation [19]. One major driver of this change is technology [3, 19].

Expectations, behaviours and wishes of social life transfer into work life. For example, the desire to have children leads to the requirement in the matching of working hours and childcare or a work-absence for the first period of the child's life. Further customers do have expectations for economical behaviour. Like a code of conduct, the use of information technology has become a prerequisite for firms. Moreover, this technology enables a direct contact via Internet between the customer and the company. Also, customers are better informed than ever and products and services become more and more substitutable. This drives rivalry among firms and also drives innovation within firms [20].

2.1.6   Politics

Political organs usually govern a nation. All stakeholders, not only political ones, in a nation can influence these governments. People express their will, culture and views for example in elections that form the government. Economic actors interact with political organs on numerous matters. As such politics are expression of societies and embedded or interlinked with every above-mentioned aspect of globalization. In this relation politics influences the public discussion or opinion.

Interaction with political organs may provide advantages for businesses and gives the opportunity to avoid threats.

In sum this aspect of globalization is highly interlinked with aspects three to five. So, influences on globalization may be a result of the globalization process of these other aspects.

2.2 Technology: Driver of Globalization

Technology has become a persistent companion. A more recent example for the impact of technology is modern technology like Telecommunications (TC) and Information Technology (IT), such as PCs, Laptops, tablets and smartphones. They are familiar to and used by many people mostly in combination with the "Internet" or "World Wide Web" [21]. Moreover, the usage of the Internet may also increase economic growth and international trade [22].

These modern technologies do also enable easier hurdle of distance. An example is the availability of information. Due to TC and IT, information is available around the globe nearly at the same time. It may also have eased the interaction with customers via web-shops and customer-services. This can on the one hand lead to decreased costs when real shops can be transferred like "virtual shops" into the Internet and do not have to be operated in a specific location. On the other hand this may increase customer value because goods can be purchased all the time in every place. Despite of all advantages a company can experience due to Technologies, there are disadvantages that may threaten the company such as reverse marketing [18]. These have to be watched cautiously so they are avoided.

In sum technology is a part of human life and drives the process of Globalization [3].

3 Related Work

This section gives an overview of existing approaches to the topic of this work. These approaches will be examined and put in perspective to the aspects of globalization. Additionally, this section will show, that existing approaches do not cover all aspects of globalization this work has identified.

Possible categories of the aspects of globalization will be discussed in a second step and measures for these categories will be defined.

In the following a sample of the existing approaches to measure the degree of globalization of a business are presented and evaluated. They are presented chronologically.

3.1 Networks' Spread 1979

The Networks' Spread was published by Vernon in 1979. It is usually referred to as Network Spread Index (NSI) [23]. This work analyzes the reasons behind a firms' decision to internationalize [24].

Vernon conducted a survey on 180 US firms to evaluate the pattern of internationalization. He measured the Networks' Spread by counting the number of countries in which the company has subsidiaries.

In conclusion Vernon's NSI is uni-dimensional and uses one level of detail.

3.2 Degree of Internationalization 1994

Sullivan was one of the first using more than a single measure for a combined or multidimensional index in 1994. He called his index *degree of internationalization scale* (DOI) [25]. To overcome the danger of a statistical Type 1 or Type 2 error and the threat to misinterpret a concept due to an abnormal characteristic of the data, he proposed to follow a multidimensional approach. In the following he proposed his index *degree of internationalization*:

He develops his index based on three attributes of internationalization: The firms' sales-performance, its structure and attitudes, which are subdivided into five measures [25].

These five measures are each weighted equal and summed thereafter. As a result, the DOI is a number between 0 and 5 where 5 signifies total internationalization.

Sullivan operationalized his index on a sample of "Forbes (…) "Most International" 100 American manufacturing and service firms on the basis of total foreign revenues" [25].

3.3 Transnationality Index 1995 and Transnational Activities Spread Index 1998

The United Nations Conference On Trade and Development (UNCTAD) proposed in their *World Investment Report 1995* the transnationality index (TI) [26]. The TI measures the degree of internationalization by combining three measures. The percentage of foreign assets, foreign sales and foreign employment are derived in combination with their equivalent in total. The normalized result is a number between zero and one, which can be interpreted as a percentage of Internationalization. Easy data- availability and accessibility is an advantage of the UNCTADs' index.

The TI earning wide criticism [23, 27]. Grazia Ietto-Gilles proposed a composite index of two previous approaches in 1998 [23]. She developed a combined index consisting of the UNCTADs' TI and Vernon's' NSI [24, 26]. The purpose of this was to overcome the home/foreign perception in previous approaches to measure the subject.

Ietto-Gilles proposes to combine them to solve the shortcomings of both indices. The result is the transnational activities spread index (TASI). This index has the three measures of UNCTADS TI weighted by Vernon's NSI.

3.4   Complex Spread and Diversity Measure 2003

The Index developed by Fisch and Oesterle in 2003 aims to measure the degree of international involvement of a company [27].

Out of the problem that arises when dividing the concepts of internationalization and globalization the authors propose their index. They state that until 2003 this was not done yet; furthermore, they propose, that previous indices did not cover the part of cultural aspects delimiting globalization from internationalization.

The index Fisch and Oesterle propose is based on a dual approach to globalization: the geographical and the cultural part.

According to the authors both parts are important and described by the indicators used. They say that until 2003 existing indices were not useful because it is necessary to measure the spread of the activities as an important aspect of globalization combined with the cultural spread. As a result, they designed the "Degree of Globalization" (DOG).

The results of Fischs' and Oesterles' work were, that until 2003 all international companies considered globalized were not thus far globalized.

3.5   Global Specialization Index 2007

The article *"How Do We Capture "Global Specialization" When Measuring Firms' Degree of Internationalization"* focuses on another view of globalization. The authors Asmussen, Pedersen and Petersen stress that previous approaches to the topic leave out the fact, that one aspect of globalization is the exploration of locations-specific advantages. Thus, a truly globalized MNC has to have their value chain spread over all countries in which the company is active and not only several mini-replicas of the home value chain structure set up abroad. They base their approach on Porters concept of the value chain [20, 28].

In their work the underlying proposal is derived that the more the company is exploring these location specific advantages, the more the company is globalized. For measuring the degree of specialization of a company the authors use the international division of labour. The concept of their index is based on the total number and the number of employees located in a specific country. Furthermore, the value-added activity of the employees was considered. With this data the percentage of employees working in each activity per country was calculated. The outcome was pooled to the *global specialization index* (G).

The G was tested on a sample of Danish MNCs. For this purpose, the authors made a survey. With this the authors derived the data for their sample by using a questionnaire. The result of the work was, that three clusters of MNC were identified in which the Danish firms are bundled. Also, the approach was the first measuring the aspect of specialization in terms of globalization.

3.6 Discussion

The sample of approaches shows that differentiation between internationalization and globalization is very important. The first three approaches either do not differentiate between them or label it internationalized but measure aspects that go beyond internationalization. The approaches that differentiate between the two concepts of internationalization and globalization do not provide a consistent definition of these concepts.

Additionally, the indices do not cover all aspects of globalization derived above. Also, the validity for the covered aspects of globalization is limited because the aspects are measured with help of maximum three calculations (Sullivan/Economical). This is problematic for the case that this item is an anomaly, on what reason so ever, compared to the rest of the data.

Moreover, the presented indices do not categorize their aspects of globalization. In some approaches a discussion of aspects related to the concept of globalization is not conducted. Therefore, a consistent basis for argumentation and evaluation is missing.

This is because Vernon, Sullivan and UNCTAD use global and international in their papers as synonyms. In conclusion they conduct no evaluation whether it is international or global and in order to what the intents of measures are undefined on the one hand.

On the other hand, Fisch & Oesterle and Asmussen, Pedersen & Petersen differentiate between these concepts, but the definitions vary and lack aspects of globalization previously found.

4 Business Globalization Index (BGI)

Based on the previous sections an index will be proposed that tries to overcome shortages of previous approaches and fulfill the criteria developed for a global business.

4.1 General Index Model

In a first step a general index model was developed. Its architecture uses the definitions of section global business in order to classify the aspects of globalization after Dicken and Scholte [3, 11] and is structured in aspects.

In the second step the index uses a modular, extendible and multilevel model inspired by a paper written by Daft and Albers [29]. The levels are *aspects, dimensions, elements* and *items* (See Table 1).

The index model will use *dimensions, elements* and *items* to derive the state of globalization of a business.

Table 1 General index model

Aspects	Dimensions	Elements	Items
Aspect 1	Dimension 1	Element 1	Item 1
			…
			Item u
		…	…
Aspect 2		Element q	Item v
Aspect 3	Dimension 2	Element r	Item w
			Item x
	…	…	…
	Dimension o	Element s	Item y
…	…	…	…
Aspect n	Dimension p	Element t	Item z

Based on this, items were compiled to measure the aspects of globalization in a third step. This step was then discussed in a dialogue with experts from the institute for border-region-studies at the SDU. On this basis the items were pooled to elements and in a last step the elements were pooled into dimensions.

As shown in Table 1, the model becomes more detailed comparing the levels from the left to the right.

4.2 Five Aspects

As derived in section global business the aspects influencing the concept of globalization are:

1. Economical
2. Geographical
3. Legal
4. Social
5. Cultural
6. Political

These aspects describe the overall scope of the concept of globalization. As this work has to focus on the globalization of a business, the aspect no. 6: political will be left out. Anyhow, the political aspect of globalization will be incorporated indirectly in the index due to the interdependency of the aspects three to five.

4.3   Dimensions

The five remaining aspects are merged into three frames: *economical frame, spatial frame* and *regulatory frame* (See Table 2). This framing of aspects will be called dimensions. Merging of aspects into three frames is conducted for reasons of clarity, applicability and ease of use. Due to content resemblance the aspects three to five are merged into one dimension. All dimensions will be normalized.

4.3.1   Economical Frame

The economical frame dimension combines all economical aspects of globalization. This dimension measures the core of a business. businesses face their economical surrounding and are engaged in economical sourcing. As this frame represents the core of a business it is indicated to use multiple data to calculate its complexity.

4.3.2   Spatial Frame

The spatial frame dimension combines all geographical aspects of globalization. As derived in the previous sections the concept of globalization is highly linked to a global spread of activities. Therefore, the businesses' spread and the firms' spread of activities and market involvements have to be assessed.

4.3.3   Regulatory Frame

The dimension regulatory frame combines the aspects legal, social and cultural of globalization, for a better understanding and usage within the derived index. As stated in paragraph spatial frame, a global business is spread over national borders. This dimension will contain the company's adaption to the requirements the geographical spread goes along with. An example is the comprehensibility of the respective company's communication. Moreover, a business faces three sides of these social and cultural requirements:

First of all, there are employees, which place requirements.

Table 2 Aspects and dimensions

Aspects of globalization	Dimensions
Economical	Economical frame
Geographical	Spatial frame
Legal	Regulatory frame
Social	
Culture	

Secondly, there are customers with their needs and their cultural background.

Thirdly, a certain way of conducting business may exist in a country where the company operates and to which it must adapt to do its business successfully.

Sometimes the aspects social and cultural are used as synonyms. Cultural and social characteristics may refer sometimes to supranational geographical forms but as the basic entity for this work is the national state, both refer here to a national state. The legal aspect of globalization is linked also directly to the national state, as every country has its own legal system.

Hence, the legal aspect of globalization adds to this frame because all three aspects refer to the same body: The national state.

4.4 Elements of the BGI

The next and more detailed level of the index is the level of elements. Elements describe all immanent values of their dimension. The dimensions are constructed out of twelve elements. Due to the preceding normalization it is not problematic to sum all elements into a number for each dimension (see Table 3).

The economical dimension consists of five elements:

1. *International Shares*: The adding value is the main purpose of a business. This element derives how much of this value adding takes place outside the home country of the company. It may show also where the company focuses on.
2. *Network*: This element shows if the company is embedded in a global cluster of production.
3. *Profits*: This element demonstrates the company's ability to profit from the activity in foreign markets.

Table 3 Dimensions and elements

Dimensions	Elements
Economical frame	International shares
	Network
	Profits
	Growth
	Innovation
Spatial frame	Spatial distribution
	Spatial distance
Regulatory frame	Legal structure
	Network
	Social diversity
	Global perspective
	Demand diversity

4. *Growth*: A global company participates in multiple marketplaces. This enables the company to grow beyond the potential of the domestic market.
5. *Innovation*: A global company has the chance for diversification of knowledge sourcing. Furthermore, a spread of innovations may have advantages in owning special knowledge.

In conclusion the normalized dimension economical frame mathematically is shaped as follows:

$$\text{Economical Frame} = [(1/5) \times \text{International shares}] + [(1/5) \times \text{Network}]$$
$$+ [(1/5) \times \text{Profits}] + [(1/5) \times \text{Growth}] + [(1/5) \times \text{Innovation}] \quad (1)$$

The spatial frame combines two elements:

1. *Spatial distribution*: This element shows the spatial spread of activities of the company and its market involvements.
2. *Spatial distance*: This element measures the widest distance between the firms' HQ and a company's subsidiary.

It is represented using two elements. In conclusion the normalized dimension spatial frame is shaped as follows:

$$\text{Spatial Frame} = \big[(1/2) \times \text{Spatial distribution}\big] + \big[(1/2) \times \text{Spatial distance}\big] \quad (2)$$

The dimension regulatory frame consists of five elements:

1. *Legal structure*: This element incorporates the company's ability to use different legal conditions to its advantage.
2. *Network*: Here the company's potential to identify, use and/or create favourable conditions is measured.
3. *Social diversity*: This element assesses the potential to understand, adapt and use cultural diversity of countries on the basis of internal structure.
4. *Global perspective*: This element assesses how a company can understand, adapt and use cultural diversity.
5. *Demand diversity*: This element assesses the degree of adaption the company practices to local markets based on the company's communication.

In conclusion the normalized dimension regulatory frame is shaped as follows:

$$\text{Regulatory Frame} = \big[(1/5) \times \text{Legal structure}\big] + [(1/5) \times \text{Network}]$$
$$+ \big[(1/5) \times \text{Social diversity}\big] + \big[(1/5) \times \text{Global perspective}\big]$$
$$+ \big[(1/5) \times \text{Demand diversity}\big] \quad (3)$$

4.5   Items of the BGI

The most detailed level in this model is the level of items. This is residing above the data compiled from the dataset of firms. The architecture is shown in Table 4. Using a relatively small number of items create the danger of an error. This is when one of these few items and its measures is an anomaly to the rest. This could lead to misinterpretations and therefore this index will try to overcome this threat by using a number of twenty-two distinct items. This ensures an allocation of a minimum of three items per dimension.

Table 4 Elements and items

Elements	Items
International shares	Foreign costs
	Share of foreign assets
	Share of foreign revenues
	Share of foreign sales
	Percentage of foreign shareholders
Network	Percentage of foreign suppliers
	Average market share of the main three markets
Profits	Profits
Growth	Foreign growth rate
Innovation	Share of foreign patents
	Average registration of patents in foreign countries
Spatial distribution	Number of market involvements (in how many markets is the company involved?)
	Number of countries with local affiliates
Spatial distance	Distance between HQ and subsidiary
Legal structure	Number of stock markets listed
	Taxation (general amount/percentage of profit)
Network	Membership in international branch associations
Social diversity	Composition of company's workforces nationalities (GINI)
	Share of foreign employees
Global perspective	Employees experience in foreign activity (years abroad)
	Top Managers (TM) foreign experience
Demand diversity	Local adaption to markets

Another circumstance may cause problems when not taken into consideration. This is exceptionally the danger of different scales within the index. So, a percentage in relation to a total number could become negligible because the percentage would supposedly be relatively small (e.g. a company has thirty-two foreign subsidiaries and a growth rate of 0.25). Taken these numbers as they are, the subsidiaries would outnumber the growth rate; whereas the meaning of both numbers in line with reality may be the other way around. For this purpose, all items will be normalized to prevent this anomaly. Generally, the higher the number the more globalized a company is in this item.

For best use of the index the correct use of data will be thoughtful. For the Items only a time-period of one year will be used. This enables the index to be used on one hand for a situational analysis and on the other hand for an analytical intent over a period of time.

All items are designed in such a way as to produce a number between zero and one.

The international shares element consists of five items. The calculation combines them on a dichotomy of home/foreign values.

1. *Foreign cost* measures the costs a company has in producing its goods and services.
2. *Share of foreign assets* measures the allocation of assets.
3. *The share of foreign revenues* measures the proportion of foreign revenues on the total revenues.
4. *Share of foreign sales* aggregates the amount of sales as a percentage of the total sales.
5. *Percentage of foreign* shareholders measures the percentage of foreign shareholders out of the total shareholders of the company.

Hence the normalized element international shares aggregates as follows:

$$
\begin{aligned}
\text{International shares} = {} & \left[(1/5) \times \text{Foreign cost}\right] + \left[(1/5) \times \text{Share of foreign assets}\right] \\
& + \left[(1/5) \times \text{Share of foreign revenues}\right] + \left[(1/5) \times \text{Share of foreign sales}\right] \\
& + \left[(1/5) \times \text{Percentage of foreign shareholders}\right] \quad (4)
\end{aligned}
$$

Two items construct the element network:

1. The proportion of foreign suppliers to the number of total suppliers.
2. The average market share of the three main markets compared to the market size of the main three markets.

Conclusively the normalized element "Network" aggregates as follows:

$$
\begin{aligned}
\text{Network} = {} & \left[(1/2) \times \text{Percentage of foreign suppliers}\right] \\
& + \left[(1/2) \times \text{Average market share of three main markets}\right] \quad (5)
\end{aligned}
$$

The element profits is measured by only one item:
The proportion of profits after tax (PAT) to the profits before tax (PBT).

The normalized element profits aggregates as follows:

$$\text{Profits} = \left[\text{Proportion of PAT to PBT}\right] \tag{6}$$

The element growth is measured by one item as well. The foreign growth rate is assessed from analytical year to previous year. Also, on a basis of a home/foreign dichotomy this measures how much the company does benefit from growth in foreign markets. To meet the requirement of the work the figures this Element can range are numbers that are bigger or equal to zero and maximally or equal to one.

The normalized element growth aggregates as follows:

$$\text{Growth} = \left[\text{Proportion of foreign growth to total growth}\{0 \leq \text{Growth} \leq 1\}\right] \tag{7}$$

Innovation combines two items.

1. The *share of foreign patents* assesses how many patents are registered outside the home country.
2. The *average registration of foreign patents* combines the registration of patents compared with the previous year in a home/foreign dichotomy.

The normalized element innovation is shaped as follows:

$$\text{Innovation} = \left[(1/2) \times \text{Share of foreign patents}\right]$$
$$+ \left[(1/2) \times \text{Average foreign registration of patents}\right] \tag{8}$$

The spatial distribution element consists of two items:

1. *The number of market involvements* compared to the total number of markets. A market involvement is the availability of the company's products and services in a specific market. As markets are bound to nations (see section global business) the total number of markets is the number of nations on this planet. For general application this item will use the number of nations represented in the UN [30].
2. *The percentage of countries in which the company has affiliates* out of the total number of countries. Here the same total number of countries is used, according to the argumentation of number one.

In conclusion the normalized element spatial distribution aggregates as follows:

$$\text{Spatial distribution} = [(1/2) \times \text{Number of market involvements}]$$
$$+ [(1/2) \times \text{Number of countries in which the firm has affiliates}] \tag{9}$$

The element spatial distance is measured by one item:

The distance between the company's headquarters (HQ) and its furthermost affiliate. This is then related to the distance maximally possible on the planet: The half of the equator: 40.075 km/2 = 20,037.51 km {12,450.73 miles} [31].

In conclusion the normalized element spatial distance is shaped as follows:

$$\text{Spatial distance} = [\text{Distance between HQ and affiliate}] \qquad (10)$$

The element legal structure consists of two items:

1. The number of stock markets on which the company is listed. The World Federation of Exchanges was chosen as representative for the total number of stock exchanges [32].
2. Using the proportion of PAT above PBT, the taxation of a company is assessed.

Hence the normalized element legal structure is shaped accordingly:

$$\text{Legal structure} = [(1/2) \times \text{Number of stock markets on which the firm is listed}]$$
$$+ [(1/2) \times \text{Taxation}] \qquad (11)$$

The element network describes the embeddedness in global sourcing clusters and is measured by the dichotomy of memberships in international branch associations as a proportion to the total number of branch associations in this branch.

In conclusion the normalized element network aggregates as follows:

$$\text{Network} = [\text{Proportion of membership in industry organizations}] \qquad (12)$$

The element social diversity is assembled out of two items:

1. Using the normalized GINI-Coefficient, the composition of workforces' nationalities is assessed [33].
2. The share of foreign employees is calculated in dividing by the total number of employees of the company.

So, the normalized element "Social diversity" is shaped accordingly:

$$\text{Social diversity} = [(1/2) \times \text{Composition of workforces nationalities}]$$
$$+ [(1/2) \times \text{Share of foreign employees}] \qquad (13)$$

Two items build the element global perspective.

1. The share of foreign experience of the company's workforce is aggregated by calculating the years spent abroad divided by the total years of work experience.
2. The top managers foreign experience measured as described in number one. The foreign experience is measured by years of work spent outside the home country of the company.

At first glance the Items may seem similar, but they measure different characteristics of this Element. So, on the one hand top manager have more influence on their company, on the other hand the total workforce produces the products and services of the company. Hence these characteristics are weighted equally.

The normalized element global perspective is calculated as described below:

$$\text{Global perspective} = \big[(1/2) \times \text{Share of foreign experience of workforce}\big]$$
$$+ \big[(1/2) \times \text{Share of foreign experience of top management}\big] \tag{14}$$

The element demand diversity is measured by one item. It assesses if products or services are adapted to local culture: The element uses the availability of the website in a specific language, which is spoken by more than 50 million people as their first language, in proportion to the total number of these languages [10].

The normalized element demand diversity calculates as follows:

$$\text{Demand diversity} = \big[\text{Availability of website in foreign languages}\big] \tag{15}$$

4.6 BGI Model

As previous approaches faced the flaws of combining their calculations directly this index uses a different approach. The index model for the BGI comes in two representations:

1. The index is spanned using three dimensions. This results in the advantage of using a three-dimensional *space* as analytical starting-point. The three dimensions out of sections economical frame, spatial frame, regulatory frame, the associated elements and items result in the BGI Model (BGI) (see Table 5).

A space may have n dimensions. These dimensions can be related in any way to each other. This work will use the three dimensions economical frame, spatial frame and regulatory frame introduced in previous sections of this work as the three dimensions of a space. This space will be used as analytical environment in which the dimensions are orthogonal to each other.

As a consequence, the three dimensions will be used as coordinates of a vector. This indicates the BGI as a normalized vector. The vector gives some advantages in the practical applicability of the BGI on companies. So, influences of one dimension can be visualized, analyzed and evaluated. Furthermore, an enhancement or adjustment of the index is possible within this theoretical framework.

Conclusively, the normalized vector BGI (V) calculates as follows:

$$\text{BGI}(V) = \langle \text{Economical Frame} \mid \text{Spatial Frame} \mid \text{Regulatory Frame} \rangle \tag{16}$$

2. Due to the general guideline for this work of condensing all findings into one final number this is accomplished by using the length of the BGI-Vector. To aggregate a final number between zero and one, all frames, elements and items

Table 5 BGI Model

Dimensions	Elements	Items
Economical frame	International shares	Foreign costs
		Share of foreign assets
		Share of foreign revenues
		Share of foreign sales
		Percentage of foreign shareholders
	Network	Percentage of foreign suppliers
		Average market share of the main three markets
	Profits	Profits
	Growth	Foreign growth rate
	Innovation	Share of foreign patents
		Average registration of patents in foreign countries
Spatial frame	Spatial distribution	Number of market involvements (in how many markets is the company involved?)
		Number of countries with local affiliates
	Spatial distance	Distance between HQ and subsidiary
Regulatory frame	Legal structure	Number of stock markets listed
		Taxation (general amount/percentage of profit)
	Network	Membership in international branch associations
	Social diversity	Composition of company's workforces nationalities (GINI)
		Share of foreign employees
	Global perspective	Employees experience in foreign activity (years abroad)
		Top Managers (TM) foreign experience
	Demand diversity	Local adaption to markets

have been normalized into a spectrum of numbers between zero and one to enable the aggregation into a final number.

According to the formula for calculating the length of a vector the un-normalized value aggregates as follows:

$$BGI(un-normalized)$$
$$= \sqrt{Economical\ Frame^2 + Spatial\ Frame^2 + Regulatory\ Frame^2} \quad (17)$$

This leaves a maximum value of:

$$BGI(un-normalized) = \sqrt{1^2 + 1^2 + 1^2} = 1.73 \quad (18)$$

To meet the requirement of the guideline the formula is divided by its maximum. This describes the form of the calculation and will be called BGI number (BGI (N)):

$$BGI\ (N) = \frac{\sqrt{Economical\ Frame^2 + Spatial\ Frame^2 + Regulatory\ Frame^2}}{\sqrt{1^2 + 1^2 + 1^2}}$$

(19)

5 Illustrating the BGI

In this chapter the operationalization of the BGI is shown according to the findings of the underlying work [34]. As stated in Sect. 1.2, it is for illustrative purposes, to show the easy applicability of the architecture developed.

For a valid operationalization it was ensured, that a minimum of three items per dimension were compiled from the dataset of the firms chosen. The sample was generated using the world's top 100 non-financial MNCs as a basis and consists of the Volkswagen Group, Toyota Motor Corporation, General Electric Co and ABB PLC [35, 36].

For calculating the BGI twenty-two measures were compiled out of a bunch of data, which relate directly to the items [34]. For illustrating purposes, the data of the four mentioned firms were compiled from available sources. By making sure every dimension was served by a minimum of three items, 55% of the items could be compiled. The items that could not be compiled missed for example one out of the two values needed. During the process of data-sourcing attention was paid to correct sourcing.

Items, where no data could be compiled are calculated with a zero. Using this form of calculation, the data show a minimum degree of globalization of the company measured. If more data would be used the degree globalization could rise only.

For the evaluation the data of the firms form a consistent database. Data that differ from the concentrated data are noted within the model. This means, that for all firms the same measures were derived to ensure comparability.

Table 6 shows the BGI (N) computed for each company and its position in the ranking.

If a company scores low on the BGI this does not imply a direct disadvantage by comparison to other firms with a higher BGI (N). Firms may for example have

Table 6 Company analysis results [34]

Company	BGI (N)	Rank
Volkswagen Group	0.3812	4
Toyota Motor Corporation	0.4620	3
General Electric Corp.	0.4656	2
ABB PLC	0.4853	1

reasons to allocate assets locally. Moreover, a focus-strategy on a few marketplaces can be reasonable for the company's purpose. Vice versa if firms score high on the BGI number this does not imply a direct competitive advantage over other firms that score low.

The sample shows that the firms score around half of the maximal possible value of the BGI (N). Consequently, a higher degree of globalization is possible.

ABB's maximum in the dimension *economical frame* of 0.7686 still leaves room for further globalization [34]. The analysis of the sample shows that the potential of ABB as the most globalized company is still nearly 0.25 in this field. Conclusively the overall potential of the entire sample for further globalization is still high. Notably ABB ranks in this sample first, judged by the value of the dimension *economical frame*, which is remarkably low. This does not imply that the aim for a company to score high on the BGI has to be in creating low economic figures. As example Toyota with the highest value of the dimension *economic frame* ranks third and therefore second last. In fact, this ranking shows the validity of the index framework. The index is shaped in order to prevent a proportional bigger influence of parts of the index than others have. This also prevents any parts of the BGI to be under-represented.

Conclusively this index provides two analytical tools: On the one hand the BGI Vector and the BGI Number on the other hand. Both can be used for analytical implications, which will be examined in the following Sect. 6 management implications.

6 Management Implications

Generally, it has to be noted, that before using the BGI as a basis for management actions in order to increase the degree of globalization of the company, the coverage of data has to be considered. In this sense the actions have to be shaped. So, management implications are at least questionable if they are based on a sample with coverage of 20% of the items only. For solid implications a higher coverage is recommended.

This section derives its implications based on the analysis of the findings in the underlying work [34]. This implies that due to the scope and aim of the underlying work more detailed implications are not yet possible. For further specific management implications a case study of each individual company should be performed. A case study will then consider multiple factors such as the financial, historical, environmental situation plus the mission and goals of the company chosen.

This may also make the BGI useful for two application areas: First it could be provided for firms to analyze themselves and derive management implications and secondly for use within science and educational purposes to analyze firms.

As the following implications are based solely on the analysis of data available and used to this work, the implications are adjusted accordingly.

First, a company who scores a high value within the dimension *economical frame* of the BGI may experience advantages because a global growth minimizes the threat of dependency on certain markets. This opens the chance to participate in the growth of markets around the world. A more globalized allocation of assets gives the opportunity to produce and service markets globally. On top of that a high value may indicate global knowledge sourcing through patents and also a global perspective because of the influence of a global shareholder community.

Secondly a company may experience competitive advantages because of the spread of market-involvements. The physical presence within a lot of different countries makes them agile and as a consequence this company may serve markets better than competitors.

Thirdly a competitive advantage may arise for a company with a high value in the dimension *regulatory frame*. This could be more and better knowledge-sourcing due to its global workforce. For example, a global workforce may use different approaches to solve an assignment. They could solve problems faster/easier or create innovations. Furthermore, a high value in the element global perspective may indicate better responsiveness of a company to requirements that for example could arise from the market demand, state regulations or the own workforce.

Hence a globalized company may have competitive advantages arising out of their state of globalization. As the BGI is constructed for measuring this degree, a high value might indicate these competitive advantages. Here the representation of the BGI as a vector can help to indicate where advantages are to be found. On this basis strategies can be developed to obtain more competitive advantages. Of course, this has to be made in consideration of the company itself (e.g. its financial situation to implement strategies or else) and its missions and goals-statement. Strategies that go in hand with this should be developed.

7   Critical Evaluation

Starting with possible disadvantages of the proposed index one thing becomes evident:

The broad spectrum of measurements needed may lead to difficulties, mistakes or extensive work in data generation for the sample.

Generally, the BGI is not designed to evaluate the future success of a company directly. It may indicate areas of possible competitive advantages and eventually possible threats to the company.

Some items of the index may be subjective or do not cover all inherent aspects. Examples are the top managers foreign experience, which could only be examined by using the web-presence of the firms. Sometimes the data presented are not detailed enough. The number of languages, which is limited to the first language speaker, also misses the fact, that a lot of people are English speaking as second language and the coverage of the websites may be higher than measured. The last example is the number of stock exchanges. This item uses the Federation of Exchanges, which does not represent all exchanges worldwide, but the biggest ones.

An advantage of the BGI is the use of multiple measurements. It does decrease the risk of measurement-errors or misleading interpretations through anomalies within the data-basis. Furthermore, the danger of the influence of a subjective, misleading or wrong item is decreased because of the index structure. Furthermore, this structure can help to identify these anomalies and indicate implications for the management to change them.

Due to the aim, sources and scope of the underlying work, the reachable coverage of some of the measurements may sometimes be not optimal. One example is, that the concept of glocalisation does cover more than the used availability of the company's web-presence in the native language of the foreign market. There are for example other possible aspects of glocalisation: Like the necessity to access the culture in terms of use of language for product names. (This would not have an unintended meaning in the foreign language.) Furthermore, other cultural aspects have to be considered in foreign markets and countries.

Another advantage of the proposed BGI is for example: The potential to extend the index by new requirements or knowledge. The architecture and its implementation make it possible to increase the number of items and even elements. This is an advantage because the validity can be improved.

Moreover, the structure of the index enables an evaluation over time. This is another aspect that may help to minimize the effect of anomalies within the data.

Additionally, the proposed index is shaped according to the aspects of globalization covering all relevant characteristics of a globalized business.

8   Indication for Further Work

The BGI uses twenty-two measures some of which are easy to retrieve and some not. Some data even require own data-generation or data transformation for use of the BGI.

There might be two areas for further work: Improvements of the BGI and general indications for the topic.

Within Sect. 4 it was identified that the three dimensions of the BGI are orthogonal. This implies that the immanent values of the dimensions don't have to be depending on each other.

Dimensions immanent are not dependently measured, but possibly the measured "things" are depending. This has to be identified and—if found—excluded out of the index in further studies of the concept of globalization.

The BGI could be improved by using more items within the spatial frame for better representation. This could also decrease the risk of influence of one of the items.

Furthermore, the item that has to be mentioned is the top managers foreign experience. It is likely to miss periods of foreign experience due to the form of publishing. Moreover, the foreign experience is measured in years spent entirely abroad. But the top management may spend during a year one or two month in a foreign country. This would increase the foreign experience of the management but is not considered in the index today. So, this item could be improved if more detailed data could be used.

Additionally, the coverage of the websites might be broader, than it is measured. This is because the number of languages considered in the index is limited to the languages spoken by more than 50 million people. In addition to that only the first language of a person is considered, this means, that other languages the person may understand as well are not measured. So, a more complex measurement could also improve the BGI.

The last item considered is the total number of stock exchanges on the planet. The global number of exchanges is higher than exchanges are members of the World Federation of Exchanges (WFE). But these other exchanges are relatively small in comparison to the WFE members considering the magnitude of global finance. So, this was chosen as a doable approach.

The sample of the section illustrating the BGI contains data for 55% of the items. Due to the guidelines and the timeframe of this work 45% couldn't be retrieved. These items may change the ranking within the index and in conclusion the strategic implications of Sect. 6 should to be revised, if the limitations posed by the guideline of using data free of charge is abolished.

The previous approaches have shown that there are merely no statements as: A company has to be considered global if it has reached a particular state of globalization. Attributes such as international, transnational and global are hard to define what state of globalization they try to describe.

In general, the definition of globalization and its boundaries to other similar concepts is still very difficult and hard to define. This area leaves a lot of potential for future research. In particular the effects of globalization on businesses could/should be examined more closely.

9   Conclusion

This chapter has found, that globalization is an autonomous concept, which can be differentiated from other concepts such as internationalization, liberalization or westernization.

Six aspects of globalization are influencing the concept of globalization. These are the social, political, geographical, economical, cultural and legal aspects. Furthermore, technology is another driver of globalization but this was identified as supporting influence.

Considering previous approaches to this topic this work has found that existing measures do not describe the entire concept of globalization. They do provide a broad spectrum of possibilities, but with them they measure only some aspects of globalization. Moreover, some underlie measurement errors. The approaches published by Vernon, Sullivan, UNCTAD and Ietto Gilles did not provide a differentiated view on globalization as a concept. The other approaches from Fisch, Oesterle and Asmussen, Pedersen, Petersen do provide a differentiated view on the concept. But with their measures they cover only one level of detail of the concept of Globalization.

Sections 5 and 6 have derived that businesses are influenced directly by five aspects of globalization. These aspects are the economical, geographical, social, cultural and legal aspect of globalization.

As deduction of the previous annotated this work suggests the BGI. Based on the aspects three dimensions where developed. The dimensions were the economical frame, spatial frame and regulatory frame. These three dimensions do consist out of twelve elements. Which in turn consist of twenty-two items. The items are intended to determine mathematically the concept of globalization for businesses.

The BGI provides two tools: The BGI (V) and the BGI (N). Both tools can be used for analytical purposes and are easy to use. The BGI measures all aspects of globalization and fulfills the criteria stated in the introduction to be modular, extendable and adaptable.

Afterwards, the BGI was illustrated shortly, using a sample of four companies. These were Volkswagen, Toyota, General Electric and ABB. Required data were retrieved and evaluated for these firms. The evaluation has shown that the most globalized company within the sample is ABB with a BGI (N) of 0.4853. The other firms in the sample were VW with a BGI (N) of 0.3812 as the least globalized, Toyota with 0.462 and GE with 0.4656. Generally, the highest value of the sample was the dimension spatial frame of ABB with a value of 0.7686. This indicates that the potential for further Globalization is still high within the sample. More precisely, this shows that the sample of the world's largest firms is half globalized with ABB BGI (N) of 0.4853 in 2013. Therefore, the potential for further Globalization is more than 0.5.

Considering the outcome of the sample-evaluation, management implications have been derived both in general and specific for the chosen sample. This has shown, that on one hand a high BGI (N) does not imply a high competitiveness of a company. On the other hand, the BGI (V) may help to identify areas for potential of globalization, which then in turn might lead to a competitive advantage for the company. The chapter also derived that before developing strategies for increasing the BGI (N) the firms have to be assessed according to multiple factors. Such a case study should be prepared before establishing the management implications. Furthermore, these decisions have to be made on a basis of valid data.

The BGI (V) identified the areas with the highest future potential for globalization within the sample. These were for all firms two dimensions: The economical frame and the regulatory frame with a minimum potential for the sample of 0.65.

Then the BGI was examined: the work found potential for further improvement and possible obstacles within the operationalization of the index. One example is that the data retrieving process can be long and difficult. This chapter concludes that the BGI is valid.

The BGI is the only index available measuring all aspects of globalization this work has identified. Furthermore, the BGI can compare firms over time and in addition to that the BGI (V) can support the easier development of strategies for the company.

Generally, the whereabouts of the concept of globalization as such and the intersection of businesses and globalization deserve further research.

Acknowledgements This chapter was evolved from a thesis, written at the University of Southern Denmark, Institut for Grænseregionsforskning under the kind and competent supervision of Prof. Dr. S. Albers [34]. The full paper may be retrieved from the author.

References

1. Oxford University Press (2014) Oxford English Dictionary. http://www.oed.com.proxy1-bib.sdu.dk:2048/view/Entry/98072?redirectedFrom=international#eid. Accessed 08 April 2015
2. Oxford University Press (2014) Oxford English Dictionary. http://www.oed.com.proxy1-bib.sdu.dk:2048/view/Entry/272264. Accessed 08 April 2015
3. Dicken P (2011) Global shift—mapping the changing contours of the world economy. SAGE Publications Ltd, London, UK
4. Kryvinska N, Gregus M (2018) Agile information business: exploring managerial implications. Springer Science + Business Media Singapore, Singapore
5. Google Inc. (2015) Google Scholar. https://scholar.google.de/scholar?q=Globalization%2BDefinition&btnG=&hl=de&as_sdt=0%2C5. Accessed 08 April 2015
6. Wöhe D, Döring D (2010) Einführung in die allgemeine Betriebswirtschaftslehre. Verlag Franz Vahlen, Munich, Germany
7. Encyclopædia Britannica Inc. (2015) Encyclopædia Britannica. http://www.britannica.com/EBchecked/topic/397067/multinational-corporation-MNC. Accessed 11 April 2015
8. Bibliographisches Institut GmbH (2015) Duden Online. http://www.duden.de/rechtschreibung/Globalisierung. Accessed 08 April 2015
9. Google Inc. (2015) Google Translator. https://translate.google.com/#en/. Accessed 13 May 2015
10. SIL International Publications (2015) Summary by language size. http://www.ethnologue.com/statistics/size. Accessed 13 May 2015
11. Scholte JA (2008) Defining globalization. World Econ 1471–1500
12. Morschett D, Schramm-Klein H, Zentes J (2010) Strategic international management text and cases, 2nd edn. Gabler Verlag/Springer Fachmedien, Wiesbaden, Germany
13. International Monetary Fund (2015) WEO groups and aggregates information. http://www.imf.org/external/pubs/ft/weo/2014/02/weodata/groups.htm#ae. Accessed 02 March 2015
14. Krugman PR, Obstfeld M, Melitz MJ (2012) International economics theory & policy, 9th edn. Pearson Education Inc., Boston, United States of America
15. Schäfer U (2008) Der Crash des Kapitalismus: Warum die entfesselte Marktwirtschaft scheiterte und was jetzt zu tun ist. Campus Verlag GmbH, Frankfurt am Main, Germany
16. Phatak AV, Bhagat RS, Kashlak RJ (2009) International management, 2nd edn. Mc-Graw Hill, Boston, United States of America
17. Holm K, Strauss C (1998) Industrial training issues in the Middle East. Ind Commer Train 30(7):242–245
18. Hollendsen S (2010) Marketing management a relationship approach, 2nd edn. Pearson Education Limited, Harlow, England
19. Ryder NB (1965) The cohort as a concept in the study of social change. Am Sociol Rev 30(6):843–861
20. Porter ME (1985) Competitive advantage—creating and sustaining superior performance. The Free Press A Division of Macmillan Inc., New York, United States of America
21. Statistisches Bundesamt Deutschland (2015) Genesis Online Datenbank. Accessed from Ausstattung privater Haushalte: https://www-genesis.destatis.de/genesis/online/data;jsessionid=5FB48A4C84BA15EE1002CC74417A02DA.tomcat_GO_1_2?operation=abruftabelleBearbeiten&levelindex=2&levelid=1430725554614&auswahloperation=abruftabelleAuspraegungAuswaehlen&auswahlverzeichnis=ordnungsstruktur&auswahlziel=werteabruf&selectionname=63111-0001&auswahltext=%

23SGUTGB2-PC-01%2CPC-STAT-01%2CPC-MOBIL-0%2CINTERNET-0&nummer=4& variable=2&name=ATG001&werteabruf=Werteabruf. Accessed 04 May 2015
22. Meijers H (2013) Does the internet generate economic growth, international trade, or both? http://link.springer.com/article/10.1007/s10368-013-0251-x/fulltext.html. Accessed 15 May 2015
23. UNCTAD; United Nations Conference on Trade and Development (1998) Different conceptual frameworks for the assessment of the degree of internationalization: an empirical analysis of various indices for the top 100 transnational corporations. Transnational Corporations, pp 17–39
24. Vernon R (1979) The product cycle hyphothesis in a new international environment. Oxf Bull Econ Stat 255–267
25. Sullivan D (1994) Measuring the degree of internationalization of a firm. J Int Bus Stud 325–342
26. UNCTAD; United Nations Conference on Trade and Development (1995) World investment report 1995 transnational corporations and competitiveness. United Nations, Geneva
27. Fisch J, Oesterle M-J (2003) Exploring the globalization of German MNCS with the complex spread and diversity measure. Schmalenbach Bus Rev 55:2–21
28. Asmussen C, Pedersen T, Petersen B (2007) How do we capture "global specialization" when measuring firms' degree of globalization? Manag Int Rev 47:791–813
29. Daft J, Albers S (2013) A conceptual framework for measuring airline businesses model convergence. J Air Transp Manag 28:47–54
30. United Nations (2015) United Nations Member States. http://www.un.org/en/members/. Accessed 19 May 2015
31. National Geographic Society (2015) Equator-National Geographic Education. http://education. nationalgeographic.com/education/encyclopedia/equator/?ar_a=1. Accessed 19 May 2015
32. World Federation of Exchanges (2015) Key information/World Federation of exchanges. http:// www.world-exchanges.org/member-exchanges/key-information. Accessed 19 May 2015
33. Springer Fachmedien Wiesbaden GmbH (2015) Definition >GINI-Koeffizient< / Gabler Wirtschaftslexikon. http://wirtschaftslexikon.gabler.de/Definition/gini-koeffizient.html. Accessed 19 May 2015
34. Näth-Sießegger LC (2015) Development of a business globalization index (Thesis). Accessed from the author
35. UNCTAD; United Nations Conference on Trade and Development (2015) The world's top 100 non-financial TNCs, ranked by foreign assets, 2011. http://unctad.org/SearchCenter/ Pages/results.aspx?k=The%20world's%20top%20100%20non-financial%20TNCs%2C% 20ranked%20by%20foreign%20assets%2C%202012. Accessed 12 May 2015
36. UNCTAD; United Nations Conference on Trade and Development (2015) The world's top 100 non-financial TNCs, ranked by foreign assets, 2013. http://unctad.org/SearchCenter/ Pages/Results.aspx?k=The%20world's%20top%20100%20non-financial%20TNCs%2C% 20ranked%20by%20foreign%20assets%2C%202013. Accessed 12 May 2015

Stress Testing Corporate Earnings of US Companies

Davide Benedetti and Rastislav Molnar

Abstract This paper defines a stress testing framework for corporations. Stress testing is well known in the financial sector, however it is not well established in enterprise risk management for corporations. We believe stress testing, if adapted for corporations, could play a key role in their risk management. We will test proposed framework on US publicly traded companies using two events of interest, the dot-com bubble of 2001 and financial crisis of 2008. We use OLS, Elastic-Net and Partial Least Squares Regressions (PLSR) to predict the change of corporate earnings during these events. Results suggest our proposed factors in combination with Elastic-Net or PLSR are able to predict movements of earnings better than standard approaches and OLS methodologies. In our opinion, such measurements should be done by all companies in order to make stakeholders aware what can happen if crisis hits the economy and prepare for such eventuality.

Keywords Stress testing · Enterprise risk management · Dot-com-bubble · Financial crisis · Earnings · Forecast

1 Introduction

Enterprise risk management or ERM is a set of methods and processes used by organizations to manage their risks and opportunities in order to achieve their objectives. There are several frameworks companies use for their risk management. For example, one defined by Committee of Sponsoring Organizations (COSO) who define ERM as a process implemented by corporates in strategy settings to identify and manage potential risks affecting the business operations [36]. In this paper we do not focus on ERM frameworks, nor we focus on a way how to manage risks. The

D. Benedetti (✉) · R. Molnar (✉)
Imperial College Business School, South Kensington Campus, London SW7 2AZ, UK
e-mail: d.benedetti@imperial.ac.uk

R. Molnar
e-mail: r.molnar12@imperial.ac.uk

© Springer Nature Switzerland AG 2020
N. Kryvinska and M. Greguš (eds.), *Data-Centric Business and Applications*,
Lecture Notes on Data Engineering and Communications Technologies 30,
https://doi.org/10.1007/978-3-030-19069-9_14

purpose of this paper is to define and test a framework for measuring financial risk of companies under stressed conditions. We take company's balance sheet and macroeconomic variables and analyze the interaction between them. In particular we are interested in predicting the shock transferred from sudden change in macroeconomic variables on corporate earnings. For exact implementation of ERM and stress testing into company's IT, please refer to Kryvinska [31].

We believe our framework is interesting from a number of perspectives. First, the literature on stress testing mainly focuses on credit and financial risks from the perspective of financial institutions,[1] while we focus on stress testing of balance-sheet from a corporate perspective. To the limited extend this was done by Frank Wolf in his contribution to Encyclopedia of IST where he discussed stress testing applied to corporations, municipalities and supply chain [17]. This contribution discusses the motivation for corporate stress testing and is very limited in terms of data and results. On the other hand, it points out the importance of stress testing and how it is present in our lives in banks through regulation, in computer software as well as in engineering applications. Despite not being required we agree with F. Wolf with the notion this voluntary process is of a lot benefit for corporations. It can help to reveal potential loss of revenues, which is the topic of this research, as well as interruptions in supply and order chains, workforce or even manufacturing capacities.

On the top of mentioned, we are considering machine learning approaches to stress testing. Over the past decade, machine learning is becoming a well established practice in an increasing number of companies, for more information see Molnar et al. [37] who discuss several methods as well as several cases how machine learning can be utilized for communication with customers. We test both Elastic Net and PLSR as well as traditional OLS regression in order to see which is the most suitable to stress test earnings. The major advantage of both Elastic-Net and PLSR is that they allow the users to include a large number of potential factors affecting earnings, even when the number of covariates is larger than that of observations. Elastic-Net achieves this by performing a penalised least square estimation [48], while PLSR reduces the dimensionality of the risk factors in a way which resembles principle component analysis [47]. From the industry perspective, a structured approach to risk measurement will support enterprise risk management and help to understand potential risks in the case of economic downturn as well as to predict impacts of such downturn better.

In this research, we propose a formal stress testing framework and discuss its individual steps as well as a number of potential methods that can be used. We have decided to test our framework on US data, by focusing on predicting the change of corporate earnings during dot com bubble of 2001 and financial crisis of 2008. Our results suggest both Elastic Net and PLSR do a good job predicting changes, while benchmark OLS regression yields inferior performance. This effectively implies the stress testing as proposed by this research can be used to predict earnings of companies under "crisis" conditions.

---

[1]For more information see Ong and Cihak [38] or other IMF papers in the Guide to IMF Stress Testing their paper introduces.

The rest of the paper is organized as follows. In the second and third section we define and discuss general stress testing and our stress testing framework respectively. In the fourth section we discuss data and variables we used. Fifth section looks at descriptive statistics as well as plots data and discusses them. Sixth section contain results and discussion about their implications, and seventh section concludes on our research and points out possible avenues of future research.

2 Stress Testing

Bank for international settlements describes stress testing as "an important risk management tool" used by banks for risk management [6]. It should alert bank management about possible unexpected outcomes related to a variety of risks. It fits well in Basel II recommendation on banking laws and regulations as it helps to assess risks. Although this paper does not focus on stress testing in banks and other financial institutions, stress testing methodology and scenario building was focus of research in this field. Our ambition is to bring the stress testing paradigm to corporations as we believe it to be the best known and systematic way of possible risk measurement under extreme conditions.

In order to establish a stress testing framework, let us first discuss existing frameworks and methods for banks. Probably the most comprehensive review of stress testing methods and models is provided by L. L. Ong and M. Cihak in the "A Guide to IMF Stress Testing: Methods and Models" [38]. They did a great job to present existing methods and models in a systematic way, categorize them as well as analyze their strengths and weaknesses. They point out stress testing is primarily used in banking sector and to some extend in corporations and households, however this is discussed in relation to financial sector and banks. We argue despite existing methodology focuses primarily on credit risk, same methods can be applied to simulate corporate earnings and other indicators under stressed conditions.

IMF guide to stress testing splits stress testing into three categories, based on the data used [38]:

- *Accounting based stress testing*
- *Market-price stress testing*
- *Macro-financial stress testing*

With accounting-based approach utilizing accounting data from financial statements of institutions, market-price approach relies on market prices of financial instruments and macro-financial approach linking financial and non-financial sectors of economy by utilizing macroeconomic factors on the top of accounting or market data. Regarding the categorization, we position our approach as macro-financial stress testing, as we combine balance sheet data with financial and macroeconomic information.

On the top of aforementioned categorization based on data sources, we split stress tests by methods they use. Please note here we don't aim to distinguish between cases in which given methods are applied, we are interested in methodology behind the stress tests.

- *Basic/spreadsheet methods*
- *Statistical methods*
- *Machine learning methods*

Basic methods are simplest spreadsheet approaches to stress testing. The advantage of these methods is they are simple to use and easy to understand. They often involve basic mathematical operations only done without any issues in Excel or with a calculator. An example of such stress testing on banks is provided by Ong et al. in their 2010 IMF paper, who used ad hoc stress testing scenarios to stress bank balance sheets [39].

"Statistical methods" is a single term used for all methods based on statistical approach. These methods often involve a statistical model in combination with simulations. For example, Bellotti and Crook [8] used Monte Carlo simulations and dynamic models to stress test credit card defaults with self-updating model to predict Value at Risk. Avesani et al. [3] used modified CreditRisk+ model to model probability distribution of losses on a portfolio of loans, for more information on CreditRisk+ please see [33].

Machine learning methods are relatively new entry in the stress testing methodology. The main idea is to overcome drawbacks of statistical methods, mainly the overfitting issue as well as improve the pattern recognition and precision of models. For example, Tobback et al. [43] used machine learning approach to predict bankruptcy for small and medium-sized enterprises.

Regarding the methodology behind stress tests, we position our research in the machine learning methods as we utilize Elastic net and PLSR methodology. We believe further tests could be performed using neural networks or other more advanced machine learning algorithms in the future.

It is clear a large body of literature about the stress testing in banks exists and it covers a number of different data sources as well as methods. We aim to apply this knowledge to define a stress testing framework for companies. Our proposed framework as well as details on methodology is provided in section below.

3 Stress Testing Framework

For a definition of a stress testing process by IMF, please refer to papers by Jones and Hilbers [27] or Cihak [15], they define it from a more complex perspective as they focus on identification of vulnerabilities, scenario construction and then mapping and interpreting results. We believe these are important parts and one should consider what variables make sense to stress test and what their impact might be, but in most common cases, when testing revenues, account balance or stock price the impact

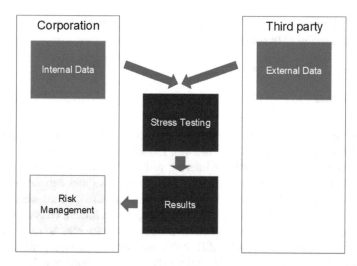

Fig. 1 Stress testing framework as proposed for corporations

is very clear. What might be not clear is the formal mechanics on how to perform the stress test itself, the scenario selection would be a topic for whole new research, while in this paper we focus on models themselves. Our aim is to define formal steps in order to successfully stress test corporate earnings. It is important to note this approach can be generalized to any company level dependent variable including other balance sheet items or stock price. On the other hand, additional factors would have to be considered, for example in the case of stock prices one would have to take asset pricing models into account as well.

Figure 1 shows a big picture stress testing framework. We can split the process into three parts, the first part is data collection, where we gather data from inside the company as well as external data. External data are in general our independent variables or factors we use to explain dependent variable of interest, in the case of this research corporate earnings. Once all data are in place, the second part is the Stress testing itself. Here we decide on a scenario we would like test as well as the method we use. After this, results are produced and outputs could be used internally in the company for their risk management.

The example usage would be if company wants to see what would happen to their earnings growth rate in the case of similar crisis as 2008 financial crisis was. In this case we would take data from recent years both factor data as well as earnings and estimate the model in the preparation phase. After model estimation, we would simulate a change in distribution function to match changes in distribution of factors during 2008 financial crisis, then we use these data to create a 90–99% confidence intervals for earnings change. Resulting number is simulated earnings change that could be used in the company for risk management. We argue by stressing multiple balance sheet items company can better prepare for shocks to economy and with potentially larger body of risk factors to other socks as well (such as natural disasters or political instability in some countries).

3.1 Process

In particular it is important to describe the process behind model estimation and calculation of stressed variables of interest. Detailed process is described in Fig. 2. It all starts with collecting data of interest, in our case we collected a number of factors, these variables are proxies of a number of risks and dependent variable quarterly earnings as described in Sect. 4. These data are used to estimate the model, in this research we consider three competing models, traditional OLS regression as well as models that automatically penalize insignificant variables, so called machine learning models, Elastic net and PLSR. As we include a large number of factors and in the future the list should expand in order to cover a larger body of risks, we believe such models are of a great benefit. For more information on regression models, please refer to the Sect. 3.2.

Once model is estimated, next steps involves applying stressed conditions to variables of interest. This is the step where stress testing scenarios play a role as different scenarios will yield different results. In general, we can split scenarios into two categories, the first category are ad hoc scenarios (or hypothetical scenarios as defined by Peria et al. [40]). These tests are created by setting stress test conditions and factors by ad hoc numbers or changing their existing distribution by some factor. The second type are historical events, these stress tests set factors to the level that existed during certain events in the past. We argue none of mentioned is superior and both types have their own advantages and disadvantages. While historical events are more familiar for everybody to understand as well as provide a good picture of what may happen to variable of interest, ad hoc stress testing scenarios are able to show

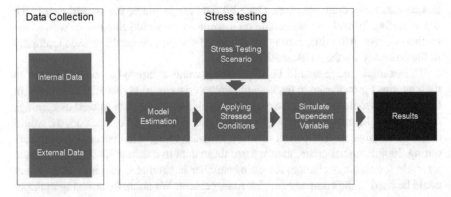

Fig. 2 Step by step stress testing process as proposed by our stress testing framework

what could happen in more extreme and unlikely, but possible cases. We select the stress testing scenario and apply its conditions to our data.

After this, the last step is to run a number of simulations, using stressed factors on model that was estimated using real data. At the end, we end up with a distribution of possible outcomes for our variable of interest and we are able to produce results we want including means and confidence intervals. Results can be later reported to the risk management of the company.

The stress testing as a process is relatively simple and straightforward, what is not straightforward are individual steps. Which variables should be included in the model, which regression model or alternative methodology should be used, which stress testing scenario selected and how results should be reported is a separate question. In this research we focus on variable selection and methodology, while we leave both, stress testing scenario selection and result reporting for future research.

3.2 Models

As mentioned above in Sect. 3.1, the selection of model used in stress testing is a big question. The simplest models used in practice are OLS regression models. Examples are models defined by IMF, for example by Peria et al. [40] or Chan-Lau [13], alternatively some other models used the vector auto regression (VAR), for example Vazquez et al. [44]. Whit recent work pushing towards machine learning approaches, for example Tobback et al. [43]. In our case, we have decided to use OLS regression as a benchmark and see how it will compare in performance with two competitive models, the Elastic Net and Partial least squares regression (PLSR).

It is important to note the Elastic Net (or combination of LASSO and ridge methods) as well as PLSR are considered to be machine learning approaches. This is mainly to the fact they try to compensate for model overfitting. We can argue the main idea behind machine learning algorithms is to be able to construct model in a way it does not learn only data, but is able to recognize patterns and predict out of sample. Both Elastic Net and PLSR compensate for this issue. This has one main implication for us, simply put, we don't have to be concerned about number of factors we include in our models as these models "get rid" (penalize) insignificant factors. Following subsections discuss individual models used in this research in greater detail.

OLS. The OLS regression is straightforward and well defined in the literature. It targets to estimate parameters of a linear functions based on a set of explanatory variables. Despite being simple it has a several key assumptions that are weaknesses. Especially the assumption of correct specification model relies on the fact the function was defined to match actual data generating process. Another weak spot are the assumption of no linear dependence of factors and normality of their distribution. In a nutshell it is relatively easy to miss-specify the model.

$$y_i = \beta\, x_i^T + \varepsilon_i \tag{1}$$

As presented in equation above, the OLS regression is a simple linear regression where we have a set of factors x and use the regression to estimate coefficients β in order to predict variable of interest (or dependent variable) y. We have decided to use this method as a benchmark for our other, more advanced methods.

Elastic Net. Elastic net, or elastic net regularization, is the method that combines LASSO (least absolute shrinkage and selection operator) and ridge (or Tikhonov regularization) regressions and it solves existing limitations of both methods. It is a convex sum of penalties presented by both underlying methods, which results in error model to look like presented in equations below. For more information on elastic net, please refer to Zou and Hastie [48]. Following paragraphs describe the elastic net further. Estimated coefficients for elastic net method are defined as:

$$\hat{\beta} \equiv \underset{\beta}{\arg\min} \left\{ \frac{1}{1N} \sum_{i=1}^{N} \left(y_i - {}'_i\beta \right)^2 + \lambda \left(\alpha \sum_{j=1}^{P} |\beta_j| + \frac{(1-\alpha)}{2} \sum_{j=1}^{P} \beta_j^2 \right) \right\} \tag{2}$$

where λ > 0 is a user supplied parameter determining the importance of the penalty, the error model then takes a form:

$$\alpha \|\beta\|_1 + \frac{(1-\alpha)}{2} \|\beta\|_2^2 \tag{3}$$

where, $\alpha \in [0, 1]$ specifies which one of the two condition (LASSO and ridge) is more important. Please note this is a tuning parameter and established methods could be used to choose this parameter correctly, for example see Hastie et al. [21]. In our case we use leave-one-out cross validation (LOOCV) to estimate tuning parameters.

In general, elastic net is known and expected to produce a model with good prediction accuracy even in the case of high dimensional data with low number of observations, which translates to our case as a lot of macroeconomic factors to which we have only limited data due to their frequency. Another big advantage is the penalties take care of overfitting with factors, this allows us to "throw" as many factors as we want to the model, without the fear of miss-specifying it.

PLSR. PLSR, or Partial least squares regression, is a regression that is somehow related to principal component regression. It was developed by Herman Wold [47] more than 50 years ago and it aims to estimate a linear regression model by predicting variables of interests as well as factors to a new space. This method is particularly useful in the cases of many factors with relatively low number of observations as well as in the case of multicollinearity between factors. For more detailed introduction to PLSR, please see Haenlein and Kaplan [20] The general model for PLSR is defined in following equations for both X and Y variables:

$$X = TP^T + E \tag{4}$$

With matrix of exogenous covariates X (N × P), matrix of exogenous latent factors T (N × K), matrix of factor loading P (P × K), matrix of errors E (N × P).

$$Y = UQ^T + F \tag{5}$$

With matrix of endogenous response variables Y (N × L), matrix of endogenous latent factors U (N × K), matrix of factor loading Q (L × K), matrix of errors F (N × L). In our case L = 1, so we get

$$Y = Uq + f \tag{6}$$

The decomposition of X and Y is made to maximize the covariance between latent components T and U. Similar to Elastic net's tuning parameters, the optimal number of principal components k for PLSR and are selected via leave-one-out cross validation (LOOCV).

4 Data

We have decided to test methodology proposed in Sect. 3.2 and answer the question whether it could be used for stress testing corporate earnings and to which success. We have decided to use two recent financial crises in order to do so. The first is the Dot-com Bubble which was more of the stock market correction rather than fully-fledged economic crisis. It affected both capital markets and way how companies are financed, with investors being more careful about their investments to young companies. More details on both macroeconomic and stock market situation around the bubble is provided by Kraay and Ventura [30], who for example point out the value of US equity increased fivefold between 1990 and 2000, especially compared to world share prices when USA is excluded which only doubled.

Second event of interest is the Financial crisis of 2008, this is the example of full-scale financial crisis that does not limited itself to stock market or a certain company type but affected all, including economy. A good discussion on 2008 financial crisis you can find in Reinhart and Rogoff [41]. Ivashina and Scharfstein [26] study the impact of crisis on lending and they found a significant decrease in 2007 and 2008. It is important to note it is in line with our data as presented in Sect. 5. An empirical view on implication of crisis on policy is presented by Taylor [42], who argued government actions did not helped the situation, but interventions prolonged and worsened the crisis.

In order to be able to capture both events of interest we have decided to consider data from 10 years prior to the dot-com bubble, this effectively means our data sample ranges from 1990 until 2017, while we do not utilize data from 2009 until 2017 in our research, we still include them in the sample in order to see current development of variables of interest. For the purpose of this research we have decided to focus on US companies and US variables used as proxies for individual risks. We have downloaded actual total quarterly revenues of whole universe of US publicly traded companies from COMPUSTAT database from 1990 until 2017. We have used FRED database to retrieve our independent variables for aforementioned periods, where possible we selected quarterly version of our variables. In the case quarterly data were not available, we lowered the frequency by including one measurement per quarter.[2]

4.1 Proxies of Risk

Kaplan and Mikes [29] described three categories of risks in organizations. First two categories focus on internal or preventable risks that are within the organization and strategy risks related to investment projects and opportunities organization follows in order to achieve its strategic goals. In this research we don't focus on those two categories, we rather focus on the third that are external risks arising from outside the company and are beyond the control.

We believe correct identification and measurement of these risks is essential building blog for risk measurement and management proposed in this paper. We follow International Country Risk Guide Methodology (ICRG) in order to identify types of risks we should consider in our model. A comprehensive description of risks is provided by Howell [23], who splits risks into three categories the political risk, economic risk and financial risk. In this paper we focus on economic risk and financial risk, some authors argue political risk is captured in the sovereign yield spread hence interest rate curves can serve as a proxy variable for political risk, for more information see Bekaert et al. [7].

As we focus on financial risk and economic risks our proxy variables are related primarily to those types of risks. We decided to split these further and define credit risk, market risk and liquidity risks as risks in financial risks category and account balance, foreign exchange risk and economic activity risk as risks in economic risks category. We did that following risk categorization as presented in ICRG [23] as well as combining it with other sources to achieve higher granularity [24]. As presented in Fig. 3 some variables capture more than one risks as well as are somehow related to more than one risks. Figure shows variables in rows and risks in columns with red color meaning no relationship with given risk, grey color somehow related and green color indicates variable being used as a proxy for given risk. Following subsections discusses individual variables, their mapping to ICRG methodology as well brief reasoning how they serve as a proxy for given risk. Please note we don't attempt an exhaustive discussion about proxies of different risk, but rather to establish a stress testing methodology for stress testing corporate earnings.

---

[2]This was done in a manner last quarterly measurement stayed, for example in the case of monthly data we take last month of the quarter as our quarterly measurement, in the case of daily data we take the last day of last month in the quarter as the measurement.

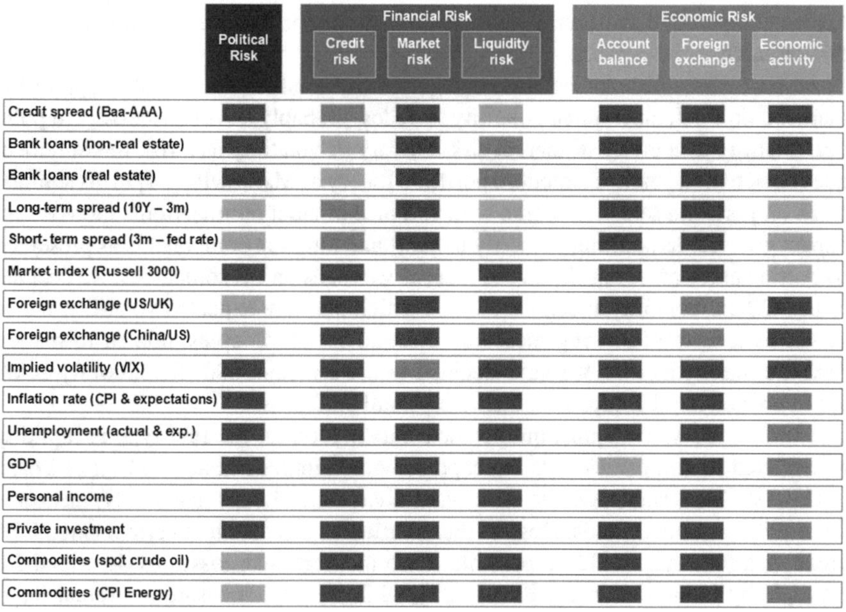

	Political Risk	Financial Risk			Economic Risk		
		Credit risk	Market risk	Liquidity risk	Account balance	Foreign exchange	Economic activity
Credit spread (Baa-AAA)							
Bank loans (non-real estate)							
Bank loans (real estate)							
Long-term spread (10Y – 3m)							
Short- term spread (3m – fed rate)							
Market index (Russell 3000)							
Foreign exchange (US/UK)							
Foreign exchange (China/US)							
Implied volatility (VIX)							
Inflation rate (CPI & expectations)							
Unemployment (actual & exp.)							
GDP							
Personal income							
Private investment							
Commodities (spot crude oil)							
Commodities (CPI Energy)							

Fig. 3 Variables considered in our framework and their mapping to risks

Financial Risk. *Credit Risk*, In banking the credit risk can be calculated for example as average net loans losses in one year divided by loss allowance in previous year as described by Imbierowicz and Rauch [25] or simply by measuring credit ratings or CDS premia as defined by Acharya et al. [1] or Hilscher and Wilson [22]. Hilscher and Wilson also point out credit risk measurement based on credit rating can be easily dominated by a simple model based on market data.

As in theory CDS should be priced as a premium over risk-free rate for credit risk he is taking, spreads should be sufficient proxies of credit risks. We have decided to use a difference between Moody's Seasoned Baa Corporate Bond Yield and Moody's Seasoned Aaa Corporate Bond Yield as a proxy for credit risk of companies (BMA). We argue the corporate bond yield spread does not contain any political risk as political risk is same for company operating within given country regardless on its internal credit rating nor it proxies economic risk.

Sovereign credit risk is in this research proxied by long term spread and short-term spread. Long term spread is defined as 10-Year Treasury Constant Maturity Minus 3-Month Treasury Constant Maturity (T10Y3MM), while short term spread is defined as 3-Month Treasury Bill Minus Federal Funds Rate (TB3SMFFM). We have also included 3-Month Treasury Constant Maturity Rate (GS3M) in our analysis as a control variable. We argue these variables (especially long-term yield) contain certain amount of political risk too as they are fully associated with government debt and yield. In our opinion political risk is reflected in long term yield to a greater extend compared to short term yield. This discussion is however out of the scope of this research.

Market Risk, We consider two categories of variables that should capture the market risk. First, the implied volatility surfaces, probably it is not necessary to point out volatility as a proxy of market risk. History of volatility as a risk measurement goes back to first years of a modern portfolio theory as Markowitz [34] in his seminal work on portfolio selection works with variance of asset returns. Some more recent papers go as far as splitting volatility to short and long run components as presented by Adrian and Rosenberg [2]. In our case, as we focus on US companies, we decided to use CBOE Volatility Index (VIXCLS) which measure expected volatility, hence serves as a forward-looking measure.[3] Second set of variables, related to market risk, are market indices, we have decided to use Russell 3000 (RU3000PR) as we are interested in all companies, including small enterprises.

Liquidity Risk, We split liquidity risk into two types, the first type is liquidity risk related to companies, this we capture with Commercial and Industrial Loans (BUSLOANS) variable. On the other hand, Real Estate Loans (REALLN) measures the liquidity of households. Despite these variables measure to a certain extend the credit risk too, this would be possible if we measure loan losses as described for example by Imbierowicz and Rauch [25], while here we measure the actual loans taken.

Economic Risk. Economic risk can be split into three subcategories. The first category is account balance risk, we don't focus on this risk. Country's current account is partially captured by GDP as GDP includes net exports. If we speak about account balance of individual companies this would be captured by balance sheet numbers. As opposed to independent variables or proxy of risk, we would be able to simulate or stress test these numbers in order to reveal what the account balance of company could be under stressed conditions. What we include in this study are foreign exchange risks and economic activity risk.

Foreign Exchange Risk, Capturing foreign exchange risk is present in the literature for many years with first papers on the topic being published more than 40 years ago, for example Westerfield [45] who tried to describe variability of foreign exchange rates using probability models or Dumas and Solnik [16] in their newer work argued stochastic changes of exchange rate result in additional sources of risk in asset pricing. Recent authors continue to study the relationship between foreign exchange risk and asset returns, for example Cenedese et al. [12]. Other authors connect it with supply chain risk management, see for example Bandaly et al. [5]. We believe research in the spirit of Bandaly et al. is related to our research in a way it's not purely focused on asset pricing, but rather aims to help with risk management of corporations, in this case of their supply chain. In contrast with them we are not interested in theoretical modeling of how foreign exchange rate is reflected in corporate earnings, we are interested to test foreign exchange variables as factors for our stress testing model.

---

[3]Existing research studying the forward-looking nature of volatility indices suggests at least some evidence in favour with it, for more information see Giot [18], Gonzalez and Novales [19] or Whaley [46]

In this case we have decided to use two foreign exchange rates, U.S./U.K. Foreign Exchange Rate (EXUSUK) and China/U.S. Foreign Exchange Rate (EXCHUS). We have decided not to include U.S./Euro exchange rate in this case as it was introduced in 1999, we would be able to use it for 2008 crisis, but as we aimed to include data since 1990 it would be hard to proxy Euro to existing currencies of individual EU countries back then.

Economic Activity Risk, Economic activity risk is a risk that is associated with "economic" activity of a country. Alternative name for it could be a "macro risk", this however would be ambiguous as it often refers to financial risk related to economy found in stocks and funds or political risk of countries. In our case we are interested in macroeconomic factors that relate to economic activity and affect the business environment and companies.

Macroeconomic risk is well defined and studied in existing literature from all angles, studying the stock markets, banks as well as companies. Some studies focused on its impact on stock market with Bali et al. [4] focused on exposure of funds to "economic uncertainty", effectively analyzing the exposure to macroeconomic variables or Lettau et al. [32] tested the stock prices and their exposure to macroeconomic risk. Another literature stream focuses on banks, it can be either as a part of stress testing as discussed before[4] or, alternatively, from a different perspective. For example, Meon and Weill [35] studied a simulated merger between EU banks and found potential gains thanks to the macroeconomic risk diversification as EU countries economies are not perfectly correlated. Third stream of literature is literature focused on macroeconomic conditions and their impact on companies. For example, Bhamra et al. [9] studied the impact of macroeconomic conditions on capital structure of firms, while Chen and Manso [14] focused on corporate debt by testing the cost of debt overhand when macroeconomic risk is taken into account, they also found out macroeconomic risk affecting the capital structure of a firm.

In contrast to existing studies, we don't aim to define our own theoretical model as for example Bhamra et al. [9] or Chen and Manso [14] did. We use macroeconomic variables as factors we consider to be able to simulate and predict the change in corporate revenues with possibility to use such model for stress testing. We have decided to use a standard set of macroeconomic indicators covering a range of areas.

The first variables are related to the inflation, we have decided to use two proxies for that, Consumer Price Index for All Urban Consumers: All Items (CPIAUCSL) and University of Michigan: Inflation Expectation (MICH) to capture both, existing inflation and expectations about future inflation. Same approach we took in the case of unemployment with Civilian Unemployment Rate (UNRATE) capturing actual unemployment rate and Natural Rate of Unemployment (Long-Term) (NROU) capturing expectations about it.

Regarding GDP we have decided to use three variables that capture economic production. Real Gross Domestic Product (GDPC1) itself as well as Industrial Production Index (INDPRO), we believe industrial production can be a good proxy

---

[4]For more information on stress testing banks see Ong and Cihak [38] and papers referenced in their study.

of economic activity of companies. We have also decided to include personal consumption and include Personal Consumption Expenditures (PCEC) in our study. In order to capture investments, we have decided to use Real Gross Private Domestic Investment (GPDIC1) and Gross private domestic investment: Domestic business (W987RC1Q027SBEA) to capture both individuals and companies. We have also included Real Disposable Personal Income (DPIC96) to capture how much money individuals have.

The last set of variables is commodities. We believe a complex approach should be applied for stress testing and as many corporations rely on commodities in their day to day operations, we have decided to capture possible development on this market too. Two commodities we have decided to include are Spot Crude Oil Price (WTISPLC) as variable capturing oil price and Consumer Price Index for All Urban Consumers: Energy commodities (CUSR0000SACE) to capture energy market prices.[5] It is important to note energy prices are in our opinion incorporating some political risk as utilities are highly regulated, for more information on regulation of utilities see for example Cambini and Rondi [11] who studied relationship between investment and regulation in EU.

4.2 Dependent Variable

In this study we have decided to focus on corporate revenues. We see this as a first step in order to establish a more complex stress testing framework for companies. Our variable is total quarterly earnings (REV) as reported and provided by COMPUSTAT. We have decided to focus on revenues as first variable as opposed to stock price as in the case of stock market other asset pricing factors may play a role in the value change besides risk factors stated above. On the other hand, testing other balance sheet values, while possibly useful would need the extension of the framework as several factors besides current economic and market situation affect investments or profits.

5 Descriptive Statistics

Following subsection reports on descriptive statistics of individual variables. As we can see in Table 1 we have in total 113 quarterly observations in our sample. This is a necessary number as we aim to study two shocks, Dot-com bubble of 2000–2002 and 2007–2009 Financial crisis. Please note all numbers reported in Table 1 are changes in variables.

---

[5]As corporations often don't trade wholesale energy on the market, we see consumer price index of energy commodities to be a good proxy of their exposure to market with energies.

Table 1 Summary of all variables (growth) over time. Table reports on all variables included in our study, please note all numbers are changes from quarter to quarter with quarterly earnings data being the average change (REV), other variables are risk factors as described in previous section

Statistic	N	Mean	St. Dev.	Min	Max
REV	113	0.047	0.242	−0.405	0.713
REALLN	113	0.016	0.016	−0.025	0.060
BUSLOANS	113	0.011	0.024	−0.065	0.073
GS3 M	113	−0.054	0.440	−1.510	0.800
T10Y3MM	113	0.007	0.463	−0.990	1.450
TB3SMFFM	113	0.006	0.188	−0.580	0.820
CPIAUCSL	113	0.006	0.006	−0.023	0.024
MICH	113	0.012	0.181	−0.615	0.900
UNRATE	113	−0.002	0.051	−0.083	0.200
NROU	113	−0.002	0.003	−0.008	0.006
GDPC1	113	0.006	0.006	−0.022	0.018
INDPRO	113	0.005	0.013	−0.064	0.028
PCEC	113	0.012	0.006	−0.024	0.025
DPIC96	113	0.007	0.009	−0.040	0.028
GPDIC1	113	0.009	0.030	−0.116	0.092
W987RC1Q027SBEA	113	0.012	0.036	−0.141	0.114
WTISPLC	113	0.025	0.178	−0.455	0.927
CUSR0000SACE	113	0.013	0.093	−0.407	0.277
EXCHUS	113	0.003	0.050	−0.035	0.503
EXUSUK	113	0.000	0.045	−0.152	0.113
BMA	113	−0.001	0.232	−1.320	1.110
RU3000PR	113	0.021	0.080	−0.206	0.248
VIXCLS	113	0.049	0.361	−0.495	1.864

Table 2 Summary of revenues (raw) over the cross-section. This table reports raw numbers in million USD as the cross section of all companies

Statistic	N	Mean	St. Dev.	Min	Max
REV	15,436	250.338	1573.268	−200.112	53147.100

Summary statistics for cross section of companies is reported in Table 2. Please note numbers are in million USD with highest quarterly revenues reported by APPLE INC. On the other hand, negative revenues were booked by AMERICAN INTER-NATIONAL GROUP (AIG) in 2008. According to reports [10], this was due to restructuring activities that took place in 2007–2008. This was the only case of substantially negative revenues booked in our sample with other lowest revenues being around 0. Despite this effectively means we include a number of small companies in

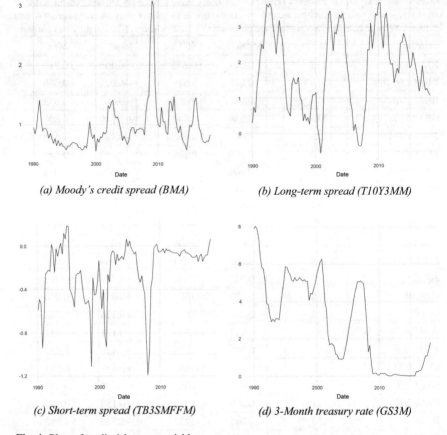

(a) Moody's credit spread (BMA)

(b) Long-term spread (T10Y3MM)

(c) Short-term spread (TB3SMFFM)

(d) 3-Month treasury rate (GS3M)

Fig. 4 Plots of credit risk proxy variables

our study, we don't set any cut-off point for them. We argue in this case issue coming from businesses being relatively small do not affect relationship between our risk factors and revenues. This would be in contrast with stock prices, which would be affected by low liquidity of cheap stock.

Besides of overall descriptive statistics which doesn't need any further comments, we present figures for individual variables split by risk they proxy. These are presented in Figs. 4, 5, 6, 7 and 8. Looking at first group of variables in Fig. 4 we can clearly see large decrease in long and short term spread around crises in 2001 and 2008, this is visible in Fig. 4b, c. The decrease in the spread is also supported by decrease in short term treasury rate as pictured in Fig. 4d. As for Moody's credit spread represented in Fig. 4a we can see decrease in credit spread in 2001, but sharp and significant increase in 2008. While sovereign spread decrease could be explained by higher demand for "safe" investment pushing rates down in the time of crisis, 2008 sharp increase (compared to 2001 sharp decrease) of credit spread could be explained by

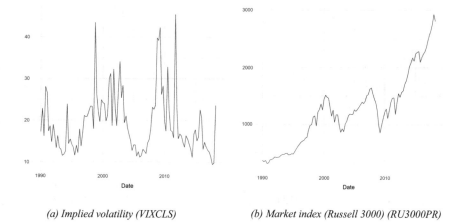

(a) Implied volatility (VIXCLS)   *(b) Market index (Russell 3000) (RU3000PR)*

Fig. 5 Plots of market risk proxy variables

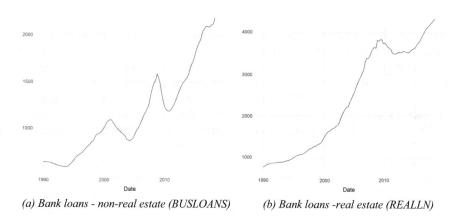

(a) Bank loans - non-real estate (BUSLOANS)   *(b) Bank loans -real estate (REALLN)*

Fig. 6 Plots of liquidity risk proxy variables

the difference between two events. While first could be considered a stock market correction, meaning investors started to switch from stock to bonds in 2001, the latter is economic crisis, and investors tried to avoid risky assets including bonds with lower rating.

In Fig. 5 we present our proxies for market risk. As visible in Fig. 5a, the volatility index is high around our events of interest, which implies a high volatility in the time of crisis, despite is behaviour being somehow different in 2001 (lower volatility compared to 2008), the big picture remains. On the other hand, Russell 3000 index described in Fig. 5b behaves as expected with decreases in value in 2001 and 2007. Our liquidity proxy variables (bank loans) shed even more light on the difference between both events. As described in Fig. 6, in the case of non-real estate bank loans (Fig. 6a), we can see a clear decrease in loans following both crises. On the other

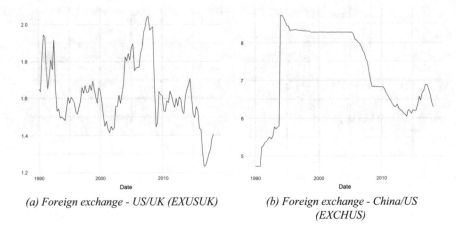

(a) Foreign exchange - US/UK (EXUSUK)   (b) Foreign exchange - China/US
                                                                        (EXCHUS)

Fig. 7 Plots of foreign exchange risk proxy variables

hand, Fig. 6b shows a different picture. In the case of real estate bank loans, we see literally no change in 2000–2002 period, however we see how previously increasing trend stopped and eventually decreased in years following 2007. This supports the notion 2001 was more of stock market correction, while 2008 was crisis affecting whole economy. Figure 7 plots foreign exchange rates. It is visible how China's Renminbi was pegged to US dollar until 2005, with correction following. In the case of USD to GBP exchange rate, we can see sharp drop in value in both 2001 and 2008, while 2008 drop was much significant as visible in Fig. 7b.

Last set of figures describes the economic activity risk proxies. As expected from those types of variables in Fig. 8 we can clearly identify several trend lines. Consumer price index shown in Fig. 8a shows almost no change in 2001, but decrease in 2008. This is another argument for 2008 being economic crisis with greater impact (in this case the deflation of prices). Inflation expectations shown in Fig. 8b, illustrate a drop in expected inflation in both 2001 and 2008. It is important to point out a relatively high expected inflation preceding the financial crisis of 2008. The unemployment rate presented in Fig. 8c behaves as expected, there is a decrease in unemployment rate before crises and increase after them as jobs are lost. On the other hand, is reported natural unemployment rate reported in Fig. 8d, here we can clearly see no change in 2001, however relatively sharper decrease in years just before the financial crisis of 2008, followed by sharp increase after that. Interesting is also the correction happening in recent years which seem to have corrected back to trend it had before 2008 crisis.

While real GDP and Personal consumption (Fig. 8e, g) are similar in terms of their behaviour over time, the third proxy for economic production, the industrial production shown in Fig. 8f displays a different pattern. It reacts much more to changes in economy or on the stock market compared to personal consumption, this makes sense as companies are much more sensitive to changes on financial markets

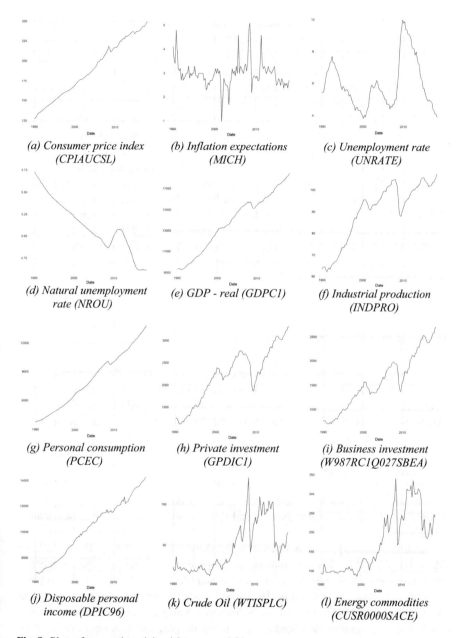

Fig. 8 Plots of economic activity risk proxy variables

and in economy in general, compared to households which in general change their behaviour only if their income changes. Speaking of disposable personal income, it is presented in Fig. 8j, it is clear year 2001 had no impact on it, and similar applies to year 2008, however it is possible to see there is a flattened trend.

Private investment and Business investment shown in Fig. 8h and i respectively show both similar patterns. It is clearly visible both periods of interest in these cases recorded a drop of investments. The last pair of variables are commodities. In this research we included both Crude oil in Fig. 8k and Energy commodities in general in Fig. 8l. Both commodities behave similarly and 2008 crisis is preceded with a period of sharp increase in commodity prices and followed by a drop in them.

As presented above, all variables behave in the way one would expect them behave, our sample includes two periods of interest. First one is between 2000 and 2002 and behaves like stock market crisis with correction in market prices, while second is between 2007 and 2008 and it behaves like an economic crisis. We use these two events to test out methodology and see whether models based on our factors are able to predict the movement in company revenues. This is an important first step to define a robust stress testing framework.

6   Regression Results

We have tested our regression models on two events of interest. We have used period from Q1 1990 until Q1 2000 to estimate the model and then predicted corporate revenues as shown in Fig. 9. In a similar fashion we estimated our models in period from Q4 2001 until Q2 2007 and then predicted corporate revenues as shown in Fig. 10. Based on our results it is clearly visible a traditional OLS regression overfits the data, given a number of factors we use as green line jumps all over in both cases.

On the other hand, both red and blue lines for Elastic net and PLSR do a great job following actual revenues (black line) in both cases. This implies both models could be considered as suitable for stress testing as they both (a) provide relatively stable predictions, (b) do not overfit the data and (c) can handle a large number of potential factors well.

Results presented above have two implications for stress testing. First of all, we have shown machine learning models, as proposed by this research perform better than traditional OLS regression. This was tested on two events with similar results. Second, it is clear our proxies of risk variables are able to predict the movement of company earnings in the time of crisis. This efficiently means stress testing scenarios based on proposed methodology and using proposed factors could be used by companies to evaluate their revenues under stressed conditions.

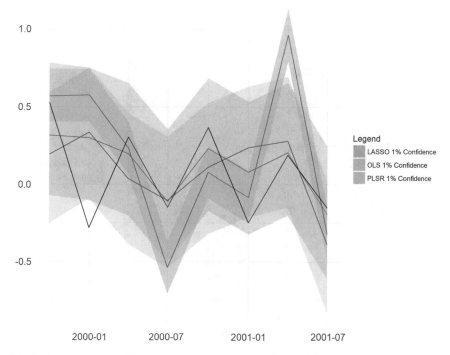

Fig. 9 Actual versus predicted earnings around dot-com bubble. Actual earnings are in black, while predicted earnings include 1% confidence interval are OLS (green), Elastic net (red) and PLSR (blue)

7 Conclusion

This paper contributes to the existing literature on a number of fronts. First of all, it discusses potential proxies of risks that can affect corporate earnings. It derives its categorization from existing literature, while applying naming and categories that make sense for corporations. It also defines factors as well as points out the rationale for selecting these factors. We understand further research would have to be done in order to test all potential factors as well as factors from number of countries would have to be included in such research.

Second contribution is the stress testing methodology in general. We contribute to the stress testing literature by testing Elastic net and PLSR as potential methodologies for stress testing. Based on our results it is clear they do a superior job compared to traditional OLS regression in predicting corporate revenues. We believe these methods are particularly useful especially in the case when a number of relatively arbitrary factors are considered. As both Elastic net and PLSR can be considered machine learning techniques in a way that they can deal with a large number of covariates by penalizing insignificant factors and reducing their number respectively.

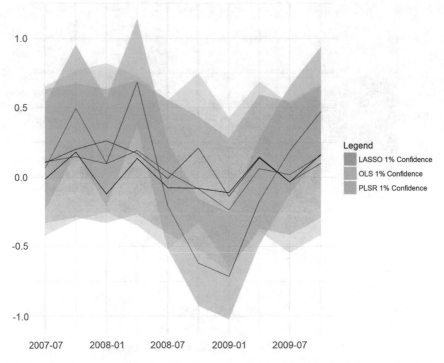

Fig. 10 Actual versus predicted earnings around financial crisis. Actual earnings are in black, while predicted earnings include 1% confidence interval are OLS (green), Elastic net (red) and PLSR (blue)

Third contribution is the application of stress testing on corporate earnings. Although stress testing for companies has always been regarded as an important activity, there is no clear definition of how stress testing should be performed and what which methodologies should be used. We addressed both issues in this research. We have defined a stress testing framework, where the only variable of choice is the method used to predict the model. We have defined three potential methods including simple OLS method widely used in stress testing today. By testing methods on actual US data for two crises we shown Elastic net and PLSR do a good job predicting revenues, while OLS provides relatively poor results.

We understand this research only touches the topic of stress testing in corporations and further work could be done on several fronts. First of all, considering factors, it would be possible to include more factors as proxies of risks not covered in this research, as well as use international data. For example, natural disasters or proxies for political risk could be added. In a similar vein, other dependent variable could be stressed, for example accounts receivable, accounts payable, or account balance itself. Alternatively, we could use our framework to predict stock prices of companies. Another type of improvement would be the inclusion of longer samples and potentially a number of crises, as well as more detailed analysis of them. Results from such stress testing can be utilized for enterprise risk management and allow companies to prepare for negative eventualities. Use of stress testing doesn't have to be limited to this, it can be utilized for business model development as well.[6]

References

1. Acharya V, Drechsler I, Schnabl P (2014) A pyrrhic victory? Bank bailouts and sovereign credit risk. J Financ 69(6):2689–2739
2. Adrian T, Rosenberg J (2008) Stock returns and volatility: pricing the shortrun and longrun components of market risk. J Financ 63(6):2997–3030. Stress Testing Corporate Earnings of US Companies 23
3. Avesani RG, Liu K, Mirestean A, Salvati J (2014) Review and implementation of credit risk models. In: A guide to IMF stress testing: methods and models, p 135
4. Bali TG, Brown SJ, Caglayan MO (2014) Macroeconomic risk and hedge fund returns. J Financ Econ 114(1):1–19
5. Bandaly D, Shanker L, Şatr A (2018) Integrated financial and operational risk management of foreign exchange risk, input commodity price risk and demand uncertainty. IFACPapersOnLine 51(11):957–962
6. Basel Committee on Banking Supervision (2009) Principles for sound stress testing practices and supervision
7. Bekaert G, Harvey CR, Lundblad CT, Siegel S (2014) Political risk spreads. J Int Bus Stud 45(4):471–493
8. Bellotti T, Crook J (2013) Forecasting and stress testing credit card default using dynamic models. Int J Forecast 29(4):563–574
9. Bhamra HS, Kuehn L-A, Strebulaev IA (2010) The aggregate dynamics of capital structure and macroeconomic risk. Rev Financ Stud 23(12):4187–4241
10. Business Wire (2009) AIG reports fourth quarter and full year 2008 loss. Business Wire. https://www.businesswire.com/news/home/20090302005619/en/AIG-Reports-Fourth-Quarter-Full-Year-2008
11. Cambini C, Rondi L (2010) Incentive regulation and investment: evidence from European energy utilities. J Regul Econ 38(1):1–26
12. Cenedese G, Sarno L, Tsiakas I (2014) Foreign exchange risk and the predictability of carry trade returns. J Bank Financ 42:302–313
13. Chan-Lau MJA (2010) The global financial crisis and its impact on the Chilean Banking System, No. 10–108. International Monetary Fund
14. Chen H, Manso G (2016) Macroeconomic risk and debt overhang. Rev Corp Financ Stud 6(1):1–38
15. Cihak M (2007) Introduction to applied stress testing, No. 7–59. International Monetary Fund
16. Dumas B, Solnik B (1995) The world price of foreign exchange risk. J Financ 50(2):445–479
17. Encyclopedia of Information Science and Technology (2008) Stress testing corporations and municipalities and supply chains, 4th edn. IGI Global, pp 6813–6823
18. Giot P (2005) Relationships between implied volatility indices and stock index returns. J Portf Manag 31(3):92–100
19. Gonzlez MT, Novales A (2009) Are volatility indices in international stock markets forward looking? RACSAM-Revista de la Real Academia de Ciencias Exactas, Fisicas y Naturales Serie A Matematicas 103(2):339–352
20. Haenlein M, Kaplan AM (2004) A beginner's guide to partial least squares analysis. Underst Stat 3(4):283–297

---

[6]For more information on business development please see Kaczor and Kryvinska [28].

21. Hastie T, Tibshirani R, Friedman J (2009) The elements of statistical learning: data mining, inference and prediction, vol 2. Springer
22. Hilscher J, Wilson M (2016) Credit ratings and credit risk: is one measure enough? Manag Sci 63(10):3414–3437
23. Howell LD (2011) International country risk guide methodology. PRS Group, East Syracuse, NY
24. Hull J (2012) Risk management and financial institutions, vol 733. Wiley
25. Imbierowicz B, Rauch C (2014) The relationship between liquidity risk and credit risk in banks. J Bank Financ 40:242–256
26. Ivashina V, Scharfstein D (2010) Bank lending during the financial crisis of 2008. J Financ Econ 97(3):319–338
27. Jones MT, Hilbers P (2004) Stress testing financial systems: what to do when the governor calls
28. Kaczor S, Kryvinska N (2013) It is all about services-fundamentals, drivers, and business models. J Serv Sci Res 5(2):125–154
29. Kaplan RS, Mikes A (2012) Managing risks: a new framework
30. Kraay A, Ventura (2005) The dot-com bubble, the Bush deficits, and the US current account. The World Bank
31. Kryvinska N (2012) Building consistent formal specification for the service enterprise agility foundation. J Serv Sci Res 4(2):235–269
32. Lettau M, Ludvigson SC, Wachter JA (2007) The declining equity premium: what role does macroeconomic risk play? Rev Financ Stud 21(4):1653–1687
33. Lutkebohmert E (2009) The CreditRisk + Model. Concentration risk in credit portfolios, pp 53–60
34. Markowitz H (1952) Portfolio selection. J Financ 7(1):77–91
35. Meon P-G, Weill L (2005) Can mergers in Europe help banks hedge against macroeconomic risk? Appl Financ Econ 15(5):315–326
36. Moeller RR (2007) COSO enterprise risk management: understanding the new integrated ERM framework. Wiley
37. Molnar E, Molnar R, Kryvinska N, Gregu M (2014) Web intelligence in practice. J Serv Sci Res 6(1):149–172
38. Ong LL, Cihak M (2014) A guide to IMF stress testing: methods and models: "Stress testing at the International Monetary Fund: methods and models". International Monetary Fund
39. Ong LL, Maino R, Duma N (2012) Into the great unknown: stress testing with weak data. No. 10–282. International Monetary Fund
40. Peria MMSM, Majnoni MG, Jones MMT, Blaschke MW (2001) Stress testing of financial systems: an overview of issues, methodologies, and FSAP experiences. No. 1–88. International Monetary Fund
41. Reinhart CM, Rogoff KS (2008) Is the 2007 US sub-prime financial crisis so different? An international historical comparison. Am Econ Rev 98(2):339–44
42. Taylor JB (2009) The financial crisis and the policy responses: an empirical analysis of what went wrong. No. w14631. National Bureau of Economic Research
43. Tobback E, Bellotti T, Moeyersoms J, Stankova M, Martens D (2017) Bankruptcy prediction for SMEs using relational data. Decis Support Syst 102:69–81
44. Vazquez F, Tabak BM, Souto M (2012) A macro stress test model of credit risk for the Brazilian banking sector. J Financ Stab 8(2):69–83
45. Westerfield JM (1977) An examination of foreign exchange risk under fixed and floating rate regimes. J Int Econ 7(2):181–200
46. Whaley RE (2008) Understanding vix
47. Wold H (1966) Estimation of principal components and related models by iterative least squares. Multivar Anal 391–420
48. Zou H, Hastie T (2005) Regularization and variable selection via the elastic net. J R Stat Soc Ser B (Stat Methodol) 67(2):301–320

From Information Transaction Towards Interaction: Social Media for Efficient Services in CRM

Vivien Melinda Wachtler

Abstract This chapter aims to illustrate the efficient services provided by CRM and social media common work. First, it provides an overview of the most important definitions that are used throughout this chapter. Second, it gives a picture about the recent changes in CRM, focusing on customer's behavior changes. Third, Sect. 4 provides the reader with information about social CRM and summarizes the new and improved services as a result of CRM and social media common work. Section 5 introduces the top 10 CRM software and their interaction to social networks. Finally, this chapter summarizes and compares the services provided by various CRM software products.

1 Introduction

At the end of 2016, 2.789 billion people were active social media users, that is approximately 37% of the total population. Compared to the previous year, the number increased by 21%. Out of 2.789 billion active social media users 2.549 billion, roughly 91% are mobile social media users. In 2016 the increase of mobile social media users was 30% [77]. In April 2017 Facebook announced that an average user spends 50 min daily on its Facebook, Instagram and Messenger platforms (WhatsApp is not included) [169].

These latest trends have led to a boost in customer demand for social customer service and therefore a transformation and reformation of CRM. Assuming a company has a CRM system and is active on social media, that does not automatically mean that both activities have a common data base and take advantage of each others data. If a company's CRM database is synced with its social media platforms, the company is in the position of capturing additional information about its customers like social profile and the content created by its audience. This is the point, where

V. M. Wachtler (✉)
Vienna University of Economics and Business, Vienna, Austria
e-mail: vivien.wachtler@gmail.com

© Springer Nature Switzerland AG 2020
N. Kryvinska and M. Greguš (eds.), *Data-Centric Business and Applications*,
Lecture Notes on Data Engineering and Communications Technologies 30,
https://doi.org/10.1007/978-3-030-19069-9_15

social CRM comes into the picture: it is an extension of the CRM system being integrated with social media platforms [72].

2 Definitions

CRM

CRM defined from a business strategy perspective: "CRM is the strategic process of selecting customers that a firm can mostly profitably serve and shaping interactions between a company and these customers. The ultimate goal is to optimize the current and future value of customers for the company" [79].

E-CRM

"Electronic customer relationship management (e-CRM); also known as CRM 1.0 is the electronically delivered set of tools that helps manage CRM. E-CRM is related to all forms of managing relationships with customers when using information technologies. It arises from the consolidation of traditional CRM with the e-business applications, and it covers the broad range of information technologies used to support a company's CRM strategy" [175].

Nevertheless, the terms CRM and e-CRM are often used interchangeably because practically all CRM solutions use some IT.

Social CRM

"Social CRM is a philosophy and a business strategy, supported by a technology platform, business rules, processes and social characteristics, designed to engage the customer in a collaborative conversation in order to provide mutually beneficial value in a trusted and transparent business environment. It is the company's programmatic response to the customer's control of the conversation" [45].

Social Media

"Social media are computer-mediated technologies that facilitate the creation and sharing of information, ideas, career interests and other forms of expression via virtual communities and networks" [110].

3 Why Is CRM More Important Now Than Ever?

First, it can be observed, that people have less time due to their busy lifestyle than before. It means, costumer service from 8 am to 5 pm is not enough anymore. Second, consumers have higher expectations considering service levels and low tolerance for failures or disappointments. Moreover, social customers also like to share their feedback with others and write reviews on social media [79]. To take a look at the most famous cases about bad customer service that went viral see Sect. 4.2.

Third, customers have access to all the necessary information about products and services on the internet. The best or cheapest product or service is just one click away with the help of search engines. To help make the best possible decision at the point of sale and avoid costly mistakes there are several apps available, like Amazon or ScanLife. The concept of the Amazon app is simple: scan the barcode and instantly check the price on Amazon. ScanLife compares the prices between providers and helps to find the best deal available [168].

On the one hand, customers are becoming less loyal to companies and brands [185]. Nowadays almost every company offers a loyalty program for its customers to get them return to the firm and purchase frequently. The customer benefits from the free beverage, coupons or other rewards depending on the type of the business [78]. According to a study conducted by Accenture in 2017, 71% of interviewed claimed that despite the companies' attempts, loyalty programs do not generate loyalty. Moreover, 61% shifted the majority of their businesses to a competitor in the last year. Furthermore, 30% of Millennials (generation between the age of 22 and 40 [172]) stated that their loyalty can not be earned [1]. On the other hand, there are some companies who still live off their loyal customer base. The most prominent example is Apple. Figure 1 shows that in 2010 about 95% of Apple's customers were counted as loyal customers. However, the average customer retention rate has declined significantly in the recent years and reached its all time low at the end of 2016 at 75%. This number may seem low but is still the highest among the competition [38].

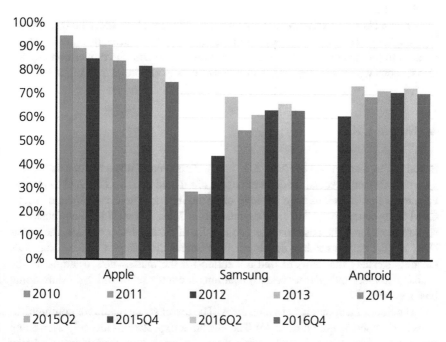

Fig. 1 Retention rate by major participants (Apple, Samsung, Android) [38]

Another reason is that consumers demand more convenience and self-service. It was already mentioned that customers have less time now than before due to peoples busy lifestyle. Consumers often go to small convenient stores, that are either close to their residence or are open 24/7 and offer limited product range for everyday needs, e.g. Tesco Express in the UK, 7-Eleven in the US [33]. Meanwhile demanding more convenience have brought us to self-service. Since consumers are often under time pressure, they do not want to wait in line but rather check out on their own in a store. Not just the customers benefit through time saving but also the stores by saving labor costs. In the recent years some supermarket chains in Austria e.g. Spar and Merkur introduced the self-checkout service in certain stores. Amazon took self-checkout into a whole new level when Amazon Go was introduced in a video in December 2016. The concept of the store is very convenient: shopping experience without lines and checkout. This new type of store is supported by the newest technology: computer vision, sensor fusion, and deep learning and it is called "Just Walk Out Shopping Technology". To shop in the store customers only need to have an Amazon account and the Amazon app installed on their smartphone [12]. Currently only one shop exists in Seattle and it is only for Amazon employees available. According to Business Insider, Amazon intends to build 2000 grocery stores in the US in the next ten years [82].

Finally, the last reason why consumers' behavior has dramatically changed in the recent years is the increased use of social media. To find out more about the most popular social network sites see Sect. 4.1 and to know more about the social customer see Sect. 4.2.

In this section only customers' behavior changes were discussed. Changes regarding consumers such as aging population in developed countries or adjustments in relation to the marketplace like difficult differentiation among competition etc. are also reasons why managing customer relationships is important to companies.

4 Social CRM

Social CRM, also known as CRM 2.0., is an extension of CRM and not a substitute. Two new elements: social media and people are joined in. It was created to involve customers in discussions with the help of social media. A relevant object of social CRM is to benefit the company (e.g. increased sales, knowledge about customer requirements) and the customers (e.g. quick fix of customers' problems, the possibility to take part in product developments). Social CRM is part of the company's business strategy that aims to find a solution for the matter: how to adjust to the social customers and their newest expectations about the seller which they are doing business with [175].

"The Social customer is not going away. The use of social media for communications in channels is not going away. The customer expectations are only ever going to get more demanding. You're foolish if you ignore it. Your wise if you [...] take action. But then you're foolish if you don't continue it" [45].

4.1   The Increased Use of Social Media

According to Social Media Today, in January 2017 2.8 billion people were active on social media websites worldwide. Facebook accounted for the most popular social network site [64]. In the last ten years the number of monthly active Facebook users has dramatically increased and in the second quarter of 2017 reached its all time high at more than two billion active users each month worldwide [166] (Fig. 2).

YouTube ended up at the second place on the list with 1.3 billion monthly users. It is also recognized as a TV substitute among young adults. People watch approximately 5 billion videos on YouTube daily [43].

Instagram has the fastest growing customer base. From December 2016 to April 2017 the number of monthly active users has increased by 100 Mio and hit 700 Mio users [167]. Besides Instagram's own account 3 singers Selena Gomez, Ariana Grande, Beyoncé and one athlete Cristiano Ronaldo has made it to the top 5 most followed users [162] (Fig. 3).

Twitter is the most popular network among world leaders with 328 Mio monthly users worldwide [161]. The former US president, Barack Obama is "the uncontested political leader of the digital world" [176] with 93.1 Mio followers [177] (Fig. 4).

Since Obama left the White House, the most followed world leader on Twitter is Pope Francis with 36.2 Mio total followers from his 9 Twitter accounts in different languages [44]. At the second place is the current US president, Donald Trump with 35.3 Mio followers. To highlight the importance of social media in politics, Donald Trump said in an interview with the Financial Times in April 2017: "Without

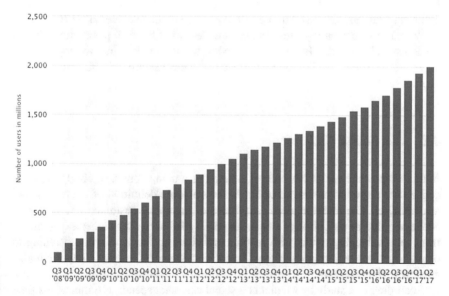

Fig. 2 Number of monthly active users on Facebook [166]

RANK	GRADE	USERNAME	MEDIA	•FOLLOWERS•
1	A++	instagram	4,641	225,670,066
2	A++	selenagomez	1,356	124,919,337
3	A++	arianagrande	3,257	112,110,799
4	A++	cristiano	1,944	108,659,940
5	A++	beyonce	1,460	105,467,364

Fig. 3 List of the most followed users on Instagram (State: 12 August 2017) [162]

Fig. 4 Barack Obama's Twitter profile (State: 12 August 2017) [177]

the tweets, I wouldn't be here... I have over 100 m [followers] between Facebook, Twitter, Instagram. Over 100 m. I don't have to go to the fake media" [176].

With 175 Mio monthly users Pinterest ended up at the fifth place on the list of the top social networks. LinkedIn and Reddit has not made it into the top 5 social networks on Social Media Today, but each of them has around 100 Mio monthly visitors [64].

4.2   The Social Customer

In this chapter the social customer is defined as a man, woman or child who is an active user on social media, since a precise, scientific definition of the term is not available yet. The most important attributes of the social customer are: very well connected due to its online presence on social media, having higher expectations from sellers, helping others with recommendations and reviews while listening to others' opinions before purchasing. The social customer is not just a buyer but also an active influencer [84].

According to a study by Mediakix a social customer spends 116 min of his time daily on social media platforms. The most time is spent on YouTube (40 min) and

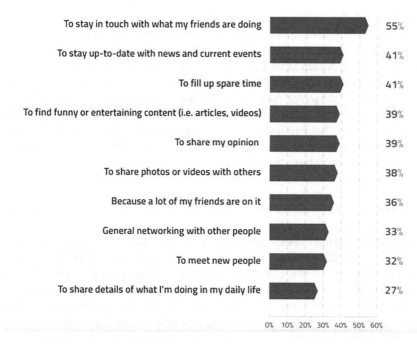

Fig. 5 Top 10 social network motivations [35]

Facebook (35 min) and the less is spent on Twitter (1 min). Mediakix also estimated the accumulated time spent on social media in a lifetime and the outcome is shocking: 5 years, 4 months [96]. In April 2017 Facebook reported that an average user spends 50 min daily on its Facebook, Instagram and Messenger platforms (WhatsApp is not included) [169].

What are people doing in that average 116 min daily? According to a study conducted by Global Web Index, the main motivator to use social media is to remain informed about friends' activities. 39% of interviewed stated that one of the motivations is to share their opinion [35] (Fig. 5). Most people also like to share their political opinion, e.g. #TrumpTrain was trending on Twitter before and during the 2016 Presidential Election in the US. The meaning behind the hashtag is the support of the Republican candidate Donald Trump [50]. The opposite meaning has #NeverTrump [49].

Writing customer recommendations and reviews falls into the category opinion sharing. Most online shops encourage customers after purchase to give feedback. For social customers it is also possible to give feedback about the company, product or service on social media, Yelp or TripAdvisor. This way customers not just share their experience with the company but also with fellow customers [124]. Jeff Bezos excellently said "If you make customers unhappy in the physical world, they might each tell six friends. If you make customers unhappy on the Internet, they can each tell 6,000" [14]. According to a research by NewVoiceMedia 59% of Millennials share disappointing customer service on social media. Nowadays it is not unusual for negative feedback to go viral and damage the trustworthiness of a company. One of the most famous cases is "United Breaks Guitars" back in 2008. The story briefly summarized: a musician Dave Caroll flow with United and his guitar got broken. Despite his constant e-mails and telephones United was not willing to pay for the repairing costs of the guitar. As a revenge he wrote a song about his bad experience with United and it has gone viral and got 3 million views in the first 10 days on Youtube [16]. Another famous incident concerning United has gone viral in April 2017 when a man was violently removed from an overbooked flight [95]. The incident is mentioned in the industry as a "public relations nightmare". The passenger sued the airline and within two weeks they reached a settlement. United had to change its policy: in the future the company will be ready to pay passengers up to $10,000 (before the incident the airline offered $800) in case they are willing to give up their seat in an overbooked flight.

Since the companies cannot control the reviews on social media, their only possibility is to provide excellent service to avoid bad criticism. Ignoring the social customer is not an option: according to a study conducted by Conversocial 45% of interviewed stated that if a company does not respond to customer complaints or questions they would feel mad, 27% claimed that they would stop buying from the company [30, 36]. Moreover, according to a study by The Social Habit, 42% of interviewed require an answer from the contacted company within an hour [13].

4.3   The Evolution of Social CRM

The evolution of CRM started with the evolution of Web 2.0. back in the beginning of 2000s. "Web 2.0 is the current state of online technology as it compares to the early days of the Web, characterized by greater user interactivity and collaboration, more pervasive network connectivity and enhanced communication channels" [126]. Some of the most important elements of Web 2.0.: wikis, software as a service programs and services, social networking sites.

Who is part of CRM? CRM 1.0. is characterized by one-way communication between the company and its customer or the company to its partner or supplier. The company generates the content that creates the value. CRM 1.0. was a task for only particular departments like sales or marketing. CRM 2.0. includes a larger community and relates numerous elements, thereby enabling the communication between customers. The focus of social CRM lies on collaborative relationship while participating in a sophisticated relationship network. The customers provide information about their expectations and needs and not the company but the conversation generates the value. In social CRM everybody participates in the company, from R&D all the way to sales [34] (Fig. 6).

As already mentioned before, the type of the communication has changed. In CRM 1.0. the channels were restricted e.g. phone, fax, email and the conversation between the company and the customer was private. The company could only collect limited

Fig. 6 How traditional CRM evolved into social CRM [34]

amount of data about its customers, based on previous interactions or customer profile data. Since the communication was one-way, the company controlled the process. In CRM 2.0. the number of channels are unlimited, starting from the traditional ones like phone or email through social networking sites, blogs, video or photo sharing to reviews and ratings etc. The company must have confidence in external sources and collect valuable data about its customers. The process in CRM 2.0. is the opposite as in CRM 1.0.: the customer is at the center of attention and the channels are dynamically guided by him. Since the conversation is controlled by the customer, it does not have to be private and it is not restricted between business hours anymore [34, 175].

The purpose of CRM has also changed: in CRM 1.0. the intension was to sell to customers, the focus of attention was only the transaction. However, in CRM 2.0. the conversation generates the value and the customer is the key feature. The main purpose of social CRM is the interaction. Brands are not just concerned about sales but satisfied customers and community building. Through communication between all employees and various customers who are also connected to each other, the customer has become the heart of the innovation cycle [47]. How to ask customers what they want? That is where crowdsourcing can be really helpful [17]. The first and most forward way is to ask the customers directly to choose between alternatives. M&M's started a vote at the same time as the Presidential Election in the U.S. in 2016 (Fig. 7). The company introduced three new flavors and the customers got to decide which flavor stays permanent in the product line. M&M's offered $100,000 for the winner and the title of "Official M&M's Taste Tester" [108]. The voting happened on the company's website but M&M's engaged its customers on social media with #MMSFlavorVote. After more than one million votes were submitted and the winner was announced, the director of M&M's said: "We are so thrilled that our fans helped to shape our brand's vibrant future. Flavor Vote was the perfect campaign to show

M&M'S U.S.A.
We've introduced three new flavors of M&M'S Peanut – Honey Nut, Coffee Nut and Chili Nut! Try
them all and vote for your favorite. Only one flavor will stay on shelves and YOU will decide which
one it is! Vote now until June 17 at http://peanutflavor.mms.com/ and follow along on our social
channels using #MMSFlavorVote!

Fig. 7 M&M's vote [41]

how much we value our fans' opinion. The brand is looking forward to adding Coffee
Nut to our peanut-flavored lineup in August" [184].

Through crowdsourcing companies can encourage their customers to be creative.
Naming new products can be really difficult, so companies often include customers
in the process and start "Name a new product" contests. One example is Sony who
brought in the customers to help find a name for their audio balls. In two weeks they
got a lot of recommendations and the best five were selected by Sony executives and
posted on Facebook [163]. However, companies have to be careful since the creativity
of the customers could turn against the brand [170]. Mountain Dew started the same
promotion for its new green apple drink, where customers could turn in suggestions
and vote for the name they like the best. People submitted and highly ranked the
names like "Diabeetus", "Soda". The promotion was useless and so offensive in the
end that it had to be shut down [125]. For creative customers the possibilities to
participate in a brand's everyday life is endless. Starbucks started the "White Cup
Contest" in 2014 and inspired customer to handle their cup as a canvas [165]. All they
had to do was take a picture of their decorated cup and post it on social media with
the hashtag #WhiteCupContest. Starbucks used the design for its reusable plastic
cups [164] (Fig. 8).

4.4   The Benefits of Social CRM

Integrating social media is about substantially more than simply likes and follows.
It includes hearing and paying attention to social discussions to help the company
understand what existing and potential customer want and how to connect with them
according to their needs. Moreover, it involves engaging with customers through
publishing appropriate and interesting content. Essential is that the content has to
be really fascinating and appealing for customers. The company can analyze the
data from all those activities and regularly develop the company's social presence to
generate more income. The essence of social CRM in a nutshell is to build trustful
relationship to customers [72].

It was already mentioned that an important goal of social CRM is to benefit both the
company and the customer. How does the company benefit? One of the main advantage is increased transparency within customers' data and sales processes. Through
centralized data from every lead and customer interaction (e.g. phone, email, chat,
social media) consistency and efficiency can be reached and everyone from the

Fig. 8 Winning design of
the Starbucks White Cup
Contest in 2014 [164]

company who has access to customer records sees the same data with real-time updates. Moreover, productivity improves through listening to social conversations and customers' needs. Social media provides a more complete view of the customers through insight into customers' demographics and behavior patterns. The extended customers' profiles provide better understanding of customer needs and wants and help the sales team target the consumers who are the most likely to buy from the company. On the one hand, through monitoring customer's conversions and feedbacks about the brand, the company is in the position to join the dialogue and be proactive. On the other hand, monitoring customer's conversation also helps to develop improved and innovative products and services tailored to customer's needs. Through customer collaboration the company can provide effective and efficient business that profits not just the company, but also the customers. Owing to customers' creation of content, social CRM provides better marketing and targeting opportunities for the company. Social CRM upgrades the reputation of the company and increases trust and loyalty. Besides the benefits that were already mentioned, the list of advantages for a company with social media presence is very convincing: increased search engine rankings, the possibility to share the company's point of view and representing the company itself in a way that creates value for its customers and not just pushes for sales [72, 158, 160, 175].

What are the advantages of social CRM for customers? First, and most importantly fast reaction and solution to customers' issues. T-Mobile USA has an outstanding customer service on Twitter, the company replies to almost every customer questions. The support representatives' signature makes T-Mobile's social media strategy unique and personalized, it also humanizes the response [48]. Second, social

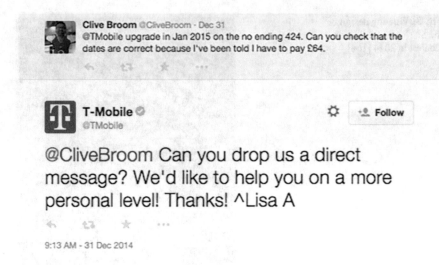

Fig. 9 T-Mobile's social customer service [182]

CRM keeps customers up to date and provides more information about products and services quickly on a more personalized level than a company's website. Finally, social CRM enhances customer satisfaction by giving customers the possibility to express their opinion and the chance for engagement using social media platforms [187] (Fig. 9).

4.5 The Challenges of Social CRM

In Sect. 4.3 it was already mentioned that social CRM is a customer-centric process. The first and most common mistake that a company can commit while adopting social CRM is the missing focus on customer centricity in the company's business strategy. If a company does not prioritize the customers in the first place, the first and most important step is to define a customer-centric business strategy for the firm, focusing on customer experience. It is not enough to define the new strategy (if it has to be defined), employees have to be trained according to it to avoid any inconvenience. Any inadequate comment or not-well chosen word can lead to customer dissatisfaction, negative WOM and damaged brand image, since in the digital age everything is documented and also public on social networks [107].

In Sect. 4.2 it was already mentioned that customers' expectations are rising and serving the social customer could be demanding for companies. Another challenge for companies is to provide customer service out of business hours. Customers expect a respond from the company within 1 h (ignoring the business hours) [13].

Figure 10 shows a customer who has been complaining about a lost suitcase and after the incident posting his opinion about the airways' customer service on

Fig. 10 British Airway's customer service [179]

Twitter. British Airways responded only after 8 h, meanwhile the tweet was seen by 76,000 users, making the customer feel ignored and angry. Of course, having a 24/7 customer service on social networks can not be expected from every company, but a large airline that operates around the clock should be prepared for such a scenario.

From the company's point of view, all the invested efforts into social CRM have to be profitable. Since social CRM is a tool just like any other to maximize the return of the company, after implementing it and putting effort into it, one should be able to measure the influence of social CRM outcome on the revenue. One of the most capable key figures to measure the result is ROI. If there is no advantage deriving from the implementation of social CRM all the invested efforts were a waste of time and money and improvements within the company are necessary [107].

All the listed challenges in this section demonstrate the most common challenges but there are several more. To sum up, although most of the companies are facing lots of challenges while implementing social CRM, it can be said that the benefits outrun the drawbacks.

4.6 Social CRM Strategies

The accomplishment or lack of success of social CRM mostly rely upon the strategy and the implementation process. The first and most important step before thinking about social CRM strategies and implementation is to choose the proper social tool (i.e. CRM software with integrated social media) that fits the company's needs. Section 5 gives an overview of the top 10 most popular CRM software including social CRM features.

There are a lot of existing social CRM strategies, a few selected ones are going to be introduced in this section. First, and most importantly the appropriate social network has to be chosen, where the company's audience is available. It depends mostly on the type of the business and the demographics of the audience. Instagram could be a great choice if the company's potential and existing customers are under the age of 35 [85].

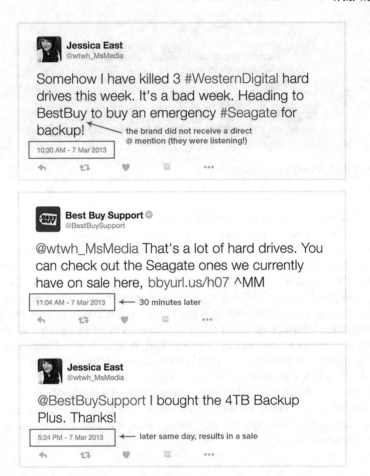

Fig. 11 Best Buy's customer service [46, 178]

Since customers both expect and benefit through fast customer service via social networks, social media managers has to act as real-time customer service representatives. They should react to customer's question and comments as soon as possible (best case scenario: within an hour) and the customer should have the feeling of the company's helping purpose instead of a transaction. Figure 11 shows the best possible customer service achievable on social networks. Even if the company Best Buy was not directly mentioned in the customer's tweet, the social CRM feature Listening made is possible for the company to see the post. Best Buy responded half an hour later and the outcome of the conversation was a purchase and a pleased and happy customer [46] (Fig. 11).

One of the essential CRM strategies is handling negative criticism. Every customer wants to be heard, not answering for angry customers just makes matters worse. The

Fig. 12 Jet Blue Airway's social customer service [46]

company's answer should be empathetic and ensuring that the company is working on a solution [18] (Fig. 12).

After mentioning some social CRM strategies, let's take a look at companies and brands that are succeeding on social networks and their recipe for success. The ambition of Innocent, the smoothie producer is to make their social media platforms an enjoyable spot that people want to check out regularly. Their method on social media is starting the conversation on subjects which excites people. Besides some promotion of the company's newest products, which they do in a funny way e.g. "Say hello to Innocent bubbles. Ignore the stares of people wondering why you're talking to a drink." [40] the most posts on Innocent's Facebook page are about everyday topics. The recipe of the company is talking to its audience like humans without jargon and not pushing for sales. Moreover, the company often posts reactions about current events like Eurovision or the solar eclipse. Figure 13 shows Innocent's Facebook post about setting back clocks in a really funny and for everyone relatable way [76].

Ellen DeGeneres, the host of the popular daytime talk show of The Ellen DeGeneres Show is the most followed TV personality on Facebook, Instagram and Twitter. She has over 28 million likes on Facebook [39], over 47 million followers on Instagram [73] and more than 73 million followers on Twitter [180] (State: September 2017). Ellen is represented on social media platforms since the beginning of her show 2003. Why are her social media posts so engaging? Just to mention the most important reasons, they are energetic, inspirational and wake emotions either making her audience laugh or cry. The content is humorous and inspiring at the same time, encouraging the audience to like, share and comment. Ellen has a remarkably active profile. She shares content more than once daily on her social media platforms [75].

Last, but not least, who would not remember the famous Oscar selfie? It can be linked to Ellen's name. In 2014 she was hosting the Oscars and asked Bradley Cooper to take a selfie with her and some other celebrities like Jennifer Lawrence, Meryl Streep and Brad Pitt from the front row. She announced during the Oscars that she

Fig. 13 Innocent's post about putting the clock back [40]

wanted the photo to be the most-retweeted photo of all time and she achieved it within 1 h [74] (Fig. 14).

4.7 Conclusion

From the company's perspective CRM integrated with social media provides a lot of new and better services. Without integrating the social platform of the company into its CRM system, capturing private information about customers from social media was not possible. Social CRM provides the company additional information about its audience's attributes like demographics, interest, needs and behavior patterns. Social CRM does not just provide more data about the customers but also what they are

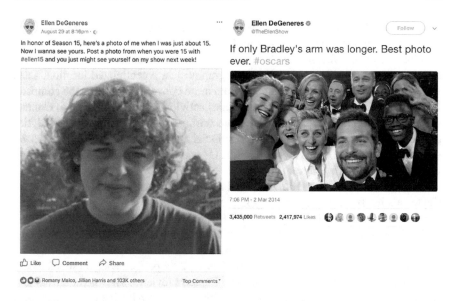

Fig. 14 Ellen's posts [39, 180]

talking about the brand, giving the company the opportunity to join the conversation and be aware of its audience's opinion. Moreover, the social CRM feature Listening makes it possible for the company to get notified when people on social media are talking about the brand, even if it was not tagged in the conversation.

From the customer's perspective social customer service is the biggest advantage of CRM and social media common work at the age of a mobile word. Social customer service is not just a new communication channel for the customer with the company but a rapid response opportunity to customers' requests.

5 Top 10 CRM Software and Their Interaction to Social Networks

CRM software is not a brand new phenomenon. The earliest versions of the software were available in the beginning of the '90s. Nowadays, the market of CRM businesses is worth supposedly $36 billion and it is still growing 27% yearly, dominated by the following industry leaders: Adobe with 26.9% of the market share, Salesforce (21.1%), Microsoft (20%), diverse other vendors combined together (13%) and Oracle (Fig. 15).

What is new, what has changed in the recent years? Ten years ago, 88% of the vendors offered only the traditional on premise CRM solutions, according to a Software Advice Study. In contrast, in 2017, 87% of CRM vendors offer cloud-based solutions for their customers. As a result of cloud-based structure mixed with progressive mobile technologies, mobile CRM usage has risen significantly recently. Most CRM vendors offer Android and iOS apps.

What are the key features of CRM software? Contact management to systematize customer data (e.g. contact details, demographics) in a centralized location. Lead management to help the sales team generate more conversion. It helps the company follow through every stages of different sales activities and is connected to Contact management. Campaign management from setting up target customers to analytics to evaluate the success of the campaigns. Social media management to track social media activity of the company's audience. Of course, every vendor offers different packages that include different features.

In this section, the top 10 most popular CRM software's vendors according to Capterra's list from July 2016 are going to be shortly introduced: including a brief introduction about the company, offered CRM packages with some selected features, social CRM functionalities and a few randomly picked customers of the company. The list from Capterra is based on a market score that consists of three criteria: 40% amount of customers, 40% amount of users, 20% performance on social media including the total number of followers and Capterra reviews. The maximum achievable score is 100. In my view, the ranking of the vendors on the list is representative, since the market score is based on more than one criteria and the market share is not taken into account, preventing the list from registering only the industry leaders at the first places (Fig. 16).

5.1 Salesforce

Salesforce is the "world's #1 CRM platform" [136] according to its self presentation, headquartered in San Francisco. The American cloud computing company was established in 1999 by Marc Benioff who is still the CEO and chairman of the company [140, 141].

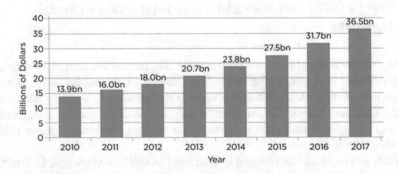

Fig. 15 CRM market from 2010 to 2017 [81]

Fig. 16 Top 10 most popular CRM software according to Capterra [19]

Salesforce offers different CRM packages for different sized companies. Every company from a start up to an enterprise in every industry can find a solution for their different needs. SalesforceIQ is the standard CRM software the company offers and the pricing looks as follows: Starter package offers out-of-the-box CRM for €25 per user per month with features like Automatic data capture and Personalised sales tracking. However, this package only fits small businesses since it only serves up to 5 users. Lightening Professional package offers complete sales CRM and fits middle-sized companies for €75 per user per month without the limitation of the number of users. Of course, more expensive packages offer more features. Lightening Professional grants features like Opportunity Tracking so Salesforce's customer can easily follow up on his deals and take a look at the details such as phase, products, competition and more. The most popular SalesforceIQ package is called Lightning Enterprise, costs €150 per user per month. It grants services like Advanced forecasting and Advanced reporting. The most expensive package costs €300 per user per month and is called Lightning Unlimited. It consists of the features of the cheaper packages and offers some exclusive features such as 24/7 customer support and Unlimited online training. Every package provides smart mobile app for iOS and

Android for free. The company offers a 14-day free trial to all its packages without obligation but with full access to Sales Cloud. This way the customer has the opportunity to watch tutorials and perform common actions [142, 145–148].

Salesforce not just offers CRM solutions, it provides a wide range of products like Service Cloud, Marketing Cloud or Community Cloud. Social Studio is offered in Marketing Cloud Einstein just like E-mail and Advertising Studio. With the help of Social Studio the customer can Listen to conversations about its brand, gain feedback and use them for creating new marketing strategies or optimizing existing ones. Of course, there are some other features like Responding to social posts, Providing customer service, Creating content on social media and Analyzing marketing performance. The price of Social Media Marketing software is not published on Salesforce's website, the company must be contacted for further information. The company can be reached via telephone in urgent cases or by filling out a form with information about the person's name, job title, e-mail. The company answers the incurred question as soon as possible [143, 144, 135].

According to Capterra, Salesforce has 150,000 customers and 7,200,000 users [27]. Canon, American Express, Toyota, Aston Martin, Dunkin' Brands among others belong to Salesforce's clients [139]. Canon, one of the world's largest producer of cameras, uses the Sales Cloud to connect with its customers [138]. Aston Martin profits both from Sales and Marketing Cloud and Salesforce helps the luxury brand to follows its goal: to create a unique and personal experience to satisfy its customers [137]. Dunkin' Brands, mostly known from its coffee and donuts, manages its social media platforms with Social Studio. The company's social aim is to inspire their audience and engage with them. The social aim follows the company's overall goal to generate loyal customers [37].

5.2   Zoho

Zoho (formerly AdventNet Inc.) was established by Sridhar Vembu and Tony Thomas in 1996. Sridhar Vembu is still a key person at the company, he is the CEO at present. Zoho Corporation holds three brands: Zoho, ManageEngine, WebNMS. Zoho is a provider of online applications. The company launched its CRM software in 2005, Zoho Social in 2015 [183, 186, 188, 201].

According to the company's self-presentation Zoho CRM is a "CRM software that lets you close more deals in less time" [195]. The company offers CRM tools from small businesses to enterprises. Zoho provides five different CRM packages: Free Edition, Standard, Professional, Enterprise, Ultimate. The Free Edition serves up to 10 users and can be the perfect solution for small businesses. It offers some basic functionalities like Accounts and Contacts that are responsible for keeping track of important information in the company. The Standard Edition costs €12 per user per month and offers limited features like Sales forecasting and Mass email. The Professional Edition is available for €20 per user per month and provides services like Email integration and Inventory management. The most popular CRM package is called Enterprise Edition and is sold for €35 per user per month. It includes ZIA, the sales assistant. Zia is data intelligent, she provides performance-based trend analysis, explanations and forecasting. The Ultimate Edition offers the most features like Advanced customization and Advanced CRM Analytics for €100 per user per month. Zoho offers a 15-day free trial period for all its packages [189–192, 194, 199].

Social CRM is included in every CRM package beginning from the Free Edition with restricted or unlimited functionalities. Every edition offers integration with Twitter, Facebook and Google+. The Professional, Enterprise and CRM Plus offer more features: Social tab, Capturing leads from Twitter or Facebook and Social interaction with leads and contacts. It is also possible to subscribe only the Social CRM functionalities without the whole CRM packages. The Free Social CRM plan offers access for one user for one brand with restricted functionalities. The Standard Social CRM costs €10 for a month and allows access for two users and one brand with extended functionalities like List of most engaged connections, Basic analytics. The Professional Social CRM costs €50 for a month and grants access for five users and three brands with unlimited functionalities like Advanced analytics and Custom reports. In all three social CRM plans are the integration with Facebook, Twitter, LinkedIn, Google+ and Instagram included [196–198, 200].

According to Capterra, Zoho has 80,000 customers and 20,000,000 users world-wide [29]. The most famous customer is Amazon and a lot of small to middle sized companies are listed on Zoho's website e.g. Selectra, a website, specialized in energy price comparison [159]. A lot of success stories can be found on the website, some of them emphasizes the benefits of Zoho after switching from another CRM provider like Salesforce or Sugar [193].

5.3   Act!

Act! is a CRM software application that belongs to Swiftpage. As a solution provider for small businesses and sales teams, the aim of the company is to help grow these businesses with revolutionary technologies combined with relevant content [7, 171].

Act! offers three different CRM packages: Act! Premium, Act! Premium Cloud, Act! Pro. Act! Premium is the most affordable option for €22 per user per month. It includes features like Contact and Activity Management that helps you keep track of the companies contact details along with email, phone numbers, documents and social media updates and Opportunity and Sales Tracking with details about the company's sales processes and Email Marketing. The only difference between Act! Premium and Act! Premium Cloud is that the cloud version does not require IT from the customer's side and Act! handles the technical support. Act! Premium Cloud costs €35 per user per month. Act! Pro offers the same features like the other two versions but is economically the better choice for individuals and small workgroups since the package costs €249 per user for a whole year. For small companies and start ups with a lesser budget Act! offers Essentials for €9 per user per month, with useful features like Email Marketing from its CRM packages. Unfortunately, besides the fact that Act! offers social media integration, detailed information e.g. social CRM features can not be found on the company's website. However, Act! provides a lot of tutorial videos with step by step instructions on its website and the company offers a 14-day free trial [2–6, 8, 10].

Act! has 71,000 customers and 3,000,000 users according to Capterra [20]. Since Act! provides CRM solution for small to middle sized businesses, under customer stories enterprises or worldwide well-known names can not be found. One of the success stories is about Reebok Sports Club, a gym in London that counts approximately 8000 members [9]. Diane Kay, the Sales and Marketing director of the club said when asked about Act!: "Act! has become our bible when it comes to monitoring and managing new memberships, as it let's us trace the entire sales chain from initial contact, so we always know who's done what" [11].

5.4 Microsoft Dynamics

Dynamics CRM belongs to the American multinational technology company Microsoft Corporation. The company was founded by Bill Gates and Paul Allen in 1975. The first CRM software from Microsoft was launched in 2003 and was called Microsoft CRM 1.0. After rebranding and renaming the software as Microsoft Dynamics to indicate the inherence to the Dynamics product family, the first version of the software was released in 2005 [101, 102, 109, 173, 174].

Microsoft offers Dynamics CRM currently in Dynamics 365. It is a cloud application that brings together ERP and CRM solutions. Dynamics 365 offers Sales, Customer Service, Field Service, Talent, Finance, Operations, Retail, Project Service Automation, Marketing and Customer Insights. Relationship Management is included in Sales. Dynamics 365 is available in two editions: Enterprise edition and Business edition. The best valued package within the Enterprise Edition is called Dynamics 365 Plan and it includes an ERP system for the customer inclusive Sales and costs $210 per user per month. It is also possible to start with only Sales for $95 per user per month and upgrade the package later. For small to medium sized businesses Microsoft offers Dynamics 365 Business Edition with full access to application functionality for $40 per user per month. The company provides a 30-day free trial period [100, 103, 104, 105].

Social media integration is possible within Dynamics 365 when connecting Dynamics 365 to Social Engagement from Microsoft. Social Engagement collects data from social media websites like Facebook or Twitter, mostly focusing on conversation about the customer's brand or products. Another option is to combine LinkedIn Sales Navigator and Dynamics 365 Sales in a package called Microsoft Relationship Sales solution at a monthly price of $135 per user [98].

According to Capterra, Dynamics CRM has 40,000 customers and 4,250,000 users [21]. Some well-known brands among others that use Microsoft Dynamics: Sodexo, Avon, Swarovski, Bayer [99, 106]. Among other educational institutes Wirtschaftsuniversität Wien also belongs to Microsoft's customers. Microsoft Dynamics CRM helps the university to meet the challenges of modern administration, organization and marketing [97].

5.5 Hubspot

HubSpot develops and sells software products for inbound marketing and sales. The company was founded by Brian Halligan and Dharmesh Shah in 2006. Both of them are still key people at the company: Brian Halligan is the CEO and Dharmesh Shah is the CTO [54, 58, 59].

Hubspot offers three software products: CRM, Marketing and Sales. The slogan of Hubspot CRM sounds: "Why pay for a CRM when this one is free?" [63]. It can be used by unlimited amount of users and it is "free forever" (i.e. it is not a trial version, it can be used as long as the customer is subscribed). The software offers some basic functions like organizing and managing leads and customers and supports the client's sales team. Moreover, the CRM software integrates with the Sales software to help the sales team to close more deals, more efficiently. The Sales software is also for free and makes Gmail and Outlook Integration or Email Scheduling possible. An extended version of Sales software is also available from Hotspot, it costs €50 per user per month and offers features like Calling i.e. making calls directly from the browser and automatically recording them. Reports are as add-ons available for €200 per month [55, 57, 60, 62].

Social Media is included in every Marketing package the company has to offer with the same features: Social Media Monitoring, Social Media Publishing, Social Media Analytics and Social Media Bookmarklet. The Basic Marketing package is the perfect solution for small companies or as a starter package for those who are new in the world of inbound marketing. It costs €185 per month (required onboarding i.e. one-time fee €550). It offers features besides the Social Media Suite like Email Marketing and Marketing Analytics Dashboards. The most popular option is the Pro Version for €740 per month (required onboarding €2760) with extended features like Marketing Automation. The Enterprise Edition is offered at €2200 per month (required onboarding €4600) and offers the most features like Revenue Reporting [56, 61].

Hubspot has 60,000 customers and 120,000 users according to Capterra [22]. On Hubspot's website under customers a lot of enterprises, small- and middle-sized businesses are listed. Probably the most well-known name is TUI Travel. The company experienced 20–50% growth of website visitors across brands after implementing Hubspot's software (inclusive the Marketing package) [52, 53].

5.6 SAP

SAP is a multinational software corporation, headquartered in Germany. The acronym SAP stands for "System Analysis and Program Development". The company was founded by five former IBM employees in 1972. It specializes in developing and selling enterprise software to midsize and large organizations to support business operations and customer relations [149, 151]. SAP released its first CRM software in 2001 [86].

SAP offers an on premise CRM software that supports Sales, Marketing, Customer Service and Customer Interaction, the price is not published on the website. The company also offers a cloud CRM software called SAP Hybris Cloud for Customer. According to the company's self-presentation "This cloud CRM portfolio brings sales, customer service, and social CRM together—to help your team form powerful personal connections that drive customer engagement across all channels" [156]. The SAP Hybris Cloud has eight different editions. The smallest and cheapest package is called SAP Digital CRM, could be the perfect solution for individuals and small teams. It includes features like Contact management, E- mail campaign automation and Analytics. SAP Digital CRM costs €21 per user per month and a free trial period of 30 days is offered. The price and a list of features of the other packages is not published on the website, the company must be contacted for individual offers [150, 154, 155].

The feature Social Engagement is offered in the package SAP Hybris Cloud for Customers (Limited Package) and in the package SAP Hybris Social Engagement Cloud. The following product features are offered in the SAP Hybris Social Engagement Cloud: Social Media Monitoring, Real-Time Facebook and Twitter Message Response, Complete Social Profile and History, Social Media Analytics [152].

According to Capterra SAP CRM has 10,000 customers and 8,000,000 users [28]. Volkswagen, The Body Shop, Henkel, Indesit among others belong to SAP Hybris's customers [157]. The roaster of coffee Royal Cup Coffee and Tea uses SAP Hybris Cloud for Sales that includes CRM solution. The company predicts a revenue growth of 30–40% in the next decade as a result of using SAP Hybris Cloud for Sales [153].

5.7 Maximizer

The company was founded in 1987 and offers CRM solutions since 1995. It is headquartered in Canada [32]. The current president of the company is Vivek Thomas since 2009 [88, 89]. "Maximizer has been successfully competing against some of the largest software companies in the world for 28+ years. Since I took over as President, my focus has been to develop and implement a new vision and growth strategy that helps our customers, existing and new, successfully achieve their business goals while delivering top line and bottom line growth." [83] said the president about the company and his goals.

Maximizer only offers two different software products: Maximizer CRM and Financial Advisor. Maximizer CRM includes Contact Management, Sales, Marketing and Customer Service components. The customer has the option to choose between cloud and on premise solution. Maximizer CRM costs €42 per user per month. The Financial Advisor costs €3 per month more than the standard CRM package. It includes the features of Maximizer CRM and offers some additional ones like Consolidated household views and Investment account and insurance policy management. The company offers a 15-day free trial for both of its packages [91, 93, 94].

On Maximizer's website under Features social CRM is not listed and under Integration social networks can not be found, therefore it is assumed that Maximizer does not offer social CRM services [87].

According to Capterra Maximizer has 57,000 customers and 1,600,000 users [24]. To its customers belong among others the Canadian National Bank, the American media company Hallmark and Glenmark Pharmaceuticals. On Maximizer's website under success stories a lot of companies can be found from different industries: tourism, healthcare, sports, publishing, logistics, manufacturing etc [90]. With the help of Maximizer, the company Langara Fishing Adventures is capable of managing 4,000 luxury fishing vacations every year and achieved that 70% of its guest come back again. Schott Mehlenbacher, the CFO of Langara Fishing Adventure said about the CRM software: "Now, we have a process for tracking every detail of our guests' stay, along with all of their personal preferences, to ensure that each client receives a high level of communication and service" [92].

5.8 Infusionsoft

The company was established by brothers Scott and Eric Martineau in 2001, in Mesa, Arizona as a start up, named eNovasys. Clate Mask joined the firm in 2002 and is still the CEO of the company. The company changed its name in 2003 and became officially Infusionsoft. The company focuses primarily on small businesses [65, 66].

Infusionsoft offers only one software that includes CRM, Marketing and Analytics. CRM provides Contact and Data management, Calendar and Task management, Lead scoring. Marketing offers the customers features like Campaign Builder, Website tracking, Statistics and Reports. The starting price is $99 per month and requires a one-time fee starting at $999 for new user training. This package only fits customers with one user and up to 500 contacts. Essential edition serves companies with maximal 3 users and 2,500 contacts for $199 per month. Both in the Starter as well as in the Essential package CRM and Marketing automation is included. The Complete package includes CRM, Marketing automation, Sales automation and E-commerce and costs $299 per month. It serves small firms with maximal 5 users and 10,000 contacts. The company offers a 14-day free trial period [68, 70, 71].

Infusionsoft has acquired the social media marketing software company GroSocial in January 2013 [67]. According to HelpCenter by Infusionsoft integrating a social media account is possible. However, connecting to Facebook is currently not available due to a missing match to Facebook's Integration terms. Furthermore, additional information about social CRM features on the company's website can not be found [51].

Infusionsoft has 33,000 customers and 110,000 users according to Capterra [23]. Since the software serves small businesses, well-known names can not be found among Infusionsoft's users. The owner of the small bed and breakfast Les Molyneaux, in the Champagne region of France said about the software: "I use Infusionsoft to keep in touch with my past guests, and they come back. We gained the number one spot in our region on TripAdvisor as a result of guests' reviews solicited by Infusionsoft" [69]. The results speak for itself: doubled income, tripled guests.

5.9   Oracle

Oracle (Oracle Corporation) is an American software developer. The company was founded by Larry Ellison in 1977 and was named Software Development Laboratories (SDL) back in the day. Oracle mainly concentrates on developing and selling database software and technology, cloud computing, ERP- and CRM software [15, 123].

The easiest and cheapest CRM solution from Oracle is the application Customer Data Management Cloud. It only offers storage for all the customer's data for €9 per thousand contacts per month and €17 per user per month. In contrast, Oracle Sales Cloud offers CRM services like Partner Relationship Management, Sales Force Automation, Unified Sales and Services, Sales Performance Management, Unified CX Platform inclusive Customer Data Management. The Sales Cloud is available in four different packages: Professional Edition for €56 per user per month, Standard Edition for €87 per user per month, Enterprise Edition for €174 per user per month and Premium Edition for €260 per user per month. The more expensive the package the more features are offered [113–120].

For social CRM services Oracle offers the Social Cloud. It includes four key components: Social Media, Social Engagement and Monitoring, Social Network and Social Data. The price of the Social Cloud is not published on the firm's website, the company must be contacted for further information [121, 122].

According to Capterra, Oracle CRM has 5,000 customers and 4,600,000 users [25]. Li Zi Jian, the Head of the IT Department of Nubia Technology said about the Sales Cloud: "With Oracle Sales Cloud, we built a unified master sales data repository and sales management platform to unify management of our sales data, and make transparent and digitalize our sales process. In addition, with mobile support, we could effectively increase sales productivity by enabling sales teams to sell anytime, anywhere" [112]. Lego is also among Oracle's customers and uses the Social Cloud: "The whole company needs to move in a more social direction. It's across the whole value chain, from recruitment through marketing and product development" [111].

5.10 Sage

Sage (Sage Group plc) is a British multinational enterprise software company. The firm was established by David Goldman in 1981 as a start up. The original purpose of the company was to develop estimating and accounting software for small businesses. Sage offers software products for different business needs: Accounting and Finances, Business Management, HR, CRM, Payment, Business Intelligence, Inventory Management etc [127, 130].

Sage offers one CRM package that provides every service from Sales, Marketing to Reporting. The CRM software is available for small to middle sized firms with a number of users between 1 and 200 for €39 per user per month (or one-time license purchase for €750). The Enterprise Edition costs €49 per user per month (or one-time license purchase for €990). The company offers a 30-day free trial period [131–133].

Social CRM is included in the Sage CRM software. It integrates with the most popular social network sites: Facebook, Twitter and LinkedIn. With the help of social CRM, the client of Sage can track brand or company mentions, generate leads and get closer to its customers. Detailed description of social CRM features can not be found on the company's website. Sage primary emphasizes the benefits of social CRM in general from the client's perspective [134].

According to Capterra, Sage has 14,000 customers and 742,000 users [26], mostly small to medium sized businesses e.g. American Pool Enterprises, Blue Sky International and Widex [129]. The Australian Grand Prix Corporation (AGPC) is in charge for Australia's Motorsport events and uses integrated Sage CRM and Sage 300 (ERP software) to collect and control customer data. Jeremy Kann, the General Manager of AGCP said when asked about the Sage CRM software "We are already seeing improvements in the way we operate due to Sage CRM. It's been a very, very positive experience" [128].

5.11 Conclusion

The companies on the Top 10 most popular CRM Software list are very diverse: large ERP suppliers, small companies developed from start ups and only CRM software providers etc. As shown in the summary table, the pricing schemes of the various CRM software providers is different. Some companies require payments per user per month, some of them a fixed monthly fee or a one-time fee or offer their basic services for free, but every provider from the list offer a free trial period and a mobile app. Most CRM software providers offer services in every industry. The suitable CRM package and therefore the price of the CRM software depends mostly on the size of the customer's business. Some software providers like Act! and Infusionsoft concentrate on small- to medium sized businesses, but some others like Microsoft

Table 1 Comparison of different CRM software providers. *per user per month, prices that were given in $ converted into € ($1 = €0.84)

CRM software	Starting price*	Free trial	Mobile	Social CRM
Salesforce	€25.00	14d	Yes	Yes
Zoho	–	15d	Yes	Yes
Act!	€22.00	14d	Yes	Yes
Dynamics	€34.00	30d	Yes	Yes
Hubspot	–	30d	Yes	Yes
SAP	€21.00	30d	Yes	Yes
Maximizer	€42.00	15d	Yes	No
infusionsoft	€84.00	14d	Yes	Yes
Oracle	€26.00	30d	Yes	Yes
Sage	€39.00	30d	Yes	Yes

Dynamics and SAP mainly on large companies and enterprises. Most providers offer cloud solution, some of them traditional on premise CRM version or both. With the exception of Maximizer, every CRM software offer social CRM services. In most of them Facebook, Twitter and LinkedIn can be integrated (Table 1).

6 Summary

The first object of this chapter was to illustrate the efficient service provided by CRM and social media common work. Since the usage of social media has risen dramatically in the recent years, this phenomenon has led to changes in customer's behavior. The average customer became active on social media platforms and grew into an active influencer from a passive buyer. While investigating the benefits and challenges of Social CRM, it was found that the benefits of Social CRM outweigh the challenges for the company and the customer only profits through social customer services. The examples shown in Sect. 4.6 illustrated some Social CRM strategies and best practices of which a lesson can be drawn of. Finally, the new and improved services provided by CRM and social network common work were summarized.

The second goal of this chapter was to give an overview about the different CRM software providers including social CRM features. The comparison showed that the selected CRM software vendors are following the latest trends: almost every provider offers cloud based CRM solutions and every vendor provides a mobile version of the software. With the exception of one CRM software, every vendor grants social media integration and social CRM services.

In conclusion, it can be said that social CRM is not just a nice option for companies but an all-important necessity for a successful business. Companies should definitely take advantage of the bright perception of social media, since most importantly it can help companies to gain more information about its customers and their needs. Therefore, improve productivity, product development in the company and the potential to develop long-term profitable relationships with their customers.

References

1. Accenture Strategy (n.d.) Seeing beyond the loyalty illusion: it's time you invest more wisely. https://www.accenture.com/t20170216T002701Z__w__/us-en/_acnmedia/PDF-43/Accenture-Strategy-GCPR-Customer-Loyalty_Infographic.pdf#zoom=50. Accessed 04 Aug 2017
2. Act! (n.d.) Act! Essentials preisinformation. https://www.act.com/de-de/preise/act-essentials. Accessed 22 Sept 2017
3. Act! (n.d.) Act! Preisinformation. https://www.act.com/de-de/preise. Accessed 22 Sept 2017
4. Act! (n.d.) Act! Premium. https://www.act.com/de-de/produkte/act-premium. Accessed 22 Sept 2017
5. Act! (n.d.) Aktivitäten- und Kontaktmanagement. https://www.act.com/de-de/produkte/act-premium/kontakt-und-aktivitaten-management. Accessed 22 Sept 2017
6. Act! (n.d.) Automatisierung und Integration. https://www.act.com/de-de/produkte/act-premium/automatisierung-und-integration. Accessed 22 Sept 2017
7. Act! (n.d.). https://www.act.com/de-de/. Accessed 22 Sept 2017
8. Act! (n.d.) Social CRM: how customer support can benefit from social media. https://www.act.com/act-blog/post/blog/2017/05/26/social-crm-how-customer-support-can-benefit-from-social-media. Accessed 22 Sept 2017
9. Act! (n.d.). Testimonials. https://www.act.com/en-uk/customers. Accessed 22 Sept 2017
10. Act! (n.d.) Vertriebsprozess und Verkaufschancen managen. https://www.act.com/de-de/produkte/act-premium/verkaufschancen-und-vertriebsmanagement. Accessed 22 Sept 2017
11. Act! (n.d.) Act! The ultimate 'fitness' tool for sales. https://www.act.com/en-uk/customers/testimonials/reebok. Accessed 22 Sept 2017
12. Amazon (n.d.) Amazon go. https://www.amazon.com/b?node=16008589011. Accessed 03 Aug 2017
13. Baer J (n.d.) 42 percent of consumers complaining in social media expect 60 minute response time. http://www.convinceandconvert.com/social-media-research/42-percent-of-consumers-complaining-in-social-media-expect-60-minute-response-time/. Accessed 21 Aug 2017
14. Baldacci K (2013) 5 reasons why the social customer is today's undeniable authority. https://www.salesforce.com/blog/2013/07/the-social-customer.html. Accessed 20 Aug 2017
15. Bora H (2017) Oracle CRM. https://www.slideshare.net/HridayBora/oracle-crm-75519161. Accessed 03 Sept 2017
16. Brookes N (2014) The multibillion dollar cost of poor customer service [INFOGRAPHIC]. https://www.newvoicemedia.com/blog/the-multibillion-dollar-cost-of-poor-customer-service-infographic/. Accessed 21 Aug 2017
17. Bunskoek K (2014) 4 ways to crowdsource product ideas using social media contests. http://www.socialmediaexaminer.com/4-ways-crowdsource-product-ideas-using-social-media-contests/. Accessed 26 Aug 2017
18. Burke S (2016) Examples of the good, the bad & the ugly of customer service on social media! http://www.getspokal.com/examples-of-the-good-the-bad-the-ugly-of-customer-service-on-social-media/. Accessed 12 Sept 2017
19. Capterra (2016) The top 10 most popular CRM software. http://www.capterra.com/customer-relationship-management-software/#infographic. Accessed 10 July 2017

20. Capterra (n.d.) Act! http://www.capterra.com/p/133171/Act/. Accessed 22 Sept 2017
21. Capterra (n.d.) Dynamics CRM. http://www.capterra.com/p/95570/Dynamics-CRM/. Accessed 23 Sept 2017
22. Capterra (n.d.) Hubspot CRM. http://www.capterra.com/p/152373/HubSpot-CRM/. Accessed 24 Sept 2017
23. Capterra (n.d.) Infusionsoft. http://www.capterra.com/p/76390/Infusionsoft/. Accessed 24 Sept 2017
24. Capterra (n.d.) Maximizer CRM. http://www.capterra.com/p/32383/Maximizer-CRM/. Accessed 24 Sept 2017
25. Capterra (n.d.) Oracle CRM. http://www.capterra.com/p/73621/Oracle-CRM/. Accessed 03 Sept 2017
26. Capterra (n.d.) Sage CRM. http://www.capterra.com/p/45968/Sage-CRM/. Accessed 24 Sept 2017
27. Capterra (n.d.) Salesforce. http://www.capterra.com/p/61368/Salesforce/. Accessed 22 Sept 2017
28. Capterra (n.d.) SAP digital CRM. http://www.capterra.com/p/144574/SAP-Digital-CRM/. Accessed 24 Sept 2017
29. Capterra (n.d.) Zoho CRM. http://www.capterra.com/p/159406/Zoho-CRM/. Accessed 24 Sept 2017
30. Conversocial (n.d.) The 2015 definitive guide to social customer service. http://www.conversocial.com/social-customer-service/challenges-of-social-media. Accessed 12 Sept 2017
31. CRM Switch Staff (2013) A brief history of customer relationship management. https://www.crmswitch.com/crm-industry/crm-industry-history/. Accessed 10 Sept 2017
32. Crunchbase (n.d.) Maximizer. https://www.crunchbase.com/organization/maximizer#/entity. Accessed 24 Sept 2017
33. CSP Daily News (2014) 10 best convenience store chains in America. http://www.cspdailynews.com/industry-news-analysis/marketing-strategies/articles/10-best-convenience-store-chains-america. Accessed 30 July 2017
34. DBA Designs & Communications (n.d.) The evolution of social CRM. https://www.dbadesigns.com/2011/ads/the-evolution-of-social-crm/. Accessed 21 July 2017
35. Despremaux G (2015) Top 10 reasons why we use social networks. http://wersm.com/the-10-top-reasons-why-we-use-social-networks/#prettyPhoto. Accessed 20 Aug 2017
36. Drennan A (2011) Consumer study: 88% less likely to buy from companies who ignore complaints in social media. http://www.conversocial.com/blog/consumer-study-88-less-likely-to-buy-from-companies-who-ignore-complaints-in-social-media. Accessed 20 Aug 2017
37. Dunkin' Brands (n.d.). http://www.dunkinbrands.com. Accessed 22 Sept 2017
38. Edwards J (2016) iPhone users are abandoning their loyalty to Apple. http://www.businessinsider.de/iphone-users-abandon-loyalty-to-apple-2016-11?r=UK&IR=T. Accessed 30 July 2017
39. Facebook (n.d.) Ellen DeGeners. https://www.facebook.com/ellentv/. Accessed 12 Sept 2017
40. Facebook (n.d.) Innocent. https://www.facebook.com/innocent.drinks/. Accessed 12 Sept 2017
41. Facebook (n.d.) M&M's USA. https://m.facebook.com/mms/photos/a.131015676956.121420.30634981956/10153351880971957/?type=3. Accessed 12 Sept 2017
42. Finances Online (n.d.) What is CRM software? Analysis of features, types and pricing. https://financesonline.com/what-is-crm-software-analysis-features-types-pricing/#benefits. Accessed 10 Sept 2017
43. Fortunelords (n.d.) 36 mind blowing YouTube facts, figures and statistics—2017. https://fortunelords.com/youtube-statistics/. Accessed 12 Aug 2017

44. Gamble R (2017) Pope Francis tops world leaders with highest Twitter following. http://www.thetablet.co.uk/news/7598/0/pope-francis-tops-world-leaders-with-highest-twitter-following. Accessed 12 Aug 2017
45. Greenberg P (2009) CRM at the speed of light
46. Gregory S (2017) 6 key elements of using social media for customer service. http://freshsparks.com/using-social-media-for-customer-service/. Accessed 11 Sept 2017
47. Growmap (2015) The dynamics behind the social CRM evolution. http://growmap.com/social-crm-evolution/. Accessed 10 Sept 2017
48. Gushue R (2015) 5 lessons from the best brands on social media in 2015. https://blog.enplug.com/5-lessons-best-brands-social-media-2015. Accessed 10 Sept 2017
49. Hackman M (2016) #NeverTrump and the coming schism in the Republican Party, explained. https://www.vox.com/2016/2/29/11135714/never-trump. Accessed 20 Aug 2017
50. Hashtagnow (2016) #TrumpTrain. https://hashtagnow.co/hashtag/TrumpTrain. Accessed 20 Aug 2017
51. Help Center by Infusionsoft (n.d.) Social media account setup. http://help.infusionsoft.com/userguides/get-started/initial-setup-checklist/social-media-account-setup. Accessed 01 Sept 2017
52. Hubspot (n.d.) Customers filtered. https://www.hubspot.com/customers#/f=enterprise/travel-and-leisure. Accessed 24 Sept 2017
53. Hubspot (n.d.) Customers. https://www.hubspot.com/customers. Accessed 24 Sept 2017
54. Hubspot (n.d.). https://www.hubspot.com. Accessed 24 Sept 2017
55. Hubspot (n.d.) Hubspot CRM. https://www.hubspot.com/products/crm. Accessed 24 Sept 2017
56. Hubspot (n.d.) Hubspot marketing. https://www.hubspot.com/products/marketing. Accessed 24 Sept 2017
57. Hubspot (n.d.) Hubspot sales. https://www.hubspot.com/products/sales. Accessed 24 Sept 2017
58. Hubspot (n.d.) Our story. https://www.hubspot.com/our-story. Accessed 24 Sept 2017
59. Hubspot (n.d.) Our team. https://www.hubspot.com/company/management. Accessed 24 Sept 2017
60. Hubspot (n.d.) Pricing packages CRM software. https://www.hubspot.com/pricing/crm. Accessed 24 Sept 2017
61. Hubspot (n.d.) Pricing packages marketing. https://www.hubspot.com/pricing/marketing#?currency=EUR. Accessed 24 Sept 2017
62. Hubspot (n.d.) Pricing packages sales software. https://www.hubspot.com/pricing/sales. Accessed 24 Sept 2017
63. Hubspot (n.d.) What is Hubspot? https://www.hubspot.com/what-is-hubspot. Accessed 24 Sept 2017
64. Hutchunson A (2017) Top social network demographics 2017 [Infographic]. http://www.socialmediatoday.com/social-networks/top-social-network-demographics-2017-infographic. Accessed 12 Aug 2017
65. Infusionsoft (n.d.) About us. https://www.infusionsoft.com/about. Accessed 24 Sept 2017
66. Infusionsoft (n.d.). https://www.infusionsoft.com. Accessed 24 Sept 2017
67. Infusionsoft (n.d.) Infusionsoft acquires social media marketing company grosocial. https://www.infusionsoft.com/about/news/press-releases/infusionsoft-acquires-social-media-marketing-company-grosocial. Accessed 01 Sept 2017
68. Infusionsoft (n.d.) Infusionsoft pricing. https://www.infusionsoft.com/pricing. Accessed 24 Sept 2017
69. Infusionsoft (n.d.) Les Molyneaux's story. https://www.infusionsoft.com/success-stories/les-molyneaux. Accessed 24 Sept 2017
70. Infusionsoft (n.d.) Marketing automation for small business. https://www.infusionsoft.com/product/features/marketing-automation. Accessed 24 Sept 2017

71. Infusionsoft (n.d.) Small business CRM. https://www.infusionsoft.com/product/features/small-business-crm. Accessed 24 Sept 2017

72. Inside-CRM (2016) 7 surprisingly awesome benefits of social CRM. http://it.toolbox.com/blogs/insidecrm/7-surprisingly-awesome-benefits-of-social-crm-72943. Accessed 10 Sept 2017

73. Instagram (n.d.) theellenshow. https://www.instagram.com/theellenshow/?hl=en. Accessed 12 Sept 2017

74. Jarvey, N. (2014) Ellen DeGeneres' Oscar selfie tops the year on Twitter. http://www.hollywoodreporter.com/news/ellen-degeneres-oscar-selfie-tops-757364. Accessed 11 Sept 2017

75. Johnson C (2017) Ellen DeGeneres: talk show host, activist, and digital content creator. https://medium.com/rta902/ellen-degeneres-talk-show-host-activist-and-number-one-digital-content-creator-9d0ed4dd690. Accessed 11 Sept 2017

76. Kasteler J (2016) 8 companies doing social media right and what marketers can learn from them. https://marketingland.com/8-companies-social-media-right-marketers-can-learn-198228. Accessed 11 Sept 2017

77. Kemp S (2017) Digital in 2017: global overview. https://wearesocial.com/special-reports/digital-in-2017-global-overview. Accessed 17 Sept 2017

78. Kolowich L (n.d.) 7 customer loyalty programs that actually add value. https://blog.hubspot.com/blog/tabid/6307/bid/31990/7-customer-loyalty-programs-that-actually-add-value.aspx. Accessed 30 July 2017

79. Kumar V, Reinartz W (2012) Customer relationship management concept, strategy, and tools, 2nd edn

80. Lazar M (2017) 2017 CRM statistics show why is's a powerful marketing weapon. https://www.ibm.com/developerworks/community/blogs/d27b1c65-986e-4a4f-a491-5e8eb23980be/entry/2017_CRM_Statistics_Show_Why_it_s_a_Powerful_Marketing_Weapon?lang=en. Accessed 10 Sept 2017

81. Lazar M (2017) These CRM statistics prove why it should be in your marketing arsenal. https://www.readycloud.com/info/2017-CRM-statistics-why-it-should-be-in-your-marketing-arsenal. Accessed 10 Sept 2017

82. Leswing K (2017) Amazon is buying whole foods—here's Amazon's vision for the grocery store of the future. http://www.businessinsider.de/amazon-go-grocery-store-future-photos-video-2017-6?r=US&IR=T. Accessed 03 Aug 2017

83. Linkedin (n.d.) Vivek Thomas. https://www.linkedin.com/in/vivekthomas. Accessed 24 Sept 2017

84. Lisi R (2014) Der social customer. http://share4you.ch/wp-content/uploads/Erfolg_Seite27.pdf. Accessed 20 Aug 2017

85. Marvin R (2016) 7 tips for a killer social CRM strategy. https://www.pcmag.com/article2/0,2817,2491682,00.asp. Accessed 11 Sept 2017

86. Maxfavilli (2009) SAP CRM release timeline. https://maxfavilli.com/sap-crm-release-timeline. Accessed 24 Sept 2017

87. Maximizer CRM (n.d.). https://www.maximizer.com/uk/. Accessed 24 Sept 2017

88. Maximizer (n.d.) Leadership. https://www.maximizer.com/about-us/our-team/. Accessed 24 Sept 2017

89. Maximizer CRM (n.d.) About. http://maximizercrm.co.za/about/. Accessed 24 Sept 2017

90. Maximizer CRM (n.d.) CRM case studies & customer success stories. https://www.maximizer.com/customers/. Accessed 24 Sept 2017

91. Maximizer CRM (n.d.) Features. https://www.maximizer.com/features/. Accessed 24 Sept 2017

92. Maximizer CRM (n.d.) Langara fishing adventures case study. https://www.maximizer.com/customers/langara-fishing-adventures/. Accessed 24 Sept 2017

93. Maximizer CRM (n.d.) Maximizer CRM integrations. https://www.maximizer.com/features/crm-integration/. Accessed 24 Sept 2017

94. Maximizer CRM (n.d.) Maximizer CRM live pricing. https://www.maximizer.com/uk/our-solutions/pricing-eur/. Accessed 24 Sept 2017
95. Mclauhhlin T (2017) United Airlines reaches settlement with passenger dragged from plane. http://www.reuters.com/article/us-ual-passenger-idUSKBN17T2WM. Accessed 20 Aug 2017
96. Mediakix (n.d.) How much time is spent on social media? http://mediakix.com/2016/12/how-much-time-is-spent-on-social-media-lifetime/#gs.KPGGA64. Accessed 12 Aug 2017
97. Microsoft (2016) Wirtschaftsuniversität Wien. https://customers.microsoft.com/en-us/story/wirtschaftsuniversitat-wien-education-dynamicscrm2016-office2016-sharepoint2013. Accessed 23 Sept 2017
98. Microsoft (2017) Connect to Microsoft social engagement. https://technet.microsoft.com/en-us/library/dn659847.aspx. Accessed 23 Sept 2017
99. Microsoft (n.d.) Customer stories. https://customers.microsoft.com/en-us/home?sq=&ff=&p=0. Accessed 23 Sept 2017
100. Microsoft (n.d.) Dynamics 365. https://www.microsoft.com/en-us/dynamics365/home. Accessed 23 Sept 2017
101. Microsoft (n.d.) Facts about Microsoft. https://news.microsoft.com/facts-about-microsoft/. Accessed 23 Sept 2017
102. Microsoft (n.d.). https://www.microsoft.com/en-us/. Accessed 23 Sept 2017
103. Microsoft (n.d.) Microsoft dynamics 365 for sales. https://www.microsoft.com/en-us/dynamics365/sales. Accessed 23 Sept 2017
104. Microsoft (n.d.) Microsoft dynamics 365 pricing. https://www.microsoft.com/en-us/dynamics365/pricing. Accessed 23 Sept 2017
105. Microsoft (n.d.) Relationship sales solution. https://www.microsoft.com/en-us/dynamics365/relationship-sales-solution. Accessed 23 Sept 2017
106. Microsoft (n.d.) Search customer stories. https://customers.microsoft.com/en-us/search?sq=&ff=story_product_categories%26%3EDynamics&p=0&so=story_publish_date%20desc. Accessed 23 Sept 2017
107. Nayak A (2014) Social CRM challenges. https://www.linkedin.com/pulse/20141014072309-8346175-social-crm-challenges. Accessed 12 Sept 2017
108. Neudorf S (2015) M&M's will let America vote for a new peanut flavor. https://www.thedailymeal.com/news/eat/mm-s-will-let-america-vote-new-peanut-flavor/101315. Accessed 26 Aug 2017
109. Niiranen J (2013) History of Microsoft's CRM software. https://community.dynamics.com/crm/b/survivingcrm/archive/2013/09/25/history-of-microsoft-s-crm-software. Accessed 23 Sept 2017
110. Obar J, Wildman S (2015) Social media definition and the governance challenge: an introduction to the special issue. Telecommun Policy 39(9):745–750
111. Oracle (2014) Oracle social cloud. http://www.oracle.com/us/products/social-cloud-ebook-2332775.pdf. Accessed 24 Sept 2017
112. Oracle (2017) Oracle customer success—Nubia Technology Co., Ltd. https://www.oracle.com/customers/nubia-5-sales-cl.html. Accessed 24 Sept 2017
113. Oracle (2017) Oracle FAQ's. http://www.orafaq.com/wiki/Oracle_Corporation#WHO. Accessed 03 Sept 2017
114. Oracle (n.d.) Customer data management cloud features. https://cloud.oracle.com/en_US/cdm-cloud/features. Accessed 24 Sept 2017
115. Oracle (n.d.) Customer data management cloud. https://cloud.oracle.com/en_US/cdm-cloud. Accessed 24 Sept 2017
116. Oracle (n.d.) Customer data management cloud pricing. https://cloud.oracle.com/en_US/cdm-cloud/pricing. Accessed 24 Sept 2017
117. Oracle (n.d.) Oracle CRM—CX cloud suite. https://www.oracle.com/applications/customer-experience/crm/index.html. Accessed 24 Sept 2017

118. Oracle (n.d.) Sales clod pricing. https://cloud.oracle.com/en_US/sales-cloud/pricing. Accessed 24 Sept 2017
119. Oracle (n.d.) Sales cloud features. https://cloud.oracle.com/en_US/sales-cloud/features. Accessed 24 Sept 2017
120. Oracle (n.d.) Sales cloud overview. https://cloud.oracle.com/en_US/sales-cloud. Accessed 24 Sept 2017
121. Oracle (n.d.) Social cloud overview. https://cloud.oracle.com/social-cloud. Accessed 24 Sept 2017
122. Oracle (n.d.) The journey to modern marketing: season 2. https://www.oracle.com/marketingcloud/journey-to-modern-marketing.html. Accessed 28 Aug 2017
123. Oracle. https://www.oracle.com/index.html. Accessed 03 Sept 2017
124. Pickard T (2015) 10 customer service stats and what they mean for your contact center. https://www.salesforce.com/blog/2015/01/ten-customer-service-stats-what-they-mean-your-contact-center-gp.html. Accessed 21 Aug 2017
125. Rosenfeld E (2012) Mountain Dew's 'dub the dew' online poll goes horribly wrong. http://newsfeed.time.com/2012/08/14/mountain-dews-dub-the-dew-online-poll-goes-horribly-wrong/. Accessed 27 Aug 2017
126. Rouse M (n.d.) Web 2.0. http://whatis.techtarget.com/definition/Web-20-or-Web-2 [10.09.2017]
127. Sage (n.d.). https://www.business-software.at. Accessed 24 Sept 2017
128. Sage (n.d.) Australian Grand Prix Corporation. http://www.sagecrm.com/customer-stories/tabs/business-growth/australian-grand-prix. Accessed 24 Sept 2017
129. Sage (n.d.) Customer stories. http://www.sagecrm.com/customer-stories/highlights. Accessed 24 Sept 2017
130. Sage (n.d.) How we started. http://www.sage.com/company/about-sage/how-we-started. Accessed 24 Sept 2017
131. Sage (n.d.) Sage CRM. https://www.business-software.at/produkte/kundenmanagement/crm/?i=02scrm. Accessed 24 Sept 2017
132. Sage (n.d.) Sage CRM in-depth. http://www.sage.com/au/products-and-services/business-management/business-management/sage-crm/crm-modules. Accessed 24 Sept 2017
133. Sage (n.d.) Sage CRM overview. http://www.sage.com/au/products-and-services/business-management/business-management/sage-crm. Accessed 24 Sept 2017
134. Sage (n.d.) Social CRM Solutions with Sage CRM. http://www.sage.com/~/media/markets/za/sage-enterprise/brochures/crm/sage-crm-social-data-sheet-new-ci.pdf?la=en-za. Accessed 24 Sept 2017
135. Salesforce (n.d.) Social studio. https://www.salesforce.com/eu/products/marketing-cloud/social-media-marketing/. Accessed 22 Sept 2017
136. Salesforce (n.d.). https://www.salesforce.com/eu/. Accessed 22 Sept 2017
137. Salesforce (n.d.) Aston Martin. https://www.salesforce.com/eu/customer-success-stories/aston-martin/. Accessed 22 Sept 2017
138. Salesforce (n.d.) Canon. https://www.salesforce.com/eu/customer-success-stories/canon/. Accessed 22 Sept 2017
139. Salesforce (n.d.) Customer success stories. https://www.salesforce.com/eu/customer-success-stories/. Accessed 22 Sept 2017
140. Salesforce (n.d.) Executive team. https://www.salesforce.com/company/leadership/executive-team/#benioff. Accessed 22 Sept 2017
141. Salesforce (n.d.) Global offices. https://www.salesforce.com/company/locations/. Accessed 22 Sept 2017
142. Salesforce (n.d.) How to select the right salesforce lightning edition. https://secure2.sfdcstatic.com/eu/assets/pdf/datasheets/DS_SalesCloud_EdCompare.pdf. Accessed 22 Sept 2017
143. Salesforce (n.d.) Marketing cloud Einstein. https://www.salesforce.com/eu/products/marketing-cloud/overview/. Accessed 22 Sept 2017
144. Salesforce (n.d.) Marketing cloud Einstein FAQ. https://www.salesforce.com/eu/products/marketing-cloud/faq/. Accessed 22 Sept 2017

145. Salesforce (n.d.) Porducts. https://www.salesforce.com/eu/products/. Accessed 22 Sept 2017
146. Salesforce (n.d.) SalesforceIQ CRM features. https://www.salesforce.com/eu/products/salesforceiq/features/. Accessed 22 Sept 2017
147. Salesforce (n.d.) SalesforceIQ CRM overview. https://www.salesforce.com/eu/products/salesforceiq/overview/. Accessed 22 Sept 2017
148. Salesforce (n.d.) SalesforceIQ CRM pricing. https://www.salesforce.com/eu/products/salesforceiq/pricing/. Accessed 22 Sept 2017
149. SAP (n.d.) Company information. https://www.sap.com/corporate/en/company.html. Accessed 24 Sept 2017
150. SAP (n.d.) Customer engagement and commerce. https://www.sap.com/products/crm-commerce.html. Accessed 24 Sept 2017
151. SAP (n.d.) History 1972–1980. https://www.sap.com/corporate/en/company/history.1972-1980.html. Accessed 24 Sept 2017
152. SAP (n.d.) Hybrid social engagement cloud. https://www.sap.com/austria/products/cloud-customer-engagement/social-media-monitoring-cloud.html. Accessed 24 Sept 2017
153. SAP (n.d.) Royal cup coffee and tea delivers exceptional buying experiences. http://www.hybris.com/en/downloads/video/royal-cup-coffee-exceptional-buying/448. Accessed 24 Sept 2017
154. SAP (n.d.) SAP customer relationship management. https://www.sap.com/products/customer-relationship-management.html. Accessed 24 Sept 2017
155. SAP (n.d.) SAP digital CRM. https://www.sapstore.com/solutions/99026/SAP-Digital-CRM#licensing-edition-section. Accessed 24 Sept 2017
156. SAP (n.d.) SAP hybrid cloud for customer subscription options. https://www.sap.com/products/cloud-customer-engagement.subscription-options.html. Accessed 24 Sept 2017
157. SAP (n.d.) Sap hybris customers. http://www.hybris.com/en/customers. Accessed 24 Sept 2017
158. Schneider M (2015) 24 statistics that show social media is the future of customer service. http://www.socialmediatoday.com/social-business/24-statistics-show-social-media-future-customer-service. Accessed 17 Sept 2017
159. Selectra (n.d.). https://en.selectra.info. Accessed 24 Sept 2017
160. Sheptoski L (2014) 5 ways social media can improve a poor company reputation. https://www.weidert.com/whole_brain_marketing_blog/bid/204792/5-ways-social-media-can-improve-a-poor-company-reputation. Accessed 10 Sept 2017
161. Smith C (2017) Amazing Twitter statistics and facts (August 2017). http://expandedramblings.com/index.php/march-2013-by-the-numbers-a-few-amazing-twitter-stats/. Accessed 12 Aug 2017
162. Socialblade (n.d.) Top 100 Instagram users by followers. https://socialblade.com/instagram/top/100/followers. Accessed 12 Aug 2017
163. Speier K (2016) 4 examples of clever crowdsourcing campaigns. http://www.mainstreethost.com/blog/four-examples-of-clever-crowdsourcing-campaigns/. Accessed 27 Aug 2017
164. Starbucks Newsroom (2014) Starbucks announces the winner of its white cup contest. https://news.starbucks.com/news/starbucks-announces-the-winner-of-its-white-cup-contest. Accessed 12 Aug 2017
165. Starbucks Newsroom (2015) Starbucks invites you to decorate its iconic white cup. https://news.starbucks.com/news/starbucks-invites-you-to-decorate-its-iconic-white-cup. Accessed 12 Aug 2017
166. Statista (n.d.) Number of monthly active Facebook users worldwide as of 2nd quarter 2017 (in millions). https://www.statista.com/statistics/264810/number-of-monthly-active-facebook-users-worldwide/. Accessed 30 July 2017
167. Statista (n.d.) Number of monthly active Instagram users from January 2013 to April 2017 (in millions). https://www.statista.com/statistics/253577/number-of-monthly-active-instagram-users/. Accessed 12 Aug 2017
168. Steele C (2016) The best shopping apps to compare prices. https://www.pcmag.com/feature/290959/the-best-shopping-apps-to-compare-prices/1. Accessed 03 Aug 2017

169. Stewart J (2016) Facebook has 50 minutes of your time each day. It wants more. https://www. nytimes.com/2016/05/06/business/facebook-bends-the-rules-of-audience-engagement-to-its-advantage.html. Accessed 13 Aug 2017
170. Sutton L (2013) Crowdsourcing product names results in a bunch of terrible product names. https://www.fastcompany.com/3008227/crowdsourcing-product-names-results-bunch-terrible-product-names. Accessed 27 Aug 2017
171. Swiftpage (n.d.). http://www.swiftpage.com/en-gb/. Accessed 22 Sept 2017
172. Team CGK (2016) How to determine generational birth years. http://genhq.com/generational_birth_years/. Accessed 03 Aug 2017
173. Thought Co. (2017) Microsoft. https://www.thoughtco.com/microsoft-history-of-a-computing-giant-1991140. Accessed 23 Sept 2017
174. Tie Seattle (n.d.) Microsoft. http://seattle.tie.org/microsoft/ [23.09.2017]
175. Turban E, Strauss J, Lai L (2016) Social commerce marketing, technology and management
176. Twiolomacy (2017) Twiplomacy study 2017. http://twiplomacy.com/blog/twiplomacy-study-2017/. Accessed 12 Aug 2017
177. Twitter (n.d.) Barack Obama. https://twitter.com/barackobama. Accessed 12 Aug 2017
178. Twitter (n.d.) Best Buy. https://twitter.com/BestBuy?ref_src=twsrc%5Egoogle%7Ctwcamp%5Eserp%7Ctwgr%5Eauthor. Accessed 12 Sept 2017
179. Twitter (n.d.) British Airways. https://twitter.com/British_Airways?ref_src=twsrc%5Egoogle%7Ctwcamp%5Eserp%7Ctwgr%5Eauthor. Accessed 12 Sept 2017
180. Twitter (n.d.) Ellen DeGeneres. https://twitter.com/TheEllenShow. Accessed 12 Sept 2017
181. Twitter (n.d.) Jet Blue Airway. https://twitter.com/JetBlue?ref_src=twsrc%5Egoogle%7Ctwcamp%5Eserp%7Ctwgr%5Eauthor. Accessed 12 Sept 2017
182. Twitter (n.d.) T-Mobile. https://twitter.com/TMobile?ref_src=twsrc%5Egoogle%7Ctwcamp%5Eserp%7Ctwgr%5Eauthor. Accessed 12 Sept 2017
183. Vembu S (2015) I am Sridhar Vembu, Founder and CEO of Zoho Corp. AMA! https://www.techinasia.com/talk/sriharvembu-zoho-ama. Accessed 22 Sept 2017
184. Watt A (2016) Coffee nut wins M&M'S flavor vote. http://www.candyindustry.com/articles/87361-coffee-nut-wins-mms-flavor-vote. Accessed 27 Aug 2017
185. Wollan R, Davis P, De Andelis F, Quiring K (n.d.) Seeing beyond the loyalty illusion: it's time you invest more wisely. https://www.accenture.com/t20170216T035010Z__w__/us-en/_acnmedia/PDF-43/Accenture-Strategy-GCPR-Customer-Loyalty.pdf#zoom=50. Accessed 04 Aug 2017
186. Yo! Success (2016) Sridhar Vembu. http://www.yosuccess.com/success-stories/sridhar-vembu/. Accessed 22 Sept 2017
187. Zajdo C (2017) 18 very effective methods to get quality customer feedback. http://www.optimonk.com/blog/15-ways-e-commerce-websites-get-customer-feedback/ [19.08.2017]
188. Zoho (n.d.) About us. https://www.zoho.eu/aboutus.html. Accessed 22 Sept 2017
189. Zoho (n.d.) Compare Zoho CRM editions. https://www.zoho.eu/crm/comparison.html?src=crmpricing-middle. Accessed 22 Sept 2017
190. Zoho (n.d.) CRM features. https://www.zoho.eu/crm/features.html. Accessed 22 Sept 2017
191. Zoho (n.d.) CRM. https://www.zoho.eu/crm/?src=zoho. Accessed 22 Sept 2017
192. Zoho (n.d.) CRM pricing. https://www.zoho.eu/crm/zohocrm-pricing.html. Accessed 22 Sept 2017
193. Zoho (n.d.) Customers. https://www.zoho.eu/crm/customers/. Accessed 22 Sept 2017
194. Zoho (n.d.) Frequently asked questions. https://www.zoho.com/zoho_faq.html. Accessed 22 Sept 2017
195. Zoho (n.d.). https://www.zoho.eu. Accessed 22 Sept 2017
196. Zoho (n.d.) Social CRM. https://www.zoho.eu/crm/social/. Accessed 22 Sept 2017
197. Zoho (n.d.) Social features. https://www.zoho.eu/social/features.html. Accessed 22 Sept 2017
198. Zoho (n.d.) Social pricing. https://www.zoho.eu/social/pricing.html. Accessed 22 Sept 2017
199. Zoho (n.d.) ZIA. https://www.zoho.eu/crm/zia.html. Accessed 22 Sept 2017

200. Zoho (n.d.) Zoho social—plan comparison. https://www.zoho.eu/social/plan-comparison. html. Accessed 22 Sept 2017
201. Zoho Corporation (n.d.). http://www.zohocorp.com/index.html. Accessed 22 Sept 2017

Assessment of eCall's Effects on the Economy and Automotive Industry

Tomas Lego, Andreas Mladenow and Christine Strauss

Abstract eCall is a Pan European automatic emergency call system for cars. All new types of vehicles used within the European Union must be equipped with hardware enabling the operation of eCall. Despite being targeted by many researchers and official authorities, there remain many questions regarding the effects of this technology. Building on the expected monetary and non-monetary benefits of eCall, as well as other projections about the system's capabilities, in this study, we used the Delphi method to address unanswered issues and to examine the logic behind earlier projections of eCall's effects on both the automotive industry and the economy. In addition to presenting our findings from three bilingual rounds of the Delphi method with 16 experts from the automotive industry, we also assess the public perception of this technology and its future development. Lastly, we present the opinions of experts regarding eCall's potential to ensure annual monetary benefits of €20 billion, or more based on the prices of cars augmented by the implementation of eCall and its future link with autonomous cars.

1 Introduction

Worldwide, traffic accidents are the number one cause of death among people aged 15–29 years, followed by suicide, AIDS, homicide and other less common causes [1]. Given the fact that 90% of road fatalities occur in low- and middle-income countries, one might assume that Europe would not be overly concerned with this problem [2]. This might be partially true, since compared to other regions Europe's percentage of deaths due to traffic accidents, as calculated based on 100,000 road

T. Lego · A. Mladenow · C. Strauss (✉)
University of Vienna, Oskar Morgenstern Platz 1, 1090 Vienna, Austria
e-mail: christine.strauss@univie.ac.at

T. Lego
e-mail: tomas.lego@univie.ac.at

A. Mladenow
e-mail: andreas.mladenow@univie.ac.at

© Springer Nature Switzerland AG 2020
N. Kryvinska and M. Greguš (eds.), *Data-Centric Business and Applications*,
Lecture Notes on Data Engineering and Communications Technologies 30,
https://doi.org/10.1007/978-3-030-19069-9_16

users, is way below the world percentage and more than 6% below that of America, which is the world region with the second lowest average in road fatalities per capita [1, 2]. However, car accidents are still a major cause of death in the European Union [3]. Although this figure has dropped dramatically over the last few years, more than 25,000 people die on European roads every year. Until 2001, this total was more than 30,000 casualties every year and in 2004 it was 43,000 [4, 5].

Similar to the actions taken by major intergovernmental organizations such as the United Nations [2], the European Union has introduced several action plans and strategies to lower this number, with the long-term goal of reducing this cause of death to zero by 2050 [2]. It is a basic aspect of human nature to seek personal safety. Maslow's well-known pyramid confirms that fulfilling basic needs such as personal security provides a basis for humans to focus on satisfying their psychological needs. In addition to following this theory, which was introduced more than half a century ago, human society, according to the model homo economicus, also focuses on economic factors. Given that traffic accidents incur costs of approximately €160 billion per year in EU member states alone [6], of which €120 billion is due to lost lives and injuries in traffic accidents [7], there is sufficient economic motivation to further reduce this trend.

Information technologies offer new solutions to well-established problems [8–11] like automobile accidents [12]. eCall is an emergency call system first implemented in April 2018 throughout the European Union [6], which allows cars to automatically and independently contact emergency services in the event of a severe car accident. This system was developed with the intention of automating car accident notifications throughout the European Union and associated countries [13]. However, the system can also be activated manually by vehicle occupants whenever they witness a car accident or encounter any other emergency [14]. eCall cannot be deactivated by the vehicle occupants [12]. It represents another compulsory safety element toward the goal of lowering the number of road casualties [6]. While it does not do so by saving lives directly, it contributes to the more rapid and efficient deployment of emergency service units (ESU) [12]. As approximately 50% of road fatalities occur in the minutes that follow an accident, an additional 30% occur in the hours following a car crash, and just 20% of traffic deaths occur in the days following an accident [15], timely help is critical.

Against this background, we conducted a Delphi study to assess eCall's effect on the economy and the automotive industry. In the next section, we describe the research method we used. In Sects. 3 and 4, we present an analysis of our results, and in Sect. 5, we offer concluding remarks.

2   Research Design

Since eCall is a new technology with expected but as yet unproven effects, the main part of this work addresses its impacts as well as the degree to which it is understood. Using the Delphi method, based on opinions of 16 automotive-industry experts, we attempt to predict the future development of eCall, discuss its potential effects and problems, and consider its public perception as well as its influence on the automotive industry. In this study, we sought a method that would enable us to obtain precise opinions from experts regarding future eCall trends. Despite initial vacillation, we conducted this study using the Delphi method, which we believe has the best potential for obtaining the opinions of a group of specialists with respect to uncertain situations [12]. We decided for Delphi questionnaires consisting of several sections with different focuses while using both open- and Likert-scale questions.

The communication with all participating experts took place online via e-mail as instructions were provided and questionnaires were distributed electronically. Still, not all experts who initially agreed to participate in the study replied to the informational e-mail, so another e-mail requesting confirmation was sent on January 16, 2018, followed by the first questionnaire on February 2, 2018. Having received 11 responses within the established ten-day window, we sent a reminder to all remaining experts on February 12, 2018, which resulted in the first round being terminated after receiving all but one questionnaire by February 15, 2018. Of these, responses on two of the received questionnaires were not fully complete, i.e., one section of the questionnaire had been assessed on the Likert scales, but included no written responses.

After evaluating the first round, we sent out the second questionnaire on March 2, 2018 to the 16 experts. This time, we sent seven reminders on March 11, 2018 and had received 14 questionnaires by March 19. The two experts who provided no answers to the second form were the same who answered only one section in the first. On April 3, 2018, we sent out the third round of questionnaires to all 16 experts. This time, three informal reminders were sent out, in response to which we received one negative answer and no reaction at all from the two other experts. We terminated the third round on April 19, 2018. Given that all three rounds yielded sufficient data and results to draw unequivocal conclusions, we concluded this Delphi study after three rounds. Altogether, we collected 13 out of 16 possible questionnaires in the last round, resulting in a 6.25% drop-out rate in every round. Thus, at the conclusion of the study, 81.25% of all the experts had participated in all three rounds. This study took 76 days to complete and resulted in 457 answers to open questions and over 1400 Likert scale answers which were evaluated.

3 Results Concerning Monetary Effects

The eCall technology is expected to yield great monetary benefits. However, the only way in which this system will be beneficial is if it can benefit both the private and public sectors. In the automotive industry, eCall might lead to the augmentation of car prices, thereby allowing higher revenues to car retailers and manufacturers. Furthermore, through the public ownership but private administration of some public safety answering points (PSAPs), further monetary potentials of this system emerge. Similarly, eCall can replace third party services, which normally incur a charge.

However, not all eCall-related monetary effects are as obvious. The system is also expected to save lives and reduce injuries, which together account for losses of €120 billion per year in the European Union alone [7]. E-Call is also expected to achieve economic savings by limiting traffic congestion and thus preventing secondary accidents. Further, by extending this system into transportation industries other than road transportation, greater benefits might accrue.

3.1 Car Prices

Based on the first round of questionnaires, it is unclear whether the introduction of the eCall system will lead to the augmentation of future car prices. Based on the assessments of experts regarding all potential developments, more than half share the opinion that the prices of cars will not change following the introduction of eCall. Interestingly, an even higher percentage of the participating specialists believe that if this price augmentation did occur, prices would increase only slightly. This opinion is further underlined by the fact that most of the experts disagree with the claim that car retailers would misuse the introduction of this system by inappropriately raising car prices.

The uncertainty of whether the introduction of the eCall system will lead to vehicle price increases is underscored by several findings. Three-quarters of the experts in this study shared the opinion that the eCall hardware will not be expensive. This strong belief follows claims made by the European Union itself, which anticipated €100 to be the eCall-related costs per vehicle in 2009 [12]. Second, even though cost estimates are not uniform and slightly vary depending on individual publications, most experts think that these costs will be covered by car manufacturers. Lastly, most experts disagree with the idea that eCall should be an add-on and hence a paid service, which could leave its development and implementation costs uncovered if not met by the manufacturers themselves.

Considering the varied opinions of the experts about how an augmentation in car prices would be accepted by customers and how this augmentation could be presented to them, we targeted the possible augmentation of car prices in more detail in rounds two and three. The most commonly expressed opinion was that even if such an augmentation occurred, it would not be presented to the buyers as being due to the introduction of eCall. Rather, car manufacturers would try to justify the increase by claiming a general improvement in the safety equipment of cars. Interestingly, despite this sentiment, the experts also agreed that it would be correct for retailers and car manufacturers to explicitly inform their customers about any rise in prices due to the installation of eCall.

If an increase in prices were to occur, the amount of this increase could be any-where from €50 to €400, with the main reason for the increase being an attempt to finance both the development and operational costs of the system. No actual agreement was reached on whether the customers should perceive this potential price rise as acceptable. Two streams of opinions emerged, i.e., that customers would either understand eCall as another step toward enhancing vehicle safety and thus accept it, or that they would not share this opinion and be upset by the increased prices. However, ultimately, they would have to accept them as there would be no other option when buying a car. The point of view argued in earlier scientific works is that safety equipment is one of the most decisive factors when buying a new car [12]. Therefore, the former rather than latter opinion expressed by the experts is most likely, which gives car retailers potential room for augmenting the prices of new vehicles.

In summary, expert opinion is that an augmentation in the prices of cars will either not occur or will be relatively insignificant (83.33% of experts). This claim is understandable, given the nature of eCall as an emergency call system. However, from an economic standpoint and given the estimated number of cars sold in the EU having been almost 15 million units in 2017 [16], major monetary gains could be made if the prices of cars were to be augmented. If the price of every car went up by the projected amount of €50–€400, additional revenues of between €750 million and €6 billion could be achieved in the car retail market in the European Union alone.

3.2 Monetary Gains

Without explicitly mentioning any monetary benefits that might be achievable through increases in the prices of cars, by the implementation of eCall in other transportation industries, and without it being linked with private assistance services, the European Union has realized that eCall can lead to massive economic benefits [7, 17, 18]. According to the experts, the degree of achievable savings will be influenced by several factors that are not easy to predict, as it is difficult to quantify the total damage caused by car accidents. However, this might not be true. eCall technology is expected to lead to monetary savings of up to €20 billion per year [6] and the costs of traffic accidents in the European Union are estimated to be €160 billion per year [18]. Thus, the potential saving attributable to the introduction of the Pan European emergency call system, as introduced in the literature, would be limited to 12.5% of the total costs.

Based on these findings and projections, in our first-round questionnaire, we included a question targeting the level of expected monetary benefits. The results from rounds one and two obtained from experts participating in this study indicated that benefits will mainly accrue through the lives saved and reduced injuries, whereas reducing material damages will be of lesser importance. These claims echo official arguments by the European Union that lost lives and injuries account for more than three quarters of the total financial losses associated with traffic accidents on European roads every year [5, 17]. However, regarding the exact magnitude of the achievable savings, there was no unanimous expert opinion in the first round of this Delphi study. According to the experts, there are three problems that make the precise estimation of these costs impossible. First, the number of factors that can be translated into monetary values is too high. Second, these costs are difficult to estimate in advance. Third, some of the damage caused by these accidents cannot be quantified. Along with the third claims is the opinion expressed by the experts that if the technology was to save just one life, its contribution would be unquantifiable. However, according to the literature, this might not be true either, as the main factors contributing to these savings have been defined and can all be expressed monetarily. These primarily include avoidable traffic jams, avoidable secondary accidents, better traffic management, saved lives, and the lowered severity of injuries [14, 18, 19].

Furthermore, according to earlier publications, these benefits are not achieved immediately, but will accrue as this technology spreads. Roughly three quarters of the participating experts suggested that the benefits would depend on the penetration rate of the technology. Given that this fact is taken for granted by other scientific papers, this number is surprisingly low. However, assuming the importance of the system's penetration rate, in the second round, we targeted the date at which a 100% market penetration would be reached. The findings in this round were that a full penetration rate could be achieved in 15–30 years, but would depend on the average age of the vehicles in individual EU countries. It was believed highly unlikely that this would occur earlier than in 2025, but would definitely be possible by 2040. The opinion was stated that full penetration would not be achieved in all individual EU member countries simultaneously, which is also backed by earlier scientific findings introducing different average ages of car fleets in individual EU member states [20]. Moreover, in this study, our findings indicate that a true 100% penetration rate in M1 and N1 vehicles might never be achieved if the existence of veteran cars within these classes is considered.

Similar to the average age of vehicles, the costs of traffic accidents are not distributed equally over the EU member countries; every individual EU member loses a different amount of its gross domestic product (GDP) due to road accidents [1, 21]. If a 100% penetration rate is assumed, the potential savings achievable through the introduction of eCall in individual EU member states can be calculated. Figure 1 shows the potential monetary savings by individual EU member states based on a World Health Organization (WHO) report. Unfortunately, the WHO reports do not include the lost GDP figures for all 27 EU countries. Therefore, the calculation could be made for only a limited number of countries.

Of the individual factors contributing to the total monetary cost of traffic accidents of €160 billion a year [6], €130 billion is estimated to be caused by lost lives and casualties alone [17]. The 2018 estimate was lowered to €120 billion, which potentially reflects the consistently decreasing number of EU road fatalities [5, 22]. Furthermore, the tendency to express human lives in monetary values does not correspond to the opinion of experts participating in this study, who were of the opinion that the benefits achievable through saving lives are unquantifiable. However, despite the humanity of such a claim, it fails to satisfy the needs of economists.

Being able to express human life in monetary values is crucial, primarily in the health services and insurance industries [23]. Therefore, the scientific literature has targeted this issue for several decades, and reported findings regarding which factors influence the value of a human life and the amounts with which these terms can be determined. This issue has been a matter of debate for centuries, as efforts were made

	WHO 2013[1]	Ref. Year[2]	GDP in USD (referred year)[3]	EU/USD rate[4]	Absolute saving (Total loss)	Expected saving (12.5%)	WHO 2015[1]	Ref. Year[2]	GDP in USD (referred year)[3]	EU/USD rate[4]	Absolute saving (Total loss)	Expected saving (12.5%)
Austria	3.90%	2006	$335,998,557,270.00	1.26	€ 10,822,013,885.61	€ 1,352,751,735.70	3.30%	2012	$409,425,234,160.00	1.28	€ 10,915,712,841.97	€ 1,364,464,105.25
Belgium	N/A	N/A	N/A	N/A	N/A	N/A	N/A	N/A	N/A	N/A	N/A	N/A
Bulgaria	2.00%	2010	$50,610,031,140.00	1.33	€ 776,584,795.76	€ 97,073,099.47	2.00%	2015*	$50,799,117,550.00	1.11	€ 933,978,995.22	€ 116,747,374.40
Croatia	N/A	N/A	N/A	N/A	N/A	N/A	N/A	N/A	N/A	N/A	N/A	N/A
Cyprus	1.00%	2008	$27,839,460,960.00	1.47	€ 191,297,058.75	€ 23,912,132.34	1.00%	2008	$27,839,460,960.00	1.47	€ 191,297,058.75	€ 23,912,132.34
Czech Rep.	N/A	N/A	N/A	N/A	N/A	N/A	N/A	N/A	N/A	N/A	N/A	N/A
Denmark	N/A	N/A	N/A	N/A	N/A	N/A	N/A	N/A	N/A	N/A	N/A	N/A
Estonia	N/A	N/A	N/A	N/A	N/A	N/A	1.00%	2011	$23,170,239,900.00	1.39	€ 168,376,134.73	€ 21,047,016.84
Finland	1,4%[5]	2010	$247,799,815,770.00	1.33	€ 2,645,455,490.23	€ 330,681,936.28	2.20%	2012	$256,706,466,090.00	1.28	€ 4,511,393,032.64	€ 563,924,129.08
France	1.30%	2010	$2,646,837,110,000.00	1.33	€ 26,212,097,439.65	€ 3,276,512,179.96	1.00%	2015*	$2,433,562,020,000.00	1.11	€ 22,145,436,527.44	€ 2,768,179,565.93
Germany	1.30%	2008	$3,752,365,610,000.00	1.47	€ 33,621,262,073.62	€ 4,202,657,759.20	1.20%	2012	$3,543,983,910,000.00	1.28	€ 33,628,389,834.26	€ 4,203,548,729.28
Greece	0.50%	2009	$330,000,252,150.00	1.39	€ 1,193,016,348.47	€ 149,127,043.56	1.5%[7]	2011	$287,797,822,090.00	1.39	€ 3,153,027,302.60	€ 394,128,412.82
Hungary	1.50%	2009	$130,593,960,610.00	1.39	€ 1,430,748,573.31	€ 178,843,571.66	1.50%	2013	$135,215,704,420.00	1.33	€ 1,548,212,332.58	€ 193,526,541.57
Ireland	N/A	N/A	N/A	N/A	N/A	N/A	0.60%	2012	$225,571,853,190.00	1.28	€ 1,063,750,565.22	€ 132,968,820.65
Italy	2.00%	2008	$2,390,729,160,000.00	1.47	€ 33,190,742,190.75	€ 4,148,842,773.84	1.80%	2011	$2,276,292,400,000.00	1.39	€ 30,017,482,453.96	€ 3,752,185,306.74
Latvia	N/A	N/A	N/A	N/A	N/A	N/A	N/A	N/A	N/A	N/A	N/A	N/A
Lithuania	N/A	N/A	N/A	N/A	N/A	N/A	1.00%	2013	$46,417,340,370.00	1.33	€ 352,527,837.55	€ 44,065,979.69
Luxembourg	N/A	N/A	N/A	N/A	N/A	N/A	N/A	N/A	N/A	N/A	N/A	N/A
Malta	N/A	N/A	N/A	N/A	N/A	N/A	N/A	N/A	N/A	N/A	N/A	N/A
Netherlands	2.10%	2007	$839,419,655,080.00	1.37	€ 13,143,020,031.37	€ 1,642,877,503.92	2.20%	2009	$857,932,759,100.00	1.39	€ 13,884,245,266.51	€ 1,735,530,658.31
Poland	2,25%[6]	2008	$533,815,789,470.00	1.47	€ 8,358,721,062.74	€ 1,044,840,132.84	1.90%	2012	$500,284,003,680.00	1.28	€ 7,569,919,143.35	€ 946,239,892.92
Portugal	N/A	N/A	N/A	N/A	N/A	N/A	1.20%	2010	$238,303,443,430.00	1.33	€ 2,176,220,907.40	€ 272,027,613.43
Romania	N/A	N/A	N/A	N/A	N/A	N/A	N/A	N/A	N/A	N/A	N/A	N/A
Slovakia	1.40%	2010	$89,501,012,920.00	1.33	€ 955,492,825.02	€ 119,436,603.13	1.40%	2010	$89,501,012,920.00	1.33	€ 955,492,825.02	€ 119,436,603.13
Slovenia	1.00%	2010	$48,013,606,750.00	1.33	€ 364,651,072.76	€ 45,581,384.09	1.50%	2012	$46,352,802,770.00	1.28	€ 551,468,941.58	€ 68,933,617.70
Spain	0.40%	2009	$1,499,099,750,000.00	1.39	€ 4,331,281,240.07	€ 541,410,155.01	1.00%	2012	$1,336,018,950,000.00	1.28	€ 10,543,078,835.23	€ 1,317,884,854.40
Sweden	0.80%	2013*	$578,742,001,490.00	1.33	€ 3,509,228,725.99	€ 438,653,590.75	1%[8]	2010	$488,377,689,560.00	1.33	€ 3,709,103,740.87	€ 463,637,967.61
UK	1.20%	2009	$2,382,825,990,000.00	1.39	€ 20,821,011,767.10	€ 2,602,626,470.89	1.00%	2012	$2,662,085,170,000.00	1.28	€ 21,007,616,556.19	€ 2,625,952,069.52

Notes: 1) The GDP loss in individual member countries according to a WHO report on road safety (WHO, 2013; WHO, 2015). Such figures were not available for all EU member countries (N/A). 2) The referred year in which the loss was incurred was not automatically the publishing year of the report. Hence, the reference year can be found in this column. All years marked with "*" are asumed to refer to the publishing year of the report, as no other indication was available. 3) The GDP in the referred year according to: https://data.worldbank.org/country . 4) The average conversion rate USD/EU in the referred year according to: https://www.statista.com/statistics/412794/euro-to-u-s-dollar-annual-average-exchange-rate/ . 5) The mathematical average of the reported value of 1.1%-1.7%. 6) The mathematical average of the reported value of the reported value of 2.0%-2.5%. 7) 5% if underreporting was taken into account. 8) The upper boundary of the reported value of "<1.0%".

Fig. 1 Potential monetary savings by individual EU member states based on WHO reports [1, 21]

to compute the price of human life in the 17th century and even earlier during the slave trade [24].

In 1954, authors considered the value of a human life to depend on one's age and gender, but primarily on the education of the individual. Later findings expanded upon this topic with claims that the value of a human life could be estimated based on the lifetime earnings achieved by a person, in addition to their age, sex, level of education and the colour of the individual's skin [24]. Under these terms, the value of a human life could be assessed on one's contribution to the society in which one lived. In 1977, the value of a life was estimated to average £39,000, and this number has risen ever since [23]. Despite the relative lack of current research data, estimating the value of a human life remains important, and in 2018, the Czech Transport Research Centre estimated this figure to be CZK19 million, or approximately €730,000 [25].

If we stick with the projections presented by the European Union and other sources, the lives of huge numbers of road users could be saved through eCall technology every year. Based on the three most commonly projected values, Fig. 2 shows the expected number of road fatalities prevented for the years 2013 to 2016. Despite being drawn from an official EU report published in April 2018 [5], the latest available data were from 2016. Nevertheless, this calculation scheme can be applied to newer data once they become available.

The second factor influenced by eCall, traffic congestion, is also of particular importance in the calculation of the savings achievable by this technology. Two types of costs are associated with this phenomenon, i.e., the direct costs of lost time and fuel and the indirect costs. The latter include the costs of doing business and may take the form of penalties or other losses due to delayed shipments [26]. In 1961 [27], and again in 1994 [28], traffic congestion was reported to be a major issue accounting for a monetary loss of $48 billion per year in the United States of America (U.S.) alone. Given costs of this magnitude, the Nobel-Prize-winning economist William Vickrey expressed the need to avoid congestion through advances in technology [29].

Its importance in the 21st century is further confirmed by looking at the U.S., where 4.2 billion hours of travel delay are caused by traffic congestion every year. Furthermore, 2.8 billion gallons of extra gas are consumed by cars stuck in traffic jams, which account for monetary losses of $124 billion a year in the U.S. [30]. Of this value, which is expected to exceed $185 billion by 2030, approximately 20% occur in Los Angeles [26], which has yearly monetary losses totaling $23 billion [31]. A differentiation between direct and indirect costs shows that $78 billion of the total congestion costs of $124 billion in 2013 were due to direct costs, whereas only $45 billion were indirect [26]. Hence, we could generalize that the magnitude of direct costs surpasses that of indirect costs.

When looking at eCall's operability, £30 billion were lost due to traffic congestion in 2010 in the United Kingdom alone [32]. Generally, the costs of congestion can reach 4% of a country's GDP. In 2011, these costs were estimated to total 1.5% of the United Kingdom's GDP, 1.3% that of France and 0.9% that of Germany [30]. Based on this clear data, the experts in our study seem to be wrong in their statement that it is impossible to determine the monetary losses associated with traffic accidents. This further confirms the achievable monetary savings presented in Fig. 1, which apart

	EU road fatalities in 2013[1]	Calculation based on:			EU road fatalities in 2014[1]	Calculation based on:			EU road fatalities in 2015[1]	Calculation based on:			EU road fatalities in 2016[1]	Calculation based on:		
		10%[2]	2000[3]	2500[4]		10%[2]	2000[3]	2500[4]		10%[2]	2000[3]	2500[4]		10%[2]	2000[3]	2500[4]
Austria	455	45.5 (46)	35.1 (36)	43.8 (44)	430	43.0 (43)	33.1 (34)	41.4 (42)	479	47.9 (48)	36.7 (37)	45.8 (46)	432	43.2 (44)	33.7 (34)	42.1 (43)
Belgium	723	72.3 (73)	55.7 (56)	69.6 (70)	727	72.7 (73)	56.0 (56)	70.0 (70)	732	73.2 (74)	56.0 (57)	70.0 (71)	637	63.7 (64)	49.7 (50)	62.1 (63)
Bulgaria	601	60.1 (61)	46.3 (47)	57.9 (58)	660	66.0 (66)	50.8 (51)	63.5 (64)	708	70.8 (71)	54.2 (55)	67.7 (68)	708	70.8 (71)	55.2 (56)	69.0 (70)
Croatia	368	36.8 (37)	28.4 (29)	35.4 (36)	308	30.8 (31)	23.7 (24)	29.6 (30)	348	34.8 (35)	26.6 (27)	33.3 (34)	307	30.7 (31)	23.9 (24)	29.9 (30)
Cyprus	44	4.4 (5)	3.4 (4)	4.2 (5)	45	4.5 (5)	3.5 (4)	4.3 (5)	57	5.7 (6)	4.4 (5)	5.5 (6)	46	4.6 (5)	3.6 (4)	4.5 (5)
Czech Rep.	655	65.5 (66)	50.5 (51)	63.1 (64)	688	68.8 (69)	53.0 (53)	66.2 (67)	734	73.4 (74)	56.2 (57)	70.2 (71)	611	61.1 (62)	47.6 (48)	59.5 (60)
Denmark	191	19.1 (20)	14.7 (15)	18.4 (19)	182	18.2 (19)	14.0 (15)	17.5 (18)	178	17.8 (18)	13.6 (14)	17.0 (18)	211	21.1 (22)	16.5 (17)	20.6 (21)
Estonia	81	8.1 (9)	6.2 (7)	7.8 (8)	78	7.8 (8)	6.0 (7)	7.5 (8)	67	6.7 (7)	5.1 (6)	6.4 (7)	71	7.1 (8)	5.5 (6)	6.9 (7)
Finland	258	25.8 (26)	19.9 (20)	24.8 (25)	229	22.9 (23)	17.6 (18)	22.0 (23)	266	26.6 (27)	20.4 (21)	25.4 (26)	258	25.8 (26)	20.1 (21)	25.1 (26)
France	3,268	326.8 (327)	251.8 (252)	314.8 (315)	3,384	338.4 (339)	260.5 (261)	325.7 (326)	3,461	346.1 (347)	264.9 (265)	331.1 (332)	3,477	347.7 (348)	271.1 (272)	338.9 (339)
Germany	3,339	333.9 (334)	257.3 (258)	321.6 (322)	3,377	337.7 (338)	260.0 (260)	325.0 (325)	3,459	345.9 (346)	264.8 (265)	330.9 (331)	3,206	320.6 (321)	250.0 (250)	312.5 (313)
Greece	879	87.9 (88)	67.7 (68)	84.7 (85)	795	79.5 (80)	61.2 (62)	76.5 (77)	793	79.3 (80)	60.7 (61)	75.9 (76)	824	82.4 (83)	64.2 (65)	80.3 (81)
Hungary	591	59.1 (60)	45.5 (46)	56.9 (57)	626	62.6 (63)	48.2 (49)	60.2 (61)	644	64.4 (65)	49.3 (49)	61.6 (62)	607	60.7 (61)	47.3 (48)	59.2 (60)
Ireland	188	18.8 (19)	14.5 (15)	18.1 (19)	193	19.3 (20)	14.9 (15)	18.6 (19)	162	16.2 (17)	12.4 (13)	15.5 (16)	186	18.6 (19)	14.5 (15)	18.1 (19)
Italy	3,401	340.1 (341)	262.1 (263)	327.6 (328)	3,381	338.1 (339)	260.3 (261)	325.4 (326)	3,428	342.8 (343)	262.4 (263)	328.0 (328)	3,283	328.3 (329)	256.0 (256)	320.0 (320)
Latvia	179	17.9 (18)	13.8 (14)	17.2 (18)	212	21.2 (22)	16.3 (17)	20.4 (21)	188	18.8 (19)	14.4 (15)	18.0 (18)	158	15.8 (16)	12.3 (13)	15.4 (16)
Lithuania	256	25.6 (26)	19.7 (20)	24.7 (25)	267	26.7 (27)	20.6 (21)	25.7 (26)	242	24.2 (25)	18.5 (19)	23.2 (24)	192	19.2 (20)	15.0 (15)	18.7 (19)
Luxembourg	45	4.5 (5)	3.5 (4)	4.3 (5)	35	3.5 (4)	2.7 (3)	3.4 (4)	36	3.6 (4)	2.8 (3)	3.4 (4)	32	3.2 (4)	2.5 (3)	3.1 (4)
Malta	17	1.7 (2)	1.3 (2)	1.6 (2)	10	1.0 (1)	0.8 (1)	1.0 (1)	11	1.1 (2)	0.8 (1)	1.1 (2)	23	2.3 (3)	1.8 (2)	2.2 (3)
Netherlands	476	47.6 (48)	36.7 (37)	45.8 (46)	477	47.7 (48)	36.7 (37)	45.9 (46)	531	53.1 (54)	40.6 (41)	50.8 (51)	533	53.3 (54)	41.6 (42)	51.9 (52)
Poland	3,357	335.7 (336)	258.7 (259)	323.3 (324)	3,202	320.2 (321)	246.5 (247)	308.2 (309)	2,938	293.8 (294)	224.9 (225)	281.1 (282)	3,026	302.6 (303)	235.9 (236)	294.9 (295)
Portugal	637	63.7 (64)	49.1 (50)	61.4 (62)	638	63.8 (64)	49.1 (50)	61.4 (62)	593	59.3 (60)	45.4 (46)	56.7 (57)	563	56.3 (57)	43.9 (44)	54.9 (55)
Romania	1,861	186.1 (187)	143.4 (144)	179.2 (180)	1,818	181.8 (182)	140.0 (140)	175.0 (175)	1,893	189.3 (190)	144.9 (145)	181.1 (182)	1,915	191.5 (192)	149.3 (150)	186.6 (187)
Slovakia	251	25.1 (26)	19.3 (20)	24.2 (25)	295	29.5 (30)	22.7 (23)	28.4 (29)	310	31.0 (31)	23.7 (24)	29.7 (30)	275	27.5 (28)	21.4 (22)	26.8 (27)
Slovenia	125	12.5 (13)	9.6 (10)	12.0 (13)	108	10.8 (11)	8.3 (9)	10.4 (11)	120	12.0 (12)	9.2 (10)	11.5 (12)	130	13.0 (13)	10.1 (11)	12.7 (13)
Spain	1,680	168.0 (168)	129.4 (130)	161.8 (162)	1,688	168.8 (169)	130.0 (130)	162.5 (163)	1,689	168.9 (169)	129.3 (130)	161.6 (162)	1,810	181.0 (181)	141.1 (142)	176.4 (177)
Sweden	260	26.0 (26)	20.0 (21)	25.0 (26)	270	27.0 (27)	20.8 (21)	26.0 (26)	259	25.9 (26)	19.8 (20)	24.8 (25)	270	27.0 (27)	21.1 (22)	26.3 (27)
UK	1,770	177.0 (177)	136.4 (137)	170.5 (171)	1,854	185.4 (186)	142.7 (143)	178.4 (179)	1,804	180.4 (181)	138.1 (139)	172.6 (173)	1,860	186.0 (186)	145.0 (146)	181.3 (182)
Total	25,956	2,596	2,000	2,500	25,977	2,598	2,000	2,500	26,130	2,613	2,000	2,500	25,651	2,565	2,000	2,500

Notes: 1) The number of casualties in individual member states in individual years is based on an official communication published by the European Commission in April 2018 (European Commission, 2018a). 2) 3) 4) The calculations are based on three commonly communicated assumptions of eCall's contribution to saving lives. These are a) It will save approximately 10% of potential casualties every year (Newmobility.news, 2018b; European Parliament, 2015), b) It will save 2,500 lives every year (European Commission, 2005; European Commission, 2006; European Commission, 2009; Cabo et al. 2014; dan Barca, 2016; Chariete et al., 2016) and c) The lower boundary of casualties saved through the eCall system within the European Union is 2,000 lives every year (Chochliouros et al., 2005). The numbers in brackets present rounded up values of the expected figures.

Fig. 2 Number of road fatalities that could be avoided in individual EU member states [4–6, 18, 52–57]

from the lives lost and injuries caused, are also greatly influenced by traffic congestion. However, regardless of the decisive factor of the monetary benefits, 100% of
the experts agreed that massive monetary benefits are achievable when the system is
being used as an active safety element. Hence, only when in-vehicle systems (IVSs)
can communicate with each other or PSAPs actively use the information generated
by the eCall system to manage traffic can greater financial advantages be realized.
This claim is supported by earlier scientific findings regarding vehicle-to-vehicle and
vehicle-to-infrastructure communication as well as autonomous vehicles. An important basis for this claim is that in the U.S., 96% of traffic incidents are assumed to be
caused by drivers. Therefore, reducing the human factor could eliminate this critical
aspect of the cost equation to make possible significant monetary benefits. Consequently, the benefits to be achieved through the introduction of both self-driving and
interconnected cars are predicted to reach $51 billion in the U.S. [33]. Furthermore,
these technologies are expected to reduce the number of traffic accidents by 90% by
2050, leading to the potential for reducing their corresponding losses by $190 billion
[12]. The experts in this study also expected a reduction in the costs associated with
traffic accidents in Europe. However, the potential for this reduction reaching €20
billion per year is disputable.

3.3   eCall in Other Industries

Based on the promising payoffs of eCall, its possible expansion to other industries
has been projected by both the experts in this study and the authors of earlier studies.
Although the question regarding this phenomenon was not answered unanimously,
the second round brought clarity to the possible expansion of eCall to other industries,
with a majority of the experts expressing the opinion that eCall could be adopted by
other industries and that it will be extended to services without any direct connection
to emergency calls [12]. These claims are supported by existing publications and
releases as the European Union has foreseen the use of eCall in other fields and the
extension of eCall to buses and/or trains [34].

According to the opinion of our experts, eCall could primarily be used for planning
and coordinating emergency services capacities. Later, it might also play a major
role in helping police forces to search for stolen cars and providing data for better
developing the transportation infrastructure. However, eCall is expected to not only
benefit the public sector, but the private sector as well, e.g., the insurance industry.
Nevertheless, to be used for some of these purposes, two conditions must be met.
eCall must be able to store data and IVSs must also be remotely accessible from
outside the vehicle. We targeted these two assumptions in the third round of our
Delphi questionnaire.

Not only did most experts agree that both assumptions were technically possible,
they also agreed that, to be realistic, it is essential that a legal norm be established to
regulate the use of this technology for applications such as those noted above, i.e.,
locating stolen cars or car fleet management. A slight majority of the experts noted

that a nonstop activation of the eCall system might discourage customers. However, despite there being no unanimous expert opinion about the effect on the customer of permanent IVS activation, this permanent activation would be inevitable. Otherwise, eCall could never become an active safety element, which the experts earlier predicted to be the almost certain direction of eCall's development. Furthermore, other devices in daily use pose similar threats, but remain very popular among consumers. Therefore, it could be expected that, much like smartphone users being unwilling to forsake the use of their phones, drivers would not give up their vehicles because they had been equipped with eCall. Hence, though not unanimously confirmed by the results of this study, we can expect eCall to be used as a permanently activated device, and hence an active safety element in the future.

Looking at the existing literature regarding eCall's extension to other industries, rail transportation is one example [34]. There, the minimum set of data (MSD) could be broadened by the number of carriages, the type of cargo or the number of passengers. By eliminating the losses or reducing the number of fatalities and the severity of injuries after train accidents, the monetary benefits achieved through eCall would increase. Whereas this development was anticipated by the experts, they did not surmise about whether this technology would also be extended to naval transportation.

Another potential use of the technology exists, but was not unanimously acknowledged by the experts. One possible explanation might be the fact that it is not possible to guarantee the privacy required by the EU. However, data retrievable from IVSs could be of great importance to marketers, in that it would allow for a very detailed classification of drivers based on the cars they own and their driving patterns. As such, the potential contribution the eCall system might make to the private sector is as yet unquantifiable.

3.4 Co-existence with Private Services

Following the idea of cooperation between private and state services, the consensus reached in the first round was that private- and state-owned assistance systems might co-exist and actually do so today. This opinion is not unfounded and actually represents today's status quo regarding in-vehicle emergency call systems. Therefore, it is not surprising that 100% of the experts agreed on the technical possibility of their co-existence. However, they shared the opinion that there should be only one emergency call system, and that the merit of two systems is disputable. We targeted this issue in greater detail in the latter rounds of the Delphi questionnaire.

According to most of our participating experts, the co-existence of a third party service (TPS) and eCall is possible due to the different focuses of these systems. The same percentage of experts also understood TPS to be an extended service for which there could be a charge. Nevertheless, in contrast, more than three quarters of our specialists also expressed the opinion that these systems will be linked in the future. One possible explanation for this contradiction, which was further supported by their answers to the second questionnaire, is the fact that it makes no sense to have two assistance systems in one vehicle. This logic is supported by the experts defining the preferences of potential customers as moving towards an all-in-one solution rather than two different systems. However, given their contradictory ideas about what should and will happen in the future, only half of the experts prefer a link between these services and one third agreed that these systems cannot be linked in the future. Given these different beliefs and projections about future eCall developments, we can make no clear conclusion regarding future system co-existence. However, we might predict that no link will be established because eCall has been guaranteed to be free of charge and linking it with chargeable services would possibly disrupt its business model, and thereby lead to a reduction in achievable monetary benefits.

4   Further Results

4.1   Data Availability and Security

According to previous studies, approximately 75% of all vehicle users are worried about and discomforted by the idea of the uncontrollable transmission of data generated by their own vehicles [35]. Official information provided by the European Union has attempted to address this issue as the European Union has guaranteed that the system will be allowed to collect limited amounts of data, which will only be transmitted if triggered [36]. The Pan European emergency call system is officially considered to be safe with regards to the data it holds [3, 37]. Despite such official reassurances, most experts believe that the system can be hacked and a legal prohibition regarding the system's use without previous activation would ensure the prevention of eCall misuse.

The answers related to data security collected in the first round of questionnaires did not yield a unanimous opinion. Rather, there were two lines of thought, with approximately a half of the experts claiming that eCall is a secure system and the other half expressing mistrust about eCall's promised data security. As such, despite the European Union's assurances that only a limited amount of data is collected by the IVS and transmitted to PSAPs when triggered, a minority of experts believes that this risk is truly eliminated by such regulations. Moreover, three quarters of the specialists expressed the opinion that the danger might be reduced, but not eliminated. More than 70% of the experts also claimed that although eCall might not be the safest system, compared to other modern technologies, it does not represent any excessive danger.

Despite not all experts sharing this opinion, the scientific literature accepts the fact of the security of eCall as a given. Guens and Dumortier claimed that it would be easier to track someone using their smartphone than his or her vehicle [38]. Despite this sentiment and the EU's early realization that a legal regulation would be needed, the two most relevant threats perceived by the experts are the possibility of personal tracking and the wiretapping of vehicles. A possible explanation for this concern might be the fact that legal restrictions present no physical barrier against the misuse of eCall. The experts did not all share the opinion that the technology could be used for criminal purposes, but some experts believe that this development is possible. Further, despite a lack of belief in its power, most participating specialists agreed that the wording of a legal regulation regarding eCall might be a decisive factor in managing its negative aspects.

With respect to other possible downsides of the eCall technology, the study results yielded no unanimous opinions apart from a denial that road users' health might be affected by this technology or the wave frequencies it emits. One very contradictory issue is the General Data Protection Regulation (GDPR) imposed by the European Union. Despite only a small majority of experts believing that the GDPR might be problematic, the issue was addressed in the third round of questionnaires based on its topicality. Our study results indicate that it is difficult to estimate the impacts of the GDPR on eCall, as it is new. However, the experts do not expect the regulation to influence the implementation of eCall or to lead to its postponement.

4.2 Perception

Considering the novelty of eCall, in the initial round we asked about its future perception by the public as well as the public's understanding of eCall, compared to other vehicle safety features. Regarding the public's general understanding of the technology, the experts showed good agreement in the first round. As confirmed later, 85.7% of the experts shared the opinion that the system will be perceived positively. In the third round, the experts reported that the main reason for this positive perception of eCall is the fact that it is a safety element primarily focused on saving lives. Nevertheless, more than 90% of the experts also realized that any potential misuse of the system might generate a negative perception of the technology. As later confirmed, the experts also believed that misuse might be the only reason for a negative perception.

A unanimous opinion was expressed that the motorists' perceptions of eCall will depend on the age of motorists in individual EU member countries and the average age of vehicles in these states. This opinion is backed by existing scientific literature in which authors state that levels of understanding regarding the benefits brought by this technology may also depend on geographical location. It is expected that eCall may be of greater importance in countries in which cars have a higher average age. As such, the perceived importance of the system is assumed to be greater in Eastern European countries where this is the case [20].

There was no shared opinion in the first round regarding the perception of eCall as compared to other safety elements. However, a consensus was reached in the second round wherein experts agreed about the greater importance of eCall in less busy locations. As such, this is a phenomenon that does not occur with any other safety element. The third round also offered an explanation for this claim, in which there was a unified opinion regarding the lower probability of help being provided by those not involved in the accident in such locations. This claim has been supported in earlier scientific literature. On average, it takes longer for an ESU to be dispatched following an accident on a rural road, compared to one that occurs in an urban area. Whereas 95% of car crashes in urban areas are reported within 20 min and 90% in less than 10 min, in traffic accidents on rural roads, only 70% notify the ESU in less than 10 min. This percentage reaches 80% when the time horizon is doubled, but still lags significantly behind that in urban areas [39]. By limiting the dependence on random witnesses or other drivers, the experts in our study believe that eCall's greater importance will be proven on suburban roads in the future. Moreover, 55% of all accidents within the European Union occur in rural areas, which further highlights the importance and role of an eCall-like system [40].

The Delphi questionnaire results also provide evidence that it is not easy to compare eCall to other vehicle safety elements. This belief is based on the fact that most participating experts view the system as atypical. Compared to other safety equipment, eCall is "only" capable of triggering an emergency call, which makes it very difficult to compare to other safety features, such as the robustness of the vehicle chassis, its airbags or the safety belts. Perhaps due to the novelty of the technology, as yet, there are no published studies that compare this modern safety feature to the older well-established features. Given the gravity and complexity of these problems, no statement can be made regarding the future relative efficiency of eCall. Still, more than half of the experts rather shared the opinion that eCall will not save more lives than are already saved by existing safety elements. A possible explanation for this opinion might be the relatively low number of road fatalities in traffic compared to their much higher density in the years before these older safety elements were introduced.

A unified opinion was expressed regarding the fact that eCall will most probably be considered to be an extension of other safety elements, since the vehicle occupants need not engage with it on a daily base. This type of routine use can be observed with other safety features, such as the fastening of safety belts before driving. In contrast, eCall will be dormant until engaged following an accident, much like airbags. Following this logic, eCall is essentially a passive safety element to reduce the consequences of traffic accidents [41]. Nevertheless, this opinion was not clearly held by the experts in the first round, who offered a split opinion on whether eCall could be characterized as an active safety system.

Following feedback to individual arguments, we can state that none of the experts considers eCall to be primarily an active element, but rather a passive one. Still, more than 85% expressed the opinion that eCall becomes an active safety element when triggered, which is supported by previous research. As introduced by the European Union, one of the achievable goals of this system is the reduction of secondary accidents [19]. Following the general definition of an active safety element, this would classify eCall as active since it would no longer only moderate the impact of road tragedies, but would help to prevent them. The importance of this shift in the perception of eCall functionality is crucial as this issue has been targeted in numerous scientific works [13, 15, 42]. However, none of these designations introduces a date on which this feature will be fully enabled. Therefore, building on the findings from both this study and earlier scientific work, in the third round, we asked the experts to predict this date. Apart from learning that this feature might prove to be very important in the future, a consensus of more than 75% of the experts could not be reached. Still, the main statements were that we must wait until the technology is installed in more cars, which may take more than five to ten years.

4.3   Public Awareness

In this section, we address the lack of public awareness about eCall, the potential for change of this fact and the influence eCall technology will have on the automotive industry.

The experts agreed in the first round that there was almost no public awareness of eCall (as of Q1 2018). Also, in the feedback from round two, all agreed that car lovers, car experts and the clients of private assistance services were almost the only people aware of this technology. Consequently, the experts agreed that it was essential to promote this system as, otherwise, the majority of potential customers would be unfamiliar with eCall until they were buying a new car. A majority of the participating specialists stated that, if no public awareness campaign is conducted, the awareness of eCall would only begin to increase when no other cars than those equipped with this technology were available. Interestingly, the European Union realized a similar lack of knowledge about the Pan European emergency call system years ago and recommended that a promotion campaign be conducted to explain the technology to the public. This campaign was meant to be supported by the European Commission and should have had numerous goals. The public was meant to be guaranteed that there was no threat to their privacy by the implementation of this technology, whose functionality and technicality was also to be communicated to EU citizens [19]. This grand campaign never took place.

In attempting to specify the reasons that no such campaign had yet taken place, the automotive industry experts agreed that the main reason was that not all the monetary benefits of the systems have been identified. Apart from this reason, slightly more than half of the experts shared the idea that a promotion campaign would not be a decisive factor in the implementation of eCall, since the system is obligatory and has not yet lead to a campaign being carried out on the Pan European scale.

Despite the expert opinions reflecting the reality relatively well, the EU has taken official steps towards the public promotion of eCall. The Journal of the European Union reported that car manufacturers must inform their customers about the presence of the eCall system in their vehicles [43]. Information about the existence of eCall, the right of customers to choose between the Pan European solution and a TPS, as well as a description of the set of data processed by the system and potentially communicated to a 112 answering point must be included in both the technical documents provided with a new car and available on the Internet. In 2017, the EU further refined this regulation regarding which car manufacturers must inform their customers about the data being collected by the IVS and how long it is being retained. Also, citizens must be informed that the system does not track their positions throughout their travels and the PSAPs do not transmit acquired information to a third party without having received permission by the car owner [44].

Hence, there is as yet no obligation regarding what information must be offered by the car manufacturers or retailers to the general public. This reality is consistent with the claims made by the experts in this study. Unless an individual wishes to purchase a new car, there is no communication regarding the system's specifics or its presence in vehicles. However, this is likely to change, as an EU-wide marketing campaign informing the EU population about the presence of this technology in cars is to be carried out. All but one expert agreed that the authorities responsible for carrying out this public education campaign are the Ministries of Transportation of individual EU member states. Furthermore, these entities should receive support from car manufacturers who should not only inform their new customers, but also participate in a general public education campaign. The experts also agreed that such a campaign should have already been launched, but most believed it was unlikely to occur on a large scale. Rather than a massive campaign, a smaller one will likely take place. This is expected to be initiated when the first real data on the system's performance are available, which is expected towards the end of 2018 or in the first quarter of 2019. This projection is consistent with the information on eCall-related data availability acquired from the General German Automobile Club (ADAC).

These conclusions are not to say that the introduction of eCall would have been totally ignored by the media, but that a long-term information campaign financed by official authorities has not yet taken place, which is the basis for the group opinion expressed by experts participating in this study. A possible explanation for the lack of awareness about a campaign related to eCall might also be that both smaller and EU-member-state local projects are focused on lowering the number of road fatalities and car accidents in individual EU countries. eCall is not the only EU initiative targeting road safety as projects like Road Safety in South East European regions (ROSEE8) exist. This project is aimed at South Eastern European countries which are proven as well as implied by the experts in this study to underperform in road safety when compared to the European average [45]. Hence, a possible explanation might be that such local programs are more promoted or generally better known than any Pan European emergency call system.

Should a general awareness campaign regarding the eCall technology be conducted, it could not be expected to change public perception of the automotive industry. Building on the opinions expressed in the first round, three quarters of the experts agreed on this point in round two, and also stated that any possible change in the perception of this industry could be related to the quality of the information campaign. Given the different orientations of eCall and TPS, it will be crucial to differentiate between these two emergency call solutions in the information campaign. 100% of participating experts agreed on this point, and claimed that it will be of particular importance to explain the nature and content of the MSD to all road users. This is reasonable since previous scientific publications have shown that the nature of these two systems differs and the public understanding of eCall may be heavily biased if its functionality is assumed to be the same as TPS services.

Altogether, the experts believed that one of the main reasons for the lack of change in the perception of the automotive industry through the introduction of eCall might be the fact that today innovations are taken for granted and safety is expected to be continuously improved. As such, recent safety improvements in the automotive industry are some of the most important. However, the experts also agreed that safety improvements are not the only focus of development in the automotive industry, and that other emphases include the ecology (primarily focused on emissions or ecological propulsion types), connectivity and driver comfort. In any case, the opinion of most experts is that efforts to further boost vehicle safety are likely to be seen positively.

4.4 Future of eCall

Regarding projections about the future development of the eCall technology, the first round yielded no shared opinion. However, a consensus did emerge in the second round. According to more than 85% of the experts, the security of the technology will be of crucial future importance and will thus continue to be enhanced. One of the reasons for targeting the security of the system is that three quarters of the specialists believe that eCall is a system that might be hacked. The truth of this suspicion was confirmed in the U.S. in 2015 when it was shown that a car can be hacked remotely without anyone having to personally engage with the vehicle. Hence, no devices should be installed that would enable a virtual access to a vehicle's computer. Even though cars have been targeted by hackers for many years and some have been successfully hacked, the attack in the U.S. was ground-breaking since in all the earlier cases, the attackers needed physical access to the car [46]. Eliminating the need for physical access to achieve a successful cyber-attack dramatically increases the need for security for eCall with respect to such threats.

Even though the experts believe that the technology will primarily evolve towards saving lives and assets, this technology will very likely be linked with another. Of particular importance is its possible connection to the infotainment system installed in modern vehicles. Disregarding the personal beliefs of experts regarding whether such a development should take place, most foresaw this development as inevitable. However, given claims that no such development may occur in the near future, we addressed the possible time horizon for this development in the second round. Based on the answers received and the feedback regarding these arguments acquired in the third round, the general conclusion was that it might take between five to ten years for eCall to start being interconnected with other car systems. While this may even take up to 15 years, it is unlikely to be far in the future; most experts disagreed that this development would take 30 or more years to be initiated.

The idea that eCall will primarily evolve towards saving lives is evident in the numerous projects that are focused on ameliorating the system's functions. A variety of future versions of eCall are emerging. One of these, "eCall+", is an advanced version of the emergency call technology that allows video connection as well. However, the most highly discussed development, often referred to as "Next Generation eCall" (NGeCall), is expected to have a larger MSD and allow for video calls as well as simultaneous data and voice exchange. Once these new standards are formulated, all cars equipped with the eCall technology should be capable of handling both the older and enhanced versions of eCall. Also, mobile network providers should inform their customers whether their networks are capable of handling NGeCall since it is expected to be more demanding with respect to network resources [47]. Hence, given these projected developments, rather than connecting eCall to other systems from the beginning, its safety functions should be enhanced as a matter of priority.

A unanimous opinion about the future of eCall, which is not backed by existing literature and hence represents a new finding, is its being linked to self-driving cars. The results of the Delphi questionnaire reveal that experts believe eCall will inevitably be linked with self-driving vehicles. In their third round answers, experts reported that eCall's connection to autonomous vehicles should be an active safety element for sharing traffic information with other cars. To some degree, this development serves as a reminder of the EU-projected introduction of vehicle-to-vehicle and vehicle-to-infrastructure communication, which may result in massive benefits [48]. Hence, despite there having been no explicit discussion of eCall's linkage with autonomous vehicles, projections have been made regarding active safety elements to be used in self-driving cars. These developments may help the European Union reach its vision of zero road fatalities on EU roads [7].

According to the Delphi questionnaire results, the eCall system is likely to be used to record drive data in the future. Although this opinion contradicts the claims made by the European Union about eCall, it is technically possible, as demonstrated by the so-called event data recorder (EDR) introduced in the U.S. in September 2014. The EDR can be considered to be a drive data recorder that is not connected to any PSAP online, but works offline to collect telematics data that can be retrieved in case of car accidents to help authorities investigate such cases. Despite having been developed for the U.S. market, it is estimated that 4–6% of European vehicles are currently equipped with an EDR system [20]. To date, eCall has no similar EDR functionality, and collects almost no data. However, eCall's IVS can be expected to adopt this role in the future.

Despite having conducted three rounds of the Delphi questionnaires, no consensus was reached regarding whether the eCall technology would be used to restrict criminal behaviour, with only a slight majority of the experts sharing this opinion. This verdict does not follow that of earlier introduced scientific findings as the literature suggests possible use of the technology in the insurance industry and in restricting criminal behaviour. Hence, despite not having been unequivocally projected by the experts, we again addressed the issue of eCall enabling a reduction of criminal behaviour in our third round of questionnaires. As perceived by the experts, the main potentials of eCall would be locating stolen cars and possibly better proving the guilt of individual parties in car accidents. Both these effects could be achieved based on information acquirable through eCall, which might help to prevent or reduce insurance scams [34].

Similar to eCall's effect on reducing criminal behaviour, no uniform opinion was expressed on the issue of eCall being used to monitor the health conditions of vehicle occupants. Only a few of the experts shared the opinion that eCall can actively monitor the health of road users. However, this finding is supported by earlier research, which suggests an enrichment of the MSD by health-related information. Primarily, the heart rate of the driver and the breathing patterns of the remaining occupants could be added to this data set. Systems capable of determining and measuring such health conditions are currently being developed, and are already being tested by some car manufacturers. Sensors measuring the driver's heart rate can be easily installed in the steering wheel and breath-measuring sensors installed in safety belts have already been successfully tested. Furthermore, since such sensors need not be used in combination with eCall only, car manufacturers like Ford and BMW are already working on their implementation and planning to fit their vehicles with sensor-enhanced steering wheels. Apart from information about passengers' health conditions being included in the MSD, it can also be used, for example, for disconnecting the hands-free communication add-on if driver tiredness or very high stress levels are sensed [49].

The literature further predicts enhanced forms of eCall (such as NGeCall) to enable a two-way data transfer [13], thereby allowing for the PSAPs to remotely access the hardware of the car. This would allow the use of cameras installed in a vehicle and would provide PSAPs with the capability to remotely manage some of the vehicle's functions. Hence, in the future, it is expected that the ignition could be switched off by a PSAP operator, the lights of the vehicle could be switched on or off, the horn of a crashed vehicle might be sounded, and the central locking system of the car might also be remotely controlled [50]. Despite some of the experts foreseeing this situation, not all believed that the future development of eCall might go this way; less than three quarters of the experts agreed with this statement.

As we have noted, some of the conclusions drawn by our experts align with previous findings, whereas others are not supported by earlier published research. Moreover, some of the important developments suggested by the European Union and scientific publications were not identified by the experts. For example, the implementation of eCall on vehicles other than cars was only mentioned by a few experts, but can be considered to be important given the fact that 15% of all EU road fatalities are motorcyclists [51]. A newer estimate by the European Commission (2017) slightly adjusts this value and states that 46% of casualties are car users, 21% are pedestrians and 14%, the third biggest group, are motorcyclists. Hence, there are projects focused on this issue. One example is I-VITAL, a project funded by the European Union aimed at developing sensors built into motorcyclists' helmets that would enable the use of the eCall system [51].

Lastly, once eCall is routinely used as an active safety element, it might not only warn the drivers about current weather conditions or road construction sites, but may also oversee the driver obedience of traffic rules. If a speed limit is exceeded, eCall could automatically adjust and slow the vehicle [42]. Given that excessive speed is one of the biggest causes of road tragedies [2], eCall could help reduce the number of fatalities on European soil [42], thereby ensuring significant monetary benefits.

5 Conclusions

eCall is a modern safety feature whose development was made possible by recent technological advances in wireless communication and data transfer, as well as major improvements in positioning technologies in recent decades. eCall is a system aimed at automating emergency calls throughout the European Union via a single emergency call phone number 112. As such, the system will automatically trigger an emergency call in case of an accident and transmits a set of data including crucial information about the vehicle, such as its position, to a public safety answering point. After discussion for more than 13 years during which its compulsory implementation was postponed several times and three major pilot projects costing more than €47 million were conducted, the eCall system was introduced on April 1, 2018. Despite similar emergency call and assistance services being previously available in some cars, all new types of vehicles with a maximum mass not exceeding 3.5 tonnes in the European Union must now be equipped with the Pan European version of eCall.

Some existing scientific literature and official communications by the European Union, having attempted to shed light on the effects of this phenomenon, have concluded that eCall will achieve both monetary and non-monetary benefits and will eventually expand to other transportation industries. There are also claims that the system can lead to yearly savings of up to €20 billion with respect to car accident costs. However, despite being in the planning stages for more than 13 years, which allowed enough time for research, many questions remain regarding eCall's public perception, future and effects. Furthermore, no prior studies of the Pan European effects of this technology have applied the scientifically based Delphi method. Therefore, to the best of the authors' knowledge, this is the only work to date that has assessed the general impacts of eCall using the Delphi method.

As proven in this work, the future developments and effects of eCall are not easy to predict. However, based on 11 initial questions, we successfully provided unambiguous answers to a vast majority. The entire study was conducted over the course of 76 days, and its success is due to the 16 automotive-industry experts who willingly participated in this research.

This work provides both a projection of future achievements of the eCall technology, as well as an introduction to the history, functionality and technical aspects of eCall as a general overview of this emergency system. Hence, it provides a strong basis for future research on the effects of eCall and this technology in general. By offering future projections, this study also enables future researchers to confirm or disprove its findings. Since this study was conducted using the Delphi method, this report also provides the opportunity for evaluating the accuracy of this empirical method and the possible refinement of its design.

References

1. WHO (2015) Global status report on road safety 2015. World Health Organization. http://www.who.int/violence_injury_prevention/road_safety_status/2015/en/. Accessed 11 June 11 2018
2. WHO (2017) Managing speed. World Health Organization. http://www.who.int/violence_injury_prevention/publications/road_traffic/managing-speed/en/. Accessed 11 June 2018
3. EENA (2018) EENA Committees Document: eCall and open issues, 2018 revision. http://www.eena.org/download.asp?item_id=26. Accessed 15 June 2018
4. European Commission (2005) The 2nd eSafety communication bringing eCall to citizens. https://eur-lex.europa.eu/legal-content/EN/TXT/PDF/?uri=CELEX:52005DC0431&qid=1528801388090&from=E. Accessed 18 June 18 2018
5. European Commission (2018) EU road fatalities. https://ec.europa.eu/transport/road_safety/sites/roadsafety/files/pdf/statistics/trends_figures.pdf. Accessed 22 June 2018
6. Chariete A, Bakhouya M, Nait-Sidi-Moh A, Ait-Cheik-Bihi W, Gaber J, Kouta R, Wack M, Lorenz P (2016) A study of users' acceptance and satisfaction of emergency call service. Int J Commun Syst 29(15):2279–2291
7. European Commission (2018) Sustainable mobility for Europe: safe, connected, and clean. https://eur-lex.europa.eu/legal-content/EN/TXT/?uri=COM%3A2018%3A293%3AFIN. Accessed 29 June 2018
8. Mladenow A, Kryvinska N, Strauss C (2012) Towards cloud-centric service environments. J Serv Sci Res 4(2):213–234
9. Polaschek M, Zeppelzauer W, Kryvinska N, Strauss C (2012) Enterprise 2.0 integrated communication and collaboration platform: a conceptual viewpoint. In: 2012 26th international conference on advanced information networking and applications workshops (WAINA). IEEE, pp 1221–1226
10. Becker A, Mladenow A, Kryvinska N, Strauss C (2012) Evolving taxonomy of business models for mobile service delivery platform. Procedia Comput Sci 10:650–657
11. Kopetzky R, Günther M, Kryvinska N, Mladenow A, Strauss C, Stummer C (2013) Strategic management of disruptive technologies: a practical framework in the context of voice services and of computing towards the cloud. Int J Grid Util Comput 4(1):47–59
12. Lego T, Mladenow A, Novak NM, Strauss C (2017) The economic value of an emergency call system. In: International conference on research and practical issues of enterprise information systems. Springer, Cham, pp 56–66
13. Iparraguirre O, Brazalez A (2016) Communication technologies for vehicles: eCall. In: Communication technologies for vehicles. Lecture notes in computer science, pp 103–110
14. Kubát D, Weinlich P, Semerádová T (2014) Data security concerns of future eCall users. In: IDIMT 2014 proceedings, pp 21–27
15. Sihvola N, Luoma J, Schirkoff A, Salo J, Karkola K (2009) In-depth evaluation of the effects of an automatic emergency call system on road fatalities. Eur Transp Res Rev 1(3):99–105
16. Acea.be (2018) Key figures. https://www.acea.be/statistics/tag/category/key-figures. Accessed 10 July 2018
17. European Commission (2013) Proposal for a decision of the European parliament and of the council on the deployment of the interoperable EU-wide eCall. http://eur-lex.europa.eu/legal-content/EN/TXT/PDF/?uri=CELEX:52013PC0315&from=EN. Accessed 10 April 2018
18. Cabo M, Fernandes F, Pereira T, Fonseca B, Paredes H (2014) Universal access to eCall system. Procedia Comput Sci 27:104–112
19. Journal of the EU (2014) Decision No 585/2014/EU of the European parliament and of the council of 15 May 2014 on the deployment of the interoperable EU-wide eCall service. http://eur-lex.europa.eu/legal-content/EN/TXT/?uri=celex:32014D0585. Accessed 20 April 2018
20. Bonyár A, Géczy A, Krammer O, Sántha H, Illés B, Kámán J, Szalay Z, Hanák P, Harsányi G (2017) A review on current eCall systems for autonomous car accident detection. In: 40th international Spring seminar on electronics technology (ISSE). IEEE, pp 1–8
21. WHO (2013) Global status report on road safety 2013. World Health Organization. http://www.who.int/violence_injury_prevention/road_safety_status/2013/en/. Accessed 11 June 2018

22. European Commission (2016) EU road fatalities. http://ec.europa.eu/transport/road_safety/sites/roadsafety/files/pdf/observatory/trends_figures.pdf. Accessed 20 April 2018
23. Card WI, Mooney GH (1977) What is the monetary value of human life? Br Med J 2:1627–1629
24. Rice DP, Cooper BS (1977) The economic value of human life. Am J Public Health Nations Health 57(11):1954–1966
25. Idnes.cz (2018) Cena za život ztracený při dopravní nehodě: více než 19 milionů korun. https://zpravy.idnes.cz/cdv-analyza-cena-zivota-dopravni-nehody-f01-/domaci.aspx?c=A180131_101655_domaci_hell. Accessed 10 July 2018
26. Guerrini F (2014) Traffic congestion costs Americans $124 billion a year. The Forbes. https://www.forbes.com/sites/federicoguerrini/2014/10/14/traffic-congestion-costs-americans-124-billion-a-year-report-says/#594bc058c107. Accessed 11 June 2018
27. Walters AA (1961) The theory and measurement of private and social cost of highway congestion. Econometrica 29(4):676–699
28. Arnott R, Small K (1994) The economics of traffic congestion. Am Sci 82(5):446–455
29. van Woensel T, Cruz FRB (2009) A stochastic approach to traffic congestion costs. Comput Oper Res 36:1731–1739
30. de Palma A, Lindsey R (2011) Traffic congestion pricing methodologies and technologies. Transp Res Part C Emerg Technol 19(6):1377–1399
31. McKinsey&Co (2016) An integrated perspective on the future of mobility. http://www.mckinsey.com/business-functions/sustainability-and-resource-productivity/our-insights/an-integrated-perspective-on-the-future-of-mobility. Accessed 20 May 2017
32. Goodwin P (2004) The economic costs of road traffic congestion. ESRC Transport Studies Unit University College London
33. Bryans JW (2017) The Internet of Automotive Things: vulnerabilities, risks and policy implications. J Cyber Policy 2(2):185–194
34. Carutasu G, Botezatu C, Botezatu MA (2016) Expanding eCall from cars to other means of transport. J Inf Syst Oper Manag 10(2):354–363
35. Bönninger J (2015) Mobilität im 21. Jahrhundert: sicher, sauber, datengeschützt. Datenschutz und Datensicherheit, pp 388–390
36. Journal of the EU (2017) Commission implementing Regulation (EU) 2017/78 of July 2016 establishing administrative provisions for the EC type-approval of motor vehicles with respect to their 112-based eCall in-vehicle systems and uniform conditions for the implementation of Regulation (EU) 2015/758 of the European Parliament and of the Council with regard to the privacy and data protection of users of such systems. https://eur-lex.europa.eu/legal-content/EN/TXT/PDF/?uri=CELEX:32017R0078&from=EN. Accessed 15 June 2018
37. European Commission (2016) European Commission Delegated Regulation (EU) of 12.9.2016 establishing detailed technical requirements and test procedures for the EC type approval of motor vehicles with respect to their 112-based eCall in-vehicles systems, of 112-based eCall in-vehicle separate technical units and components and supplementing and amending Regulation (EU) 2015/758 of the European Parliament and of the Council with regard to the applicable standards. http://ec.europa.eu/transparency/regdoc/rep/3/2016/EN/C-2016-5709-F1-EN-MAIN-PART-1.PDF. Accessed 10 April 2018
38. Geuens C, Dumortier J (2010) Mandatory implementation for in-vehicle eCall: privacy compatible? Comput Law Secur Rev 26(4):385–390
39. Ponte G, Ryan GA, Anderson RWG (2016) An estimate of the effectiveness of an in-vehicle automatic collision notification system in reducing road crash fatalities in South Australia. Traffic Inj Prev 17(3):258–263
40. European Commission (2017) 2016 road safety statistics: what is behind the figures? http://europa.eu/rapid/press-release_MEMO-17-675_en.htm. Accessed 2 July 2018
41. Harnischmacher F, Cosyns C, Grugl KM, Moerbe M, Portouli E, Savaresi SM (2016) State of the art assessment Powered Two-Wheeler (P2W) eCall. In: 11th ITS European congress
42. Piao J, Beecroft M, McDonald M (2010) Vehicle positioning for improving road safety. Transp Rev 30(6):701–715

43. Journal of the EU (2015) Regulation (EU) 2015/758 of the European Parliament and of the Council of 28 April 2015 concerning type-approval requirements for the deployment of the eCall in-vehicle system based on the 112 Service and amending Directive 2007/46/EC. https://eur-lex.europa.eu/legal-content/EN/TXT/PDF/?uri = CELEX: 32015R0758&from=EN. Accessed 18 June 2018
44. Journal of the EU (2017) Deployment of the eCall in-vehicle system ***I European Parliament legislative resolution of 26 February 2014 on the proposal for a regulation of the European Parliament and of the Council concerning type-approval requirements for the deployment of the eCall in-vehicle system and amending Directive 2007/46/EC (COM(2013)0316—C7-0174/2013—2013/0165(COD)) (first reading). https://eur-lex.europa.eu/legal-content/EN/TXT/PDF/?uri=CELEX:52014AP0154&qid=1528801388090&from=EN. Accessed 15 June 2018
45. Laiou A, Yannis G, Milotti A, Piccoli G (2016) Road safety interventions in South East Europe. Transp Res 14:3406–3415
46. Schellekens M (2016) Car hacking: navigating the regulatory landscape. Comput Law Secur Rev 32:307–315
47. Öörni R, Goulart A (2017) In-vehicle emergency call services: eCall and beyond. IEEE Commun Mag 55(1):159–165
48. Chen L, Englund C (2018) Every second counts: integrating edge computing and service oriented architecture for automatic emergency management. J Adv Transp 1–13
49. Urbano M, Alam M, Ferreira J, Fonseca J, Simões P (2017) Cooperative driver stress sensing integration with eCall system for improved road safety. In: 17th international conference on smart technologies. IEEE, pp 883–888
50. Sdongos E, Bolovinou A, Tsogas M, Amditis A, Guerra B, Manso M (2017) Next generation automated emergency calls. In: 14th annual consumer communications and networking conference. IEEE
51. Melcher V, Diederichs F, Maestre R, Hofmann C, Nacenta JM, van Gent J, Kusic D, Žagar B (2015) Smart vital signs and accident monitoring system for motorcyclists embedded in helmets and garments for advanced eCall emergency assistance and health analysis monitoring. Procedia Manuf 3:3208–3213
52. Newmobility news (2018) eCall emergency system obligatory in EU cars. https://newmobility.news/2018/03/29/ecall-emergency-system-obligatory-in-eu-cars. Accessed 2 July 2018
53. European Parliament (2015) Automatic emergency call devices in all new car models from spring 2018. http://www.europarl.europa.eu/news/en/press-room/20150424IPR45714/automatic-emergency-call-devices-in-all-new-car-models-from-spring-2018. Accessed 2 July 2018
54. European Commission (2006) Bringing eCall back on track—action plan (3rd eSafety communication). https://eur-lex.europa.eu/legal-content/EN/TXT/PDF/?uri=CELEX:52006DC0723&qid=1528801388090&from=EN. Accessed 18 June 2018
55. European Commission (2009) eCall: time for deployment. https://eur-lex.europa.eu/legalcontent/EN/TXT/PDF/?uri=CELEX:52009DC0434&qid=1528801388090&from=EN. Accessed 18 June 2018
56. dan Barca C (2016) Graphical interface for eCall incidents. Database Syst J 7(4):47–53
57. Chochliouros IP, Spiliopoulou-Chochliourou AS, Lalopoulos GK (2005) Emergency call (eCall) Services based on approved E-112 regulations and infrastructures: a solution to improve security and release of road help. FITCE 2005 Congress: 76–84

Data-Centric Business Planning—A Holistic Approach

Valerie Busse

Abstract Based on the increasing popularity of starting an own business, the demand of having a well-prepared plan arises. Not only in order to attract investors but also to generate a solid basis for a successful start of a business. Collection, choice, analysis and presentation of numerical data are indispensable activities to make cost-effectiveness possible from the very beginning. This article provides a well-planned and structured business plan by using different approaches to fine grain essential aspects of starting a business. It not only states a detailed description of the business idea, but also considers all marketing as well as financial aspects. Approaches such as a SWOT analysis, marketing mix, competitor-, financial- and scenario-analysis, give a detailed overview of the various types of data use, interpretation and presentation.

1 Introduction

Entrepreneurship and the importance of new venture planning is constantly increasing due to an enormous growth of possible networks [8]. Starting a business is a highly complex process which includes different barriers. Especially in the beginning several scenarios have to be taken into account to generate a profitable idea [11]. It is essential to provide a well-planned concept by considering future funding methods such as venture capital, a business angel or by starting a crowdfunding campaign [4].

This article provides a structured business plan by adopting a new concept of a café/bar, which is called "The Guild".

The first section provides a detailed explanation of what the idea is about. Including the idea, legal aspects, values, principles and mission statement as well as the explanation of the logo [1].

The following section considers several marketing and market research aspects which are based on survey findings conducted within the project. In this section,

V. Busse (✉)
University of Vienna, Oskar Morgenstern Platz 1, 1090 Vienna, Austria
e-mail: valerie.busse@infinanz.de

© Springer Nature Switzerland AG 2020
N. Kryvinska and M. Greguš (eds.), *Data-Centric Business and Applications*,
Lecture Notes on Data Engineering and Communications Technologies 30,
https://doi.org/10.1007/978-3-030-19069-9_17

approaches such as target customers, competitors analysis, competitive advantage, SWOT analysis, and the marketing mix including price, product, place, promotion, physical evidence and process are used [2, 3, 5, 7, 8, 12, 13, 19, 24].

The next section gives a detailed plan about needed resources including people, premises and location, equipment and organization and administration.

Finally, a specified data driven financial analysis is provided by focusing on financial requirements and initial investment. Financial statements are given by profit and loss-, balance sheet- and cash flow-statements [4, 6, 14–16, 23]. Followed by a scenario analysis [9]. The conclusion presents a short summary of the overall plan including the financial outcomes.

2   The Business Idea and Its Framework

2.1   Concept

The core concept of The Guild is to provide people of Hamburg with a new idea and therefore a new experience while enjoying their drinks or snacks. The idea is based on the fact of a shrinking craft industry in Germany and furthermore the description of the actual generation as digital natives. The founder, which are both in their mid-twenties with an economic educational background and were inspired of similar concepts from Vienna as well as New York. This plan was started in order to give the current generation room for creativity by focusing of a non-digital concept.

The Guild is a place of creativity (Fig. 1). The Café itself will be a showroom or exhibition space for art, furniture and other ideas from local people, students and

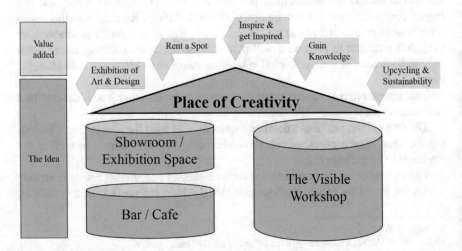

Fig. 1 The business idea

young designers. Literally, everything will be purchasable. Moreover, findings show that most of the people do not have the possibility to craft their ideas into reality because of missing tools, the lack of knowledge and/or space.

Hence, the third part of The Guild will be a visible a workshop where people can rent a spot, use the tools and get help from a carpenter or creative worker.

Concluding, the Café will be a cozy environment where open-minded, grounded and creative individuals enjoy their time by chatting, inspiring and crafting. In the evening, the place changes into a Bar, where the special environment invites the individuals to enjoy all sort of beverage of a basic range with some specials to end their days with a good feeling.

In order to provide the right materials, local sourcing is the key to a good network and small inventory. Focus is put on the creativity and the needs of every single customer to underline the possibility, which is suggested by the business idea: "Craft your dreams into reality."

The Café seats maximum 45 people, gives space for exhibiting around 15 parts of furniture and to rent 4 working-spots in the workshop. In the evening the workshop could give space for further 50 customers. The Guild will be opened from Tuesday to Sunday from 8 to 11, the opening hours on weekend evenings are longer.

2.2   Aspects of Legal Structure

The chosen legal structure for the business is a partnership [1]. According to the Partnership Act of 1890 a partnership is the relation which subsists between partners carrying on a business in common with a view to profit [1]. The two partners will run the partnership equally. The contract between the two partners will be formalised through the partnership deed. This agreement states how the partnership is run and the duties of each individual partner.

This legal structure was selected and found most appropriate for The Guild for three reasons. Firstly, low set up costs are involved. Secondly, the process is quick compared to other legal structures and thirdly more capital is available.

However, the partners are aware that this legal structure has unlimited liability and each individual partner has to submit a self-assessment tax return.

Furthermore, other legal aspects need to be taken into consideration such as the food hygiene certificate, since the main purpose of the business is to sell drinks and snacks. After getting this license there is no further licence needed for alcohol sales as all is covered in the "Schanklizens". In addition, written legal contracts need to be created with each supplier and employee. Furthermore, if the exhibition of the unique designs is successful and attracts many customers, The Guild considers setting up exclusive contracts according those designs, which should ensure the concept is unique.

2.3   Vision, Mission, Values and Behaviour Guidelines

Vision

The Guild should provide a cosy home to creativity and a community, which brings the tradition of crafting and interpersonal rules of etiquette, back to life. An idea and place to lose time and inspire as many souls as possible. A room where the diversity of personalities and ideas merge together to be cherished and supported.

Mission

In order to reach the business ideas vision, The Guild provides a unique and cosy atmosphere, which is emphasised by behaviour guidelines. These guidelines are lived throughout the organisation and therefore give people especially customers and employees, a superior significance. Thus, the foundation of interpersonal etiquette can grow from inside and make The Guild a unique venue of inspiration in any aspect. Furthermore, working with local suppliers and individuals as well as charity projects, contribute to a win-win situation for all entities.

Values

The Guild has a list of clear values which have to be lived within the business. A structured overview of the values lived in the Café/Bar will be provided at the entrance, that it is visible for customers and employees [18].

The first and most important value lived in The Guild is a *friendly attitude* towards everyone, this is underlined with the sentence "A smile never hurt nobody".

The second major value which is lived throughout the concept is the *close connection to tradition with tendency to modernity*, this will be based in Winston Churchill's quote "without tradition art is a flock of sheep without a shepherded. Without innovation, it is a corpse".

Other values such as the *imagination without barriers* is inspired by Albert Einstein's statement "Imagination is more important than knowledge. For knowledge is limited to all we now know and understand, while imagination embraces the entire world, and all there ever will be to know and understand".

Furthermore, the value of *appreciation of people* is used through John F. Kennedy's proposition "as we express our gratitude, we must never forget that the highest appreciation is not to utter words, but to live by them".

Tolerance and open-mindedness is suggested by Tom Hannah by stating: "Tolerance and celebration of individual differences is the fire that fuels lasting love".

Lastly, The Guild is mainly driven by *Innovation* which emphasised by Theodore Levitt: "Creativity is thinking up new things. Innovation is doing new things".

Behaviour Guidelines

Furthermore, several behaviour guidelines have to be taken into account by opening a new business in the sector of gastronomy and also particularly for handcrafting issues. These behaviour guidelines mirror the before mentioned aspects on operational level.

Starting with the *friendly behaviour principle*, which implies a respectful treatment of customer, co-workers and service employees.

Fig. 2 5S explanation

Moreover, all employees of The Guild will wear a *uniform*, which emphasises the aspect of tradition. A more conservative dress style and differentiation between tasks will be implemented. The bar keeper can for example be recognised by a nice shirt waist coat and bow tie while the waiters will only wear shirts and a bow tie.

Another behaviour guideline is given through *continuous improvement* through customer involvement such as personal feedback, emails, questionnaires and social media.

Employees are very important for the business and vision. Therefore, they are involved in the developing process and growth opportunities. In planning of further branches or marketing campaigns, and if financial goals are reached, a bonus in form of extra payment will be provided.

Safety instructions will be lived throughout the whole business. Safety equipment will be provided in the workshops with a strict control of usage by employees.

In The Guild particularly in the workshop, *cleanness* plays a crucial role to prevent accidents and ensure attractiveness for the customer. Therefore, lean principles and tools such as 5S are implemented in Fig. 2.

2.4 Name and Logo

The name of the business is "The Guild". The reason behind the name is that it represents the idea and feeling the business provides is easy and simple to remember, which is of vital importance to a new start up. Tradition of crafting belonging to something and quality of relationships like in the good old times will be enhanced.

Figure 3 represents the logo, which will be printed on the uniform, products and social media experience but also on the front of the Café/Bar. It is simple yet brings

Fig. 3 The logo

across the main point. The diamond represents the mentioned tendency to modernity while the inner logo the bearded man with the hat, a magician, represents the magic or imagination and creativity. It will be used as a branding, stamp and plaque on the furniture. In front of the Café/Bar itself, the logo will be carved and painted on wood to give customers a better idea of what The Guild is actually about.

3 Marketing and Market Research

3.1 Target Customers

The target customers have been selected by conducting secondary and primary research with 207 participants who live in Germany. Because of the complexity and the different aspects of the business, some additional information of prospective customers has to be mentioned.

The following gives a detailed information of demographic., psychographic-, behavioural- and geographic segmentation based on the survey findings [17, 20, 21].

- Demographic Segmentation—The main target will be students and young professionals aged between 20 and 35. Talking about gender, women and men will come to The Guild but according to our survey findings, there is slightly bigger want for supervision and tools included in a rental workshop amongst the women sector.
- Psychographic Segmentation—The target customer will be the people with a lifestyle dominated by digital media and the need to escape this environment. For that reason, the customers are provided with the possibility to rent a work-spot, get supervision and craft something with their own hands. The primary research revealed that people in that age developed this need that is not satisfied. Furthermore, the primary research implies that young designers are price sensitive when

exhibiting their design. Therefore, a win-win situation is given, by only charging a commission fee when the product is successfully sold.

- Behavioral Segmentation—The advantage of providing the whole experience in alternative area where students and young working professionals usually go enjoy a drink or snack in a Bar will be used. The primary research shows that according the purchase of furniture the majority of the survey participators is interested in uniquely designed furniture. Additionally, they are price sensitive and do not want to pay established designer prices for these unique pieces.
- Geographic Segmentation—Hamburg, Schanzenviertel area (which is situated in the west of the city of Hamburg) was chosen as an adequate site because of three main reasons: (i) the area is the main meeting point of many students and working professionals, (ii) the area is close to the city center and in vicinity of universities and offices, and (iii) the described target group with the "fitting" lifestyle tends to live or work in this area.

3.2 Competitors Analysis

The Guild shall a have no direct competitor in the chosen area. Although the competition is fierce, The Guild has a major competitive advantage to its competitors, customers will be able to create, purchase and exhibit their objects.

Direct Competitors

There are no direct competitors because the concept of The Guild is unique in the area. However, the two most comparable spots are "Salong wechsel dich" and "Haus Drei". "Salong wechsel dich" is a Café, which works as an exhibition space, but however has no workshop. "Haus Drei" has a rentable workshop in the cellar but no exhibition space.

To give an overview of those two and illustrate the difference and advantage of The Guild, the blue ocean model is applied Fig. 4.

As can be observed from Fig. 5 the indirect competition in the chosen area is fierce.

Competitive Advantage

Despite the fact that there are many cafes, bars and restaurants in Schanzenviertel, The Guild is the only one with the mixed concept of café, exhibition space and workshop.

Hence, the competitive strategy is differentiation: The Guild is a unique place in Hamburg with a personal and special environment to inspire and get inspired while enjoying a drink or snack. It can be describes as a cosy place to enjoy art or to grab a coffee to take away or sit inside. It is essential to be positioned like that in the customers' minds since the primary research showed that the atmosphere and service rather than a wide range of drinks is important to our potential customers. Furthermore, according the unique designer pieces the strategy is based on prices. While exhibition takes place, the furniture is used and will devaluate over time. Price

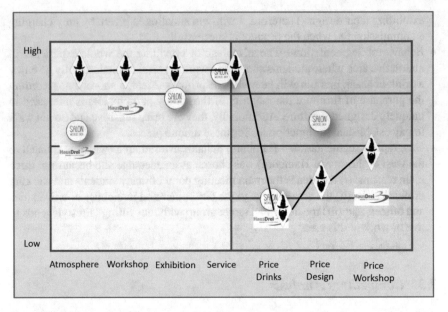

Fig. 4 Blue ocean model [10]

Inside the area			Outside the area		
Restaurants (16)	**Café (10)**	**Bars (12)**	**Workshop**	**Design Exhibition**	**Design Online**
Balutschistan	Café SternChance	Goldfischglas	Die Werkkiste	Galerie Papenhuders	Blickfang design shop
Il Cammino	Herr Max	Fritz Bauch	Tischler Akademie	Atelier Koppel 66	selekkt
Fisch Imbiss Schabi	Schmidtchen Schanze	Katze	Holzwerkstatt Ottensen	Mercedesstern-haus	Die Waescherei
Frank und Frei	Café Elbgold	Daniela Bar	Drei 21 Tischlerei	Kunstcafe	Used-design
Omas Apotheke	Less political	Galopper des Jahres	Bauer + planer	Kunstraum 17	
Kumpir Koenig	Codos	Central Park			

Fig. 5 Indirect competition

sensitive customer can buy or wait until the pieces are affordable. Moreover, if used is not what they are looking for, doing it their self or placing an order at the carpenter are the options.

3.3 SWOT Analysis

A SWOT-analysis (Strength, Weaknesses, Opportunities and Threats-analysis) identifies several strength, weaknesses, opportunities and threats which might occur during the process which is depicted in Table 1 [22].

- The service is different from the competitors as mentioned before. This is given trough the offer of the café as an exhibition space and workshop. Therefore, customers will find an atmosphere that they will not find somewhere else.
- Furthermore, The Guild, is an "alternative, creative Café/Bar" in an area, where people have an alternative, creative lifestyle. Moreover, there is a supervision in crafting provided. Additionally, there are several safety instructions provided before making use of the tools. Strengths are also give trough the fact that ideas are given to people who need to be inspired and that the whole process is highly customer focused.
- Nevertheless, there are also some weaknesses. First of all, due to insufficient kitchen material, only small snacks will be provided. These, however, could be easily prepared by the customer itself. Another weakness is given by the fact that the joinery will not provide big tools and therefore there might occur barriers by executing the customers whole creativity.
- However, there are also given several opportunities. Firstly, there is not any direct competitor in the area there is a high advantage of being the unique Bar specialized in creativity in Schanzenviertel where an alternative lifestyle is a rather a rule than a normal expectation.
- Furthermore, the area where The Guild will be located is a very well visited area of many people. Lastly, the importance of self-fulfilment amongst consumers is growing and our customers are people who want to get away from digital media.
- There might be also some threats during the concept. One might be the fierce competition, or getting competitors inspired to adopt some parts of the concept of The Guild. This would lead to a spread of the customers towards the competitors and therefore to a major loss of the overall concept. The second major threat is that if the concept is rejected itself. This treat, however, could be eliminated through several marketing concepts as listed in the following part.

Table 1 SWOT-analysis

Strength	Weaknesses
• Exhibition space • Unique atmosphere	• Insufficient kitchen • No big tools in joinery
Threats	Opportunities
• Fierce competition • Rejection of concept	• No direct competitor • Advantage of being the unique bar

3.4   Marketing Mix

The following section provides a deeper understanding of the marketing mix which is used within the concept.

Product

The customer should perceive The Guild as a unique place with a simple range of drinks and snacks. The Guild will offer a pleasant and comfortable atmosphere due to the specific store design, which is exhibited by young designers and creative people. Furthermore, the visible workshop, where a carpenter works most of the time, provides rentable spots where customers can craft under safe condition and supervision. Thus the possibility exists to relax and enjoy the product as well as the service in a friendly and unique ambiance. The major offers will be listed as followed:

Hot Drinks: Hot drinks in basic range will be provided in the Café. From Americano to Cappuccino or a Hot Chocolate. The drinks can be enjoyed inside The Guild or for take away.

Snacks: Snacks will be provided such as Panini's, sandwiches, muffins, cakes that are easier to prepare. Hence, a need for a kitchen is not present.

Alcoholic and cold drinks: In the evening alcoholic drinks will be provided and can be enjoyed in the Bar and in the workshop. While the tools are locked and the room is cleaned.

Exhibition space and Design Commission: Every person who crafts, draws and designs has the possibility to exhibit their designs within The Guild. The Café works as a showroom and furthermore the designs will be exhibited online on the website. The exhibiters will only be charged a fee in the case of successful purchase.

Rentable work-spot: There will be four work-spots every person can rent and use the tools provided by The Guild. A carpenter during the day and additionally creative will give advices and supervision.

Place

As mentioned in the previous chapters, The Guild will be located in a scene area of Hamburg in the Schanzenviertel. This location was chosen because of many different opportunities which where gained through the primary and secondary research. The Schanzenviertel is known as an alternative area, with several Bars and Cafés. However, none of them provides the opportunity of a workshop it is the Café/Bar which leads to the competitive advantage as already mentioned above. Additionally, this area directly catches the target group of 20–35 old creative and highly emotional driven people. Therefore, the chosen place can already be seen as a marketing instrument as it is directly placed where the target group is located.

Analysis of Pricing

The pricing strategy is based on the competitors. Especially the prices of drinks and snacks. The prices for the exhibiting spots as well as the workshop renting are based on the survey findings whiting 207 participators. Generally, due to the young

Table 2 Distribution of prices

Hot drinks	Average price of 2.80 €
Alcoholic drinks	Average price of 3.00 €
Snacks	Average price of 3.90 €
Design commission	15% of selling price
Workshop rent	10 € per hour

target group the prices will be kept between low and medium to generate affordable products for our target group (Table 2).

Digital and Non-digital Promotion

In order to attract potential customers and make The Guild a visible place, different advertising instruments are used.

Especially during the opening process, it is essential to draw attention to the target group and convince them of the new concept of a Café/Bar.

For winning and keeping customers a webpage is used as well as leaflets and social media. Additionally, there will be cups, bags and T-shirts provided, which are printed with The Guilds' logo.

With leaflets customers will be informed about the opening event, workshops, theme-parties, drink- and food-specials as well as live band events.

The same information as on the leaflets will be provided on the webpage as well as social media channels such as Facebook, Pinterest and Instagram. Moreover, on these channels there will be additionally do-it-yourself videos provided which give advice of creating own furniture and other unique designs. Furthermore, customers will be attracted with these videos to come and visit The Guild to use the provided work-spots and knowledge.

Besides the videos, pictures from the newest exhibition pieces including their prices will be reachable on the webpage as well as social media channels. The pictures of the designs come along with small stories of the creators and their thoughts as well as an idea explanation.

As mentioned above, another way of advertising is the webpage features such as a web designer agency, including a search engine optimisation which will be available to make it easier for customers to find us.

Concerning Facebook, Twitter and Pinterest account there will not be any additional cost. The customer will find the links for all the social media tools on our webpage which indicated an easy connection between all channels.

Furthermore, there will be customer loyalty cards which give the customer the opportunity to get a 1-h free workshop renting spot after five visits.

Physical Evidence

The physical evidence will be present through leaflets during the starting/opening process and for several special workshops. The leaflets will be handed out in the whole Schanzenviertel as well as in several universities and Café/Bar's in other areas of Hamburg. The Logo will be easily visible and recognizable. Additionally, the color concept in wooden brown and grey tones will be the same, online, on the leaflets and through the whole Café/Bar as well as the workshop. Furthermore, the logo on the stickers, cups, bags and t-shirts lead to an easy recognizable brand.

3.5 Management Hierarchy

The management work is going to be shared by the equal partners. Furthermore, there will be students as waiters which have working shifts of 8 h. The waiters work behind the Bar as well as in the service area of the Café. Additionally, a carpenter and a creative worker which provide advice to the customers in the workshop is available. The waiters are going to wear a special recognisable uniform which contains shirts, tie bow, waistcoat and braises. All this contributes a traditional and old fashion stylish environment. The management hierarchy can be seen in Fig. 6.

There will be a friendly atmosphere where the customers feel good and are likely to spend a certain amount of time in The Guild. The target group has a high affinity of handcraft.

Process-Structure
The Guild is divided into three different processes. The Café/Bar, the exhibition and the workshop. Beginning with the Café/Bar process, after showing the customer to the table the waitress explains the concept of The Guild, including the workshop and the exhibition space. After that, the customer can order coffee, soft- or alcoholic drinks and decide if he/she is going to choose something from the food range.

If the customer likes to pay, he/she blows out the candle on the table. The upcoming steam is a sign for our waitress that the customers' wish to pay.

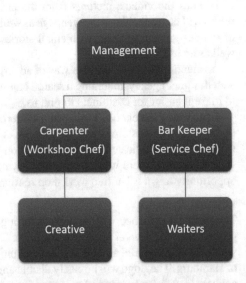

Fig. 6 Management hierarchy

The payment options are cash or the use of a credit card, to get a 10% student discount, a student ID card must be shown to the service.

Concerning the exhibiting process, customers are able to exhibit their furniture in the Café/Bar; the founders will get a provision of 15% of its selling price. While 5% go to charity projects. The exhibition items are also provided in the online shop and will devaluate over time and exchanged after a year if they are not sold. Only chosen design will be exhibited.

For the workshop the customer has to watch a 15 min' safety instruction video before entering the workshop to ensure safety. There will be safety clothes provided, such as safety cap shoe covers, protection glasses, gloves and breathing protection to support the latter.

Then the customer can start to create his/her own design. The material which will be used can be sourced either as a working kit or by bringing own material. This process happens under the supervision of either the carpenter during the day or a creative worker in the afternoon. Whereas the chosen creative worker is a retired, young designer or all-rounder with a high skills and an affinity of creative handcrafting. When the customer is finished he/she cleans his/her working spot. Then he/she can either put the uncompleted furniture in a storage-room for a limited time of two weeks, take the furniture home or provide it for exhibiting purposes in the Café/Bar.

4 Resources

Distribution of Employees

As mentioned in sections before, The Guild exists of a partnership of two. The power is divided by the two and one will have the functions which are explained in further detail in the section below.

Both partners have finished their master in business management and have different skills which will help for a successful business idea. One founder has worked in different parts of the gastronomy sector and has seen what it takes to make a Café/Bar work successful. The other one has worked in the logistic sector, and as a business consultant which means he knows the concept of supplier management as well as leadership of teams and process optimisation.

Table 3 shows the task and amount of necessary employees in more detail. Furthermore, it shows the different phases of the business and the varying amount of the service aligning to the demand.

Premises and Location

As mentioned in previous chapters, The Guild will be located in Schanzenviertel, Hamburg. This location was chosen because of its many opportunities that were found through secondary research and own experience. Schanzenviertel is home to many Bars, Restaurant and Cafés. The premises chosen will be rented because of financial reasons. Research showed that the average price of a premise about 15 €

Table 3 Amount of employee

Task	Amount employees		
	Beginning	Low season	High season
Service	4	7	10
Manager	2	2	2
Carpenter	1	1	1
Creative	1	1	1

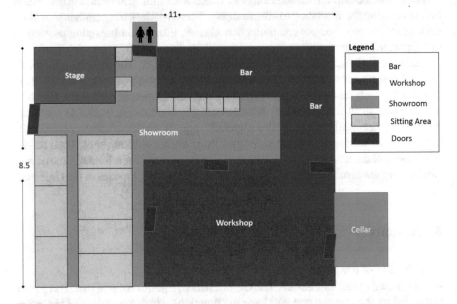

Fig. 7 The layout

per square meter, therefore the rent of The Guild is around 2100 € per month. The layout of The Guild is illustrated in the Fig. 7.

Equipment

In order to operate the Café/Bar there is Bar equipment needed in order to prepare both, alcoholic and non-alcoholic drinks. Additionally, the Café/Bar has main areas where small as well as big tables are provided. Furthermore, the Bar itself has five additional seats. Moreover, there will be a big range of working equipment provided including safety equipment in the workshop.

Organization and Administration

The Guild is open every day except Mondays. The Café/Bar is open from 8 am to 11 pm, during the week. Sundays, The Guild will close at 10:00 pm due to less demand. The Guild will have extended opening hours until 3:00 pm on Friday and Saturday. The workshop itself is open from 10:00 am to 7:00 pm during the week and 10:00 am to 7:00 pm on the weekend. However, as the workshop is used as an

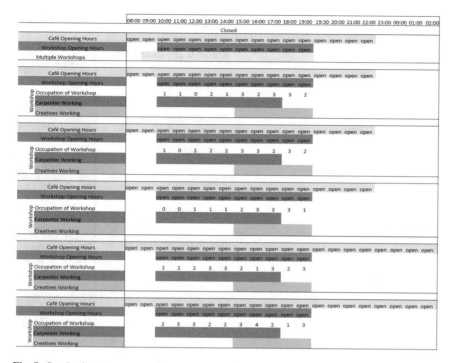

Fig. 8 Service hours

additional place for customers during the evening, it has the same opening hours as the Café/Bar just without the function of using the workplaces for creating reasons.

Both founders will work 40 h per week in the management and organisation area on a 3600 € basis. This works includes, hiring employees, financial organisation, maintenance- and repair administrations, as well as diverse other administrative operations. Furthermore, there will be students working in 8 h shifts with a salary of 10 € per hour. Additionally, there will be a creative worker who gives advice to the workshop users in the morning on an hourly basis of 10 €. Concerning the evening hours, a qualified carpenter will provide helping sessions and further advice for customers. Figure 8 represents an overview of the opening hours as well as the working hours of carpenter, creative and the occupation of the workshop.

5 Financial Analysis

Financial Requirements and Initial Investment

To cover the first expenses, and initial acquisition, a starting capital of around 75,000 € is needed. The initial cost include all expenses of the first month, such as salaries, single commission fee to the property owner (7000 €), the rent (2100 €) as well as

Fig. 9 Initial requirements

Initial Investment	Costs in Euro
Landlord Commission Fee	7000
Licence and other	750
General Renovation	35625
Marketing Expenses	3031
Tools	5095
Equipment Café/Bar	1279
Electronic Devices Bar	12150
Bar Furnishing	6000
Hygiene Settings	1980

initial investment in renovation (35,625 €), purchase of tools for the workshop (5095 €), equipment (7279 €) and further marketing expenses (3031 €). The required capital will be provided through three sources. Initial private investment of the founders (15,000 €), additionally support of a business angel where the contact already exists (10,000 €) and the major source of a Bank Loan (50,000 €) by the Postbank with an interest rate of 3.49% and a duration of 36 month. To get a deeper understanding of the necessary equipment as well as an overview of the expenses mentioned above please see Table 3. Hence, furniture will be exhibited through the mentioned process in one of the prior sections; these will not affect the financial statement in terms of costs [6].

Financial Statements
In the following sections, financial statements such as profit and loss, balance sheet and the cash flow of The Guild will be illustrated and explained [6].

Profit and Loss
As you can see in the Fig. 10, the profit and loss forecast demonstrates the sales and costs of The Guild. Where the sales consist of other sales such as design commission and workshop rent, alcohol sales, food sales and coffee sales. The cost on the other hand are the sum of direct costs and overheads such as salaries, marketing expenses as well as the initial investments mentioned before.

There will be a definition between three phases, the beginning phase, which is characterized through gradually growing sales as well as initial investment in the first year, the transition phase in the second year where salary rise aligning to increasing

Fig. 10 Profit and loss forecast

sales and the mature phase at September of the second year where we reach our peak in sales and costs.

In the beginning phase the staff members are hold rather low in order to keep the cost low as the sales rise slow in the beginning. Therefore, the founders will have multiple tasks such as service and general management.

Marketing cost will be higher in the starting month in order to boost the appearance of The Guild and attract potential customers. Especially, website and social media will be a constant investment. Other marketing costs such as flyers will occur due to promoting special events and workshops. This is illustrated in the yellow graph in the diagram underneath (Figs. 9 and 10).

In the transition phase service costs will rise aligning to increasing sales and the founder will concentrate more on managing stuff and the business itself rather than operating the Café. The marketing costs don't vary and are calculated as an investment four times a year and constant costs to ensure an up to date website and social media appearance (Fig. 11).

Every year is separated into a high and low season because of changing customer demand and behavior.

In the high season which include the winter months where the temperature is colder outside and customers seek the warmth and cozy atmosphere to enjoy hot drinks and are likely to use the workshop. Furthermore, during the Christmas period, people go shopping and hence the streets are crowded. The Christmas shopping and buying presents, will show in the increasing income of design commission as well as in the workshop rent. Therefore, the costs of service reach a peak, however, they will be covered by high sales of coffee, and alcoholic drinks.

In the low season, which includes the summer months with a high temperature outside, customers are more willing to spend their time outside than in a Café. However, due to long evenings there is an expectation of a high demand in the Bar.

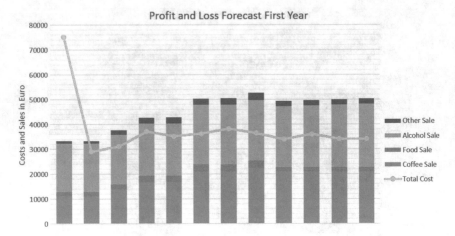

Fig. 11 Profit and loss forecast first year

Therefore, cost for service will decrease as well as sales of coffee and the workshop renting.

Overheads like depreciation, insurance, gas/electricity/water are calculated as constant prices due contracts payed half a year or yearly and contribute to a more stable total cost graph despite the marketing costs.

Furthermore, it is to mention that the described seasons can be easier obtained in the later years due to gradually increasing sales in the beginning phase (Table 4).

Balance Sheet

As can be obtained of the Table 5 the balance sheet of The Guild can be analyzed as follows [6].

The opening balance sheet shows fixed assets of 26,130 €, caused by the initial investments. The current assets of 7,260 €, creditors due within one year 16,282 €, net current assets −9,022 €, total net assets −16,802 € and the capital and reserves—16,802 € illustrates the impact of the high costs in the beginning and show an instable and risky beginning phase. This is caused by low sales assumptions and investments to develop a reputation among the customers.

Nevertheless, the following years display a steady growth in current assets and therefore in capital and reserves. This implies a stable as well as flexible company and may give the opportunity to pay loans earlier than estimated. Moreover, the fixed assets loose in value through depreciation and should be renewed after a certain period of time. This relative small amount of fixed assets again is result of the missing investment in furniture and therefore less devaluation of fixed assets. The balance sheet is characterised through liquidity. Hence, the credibility according to investors is high and gives opportunities to future growth.

Cash Flow

The cash flow shows payments and receipts of the business [3]. First of all it illustrates that the initial costs are covered by the loan and private investments, which is already

Table 4 Profit and loss forecast

The Guild

Multi-year profit and loss forecast

	Sep 18 €	Oct 18 €	Nov 18 €	Dec 18 €	Jan 19 €	Feb 19 €	Mar 19 €	Apr 19 €	May 19 €	Jun 19 €
Sales										
Coffee	11,200	11,200	14,000	16,800	16,800	21,000	21,000	22,400	21,000	21,000
Food	1,560	1,560	1,755	2,535	2,535	2,730	2,730	2,925	1,560	1,560
Alcohol	19,500	19,500	20,100	21,000	21,000	24,000	24,000	24,300	24,600	24,900
	32,260	32,260	35,855	40,335	40,335	47,730	47,730	49,625	47,160	47,460
Direct cost										
Coffee	1,120	1,120	1,400	1,680	1,680	2,100	2,100	2,240	2,100	2,100
Food	468	468	526	761	760	820	818	878	468	468
Alcohol	3,900	3,900	4,020	4,200	4,200	4,800	4,800	4,860	4,920	4,980
	5,488	5,488	5,946	6,641	6,640	7,720	7,718	7,978	7,488	7,548
Gross profit	26,772	26,772	29,909	33,694	33,695	40,010	40,012	41,647	39,672	39,912
Overheads										
Salaries/Wages	18,960	18,960	20,660	23,960	23,960	23,960	23,960	23,960	21,960	21,960
Gas/Electricity/Water	394	395	394	395	394	395	394	394	395	394
Insurance	1,300	1,300	1,300	1,300	1,300	1,300	1,300	1,300	1,300	1,300
Rent	2,100	2,100	2,100	2,100	2,100	2,100	2,100	2,100	2,100	2,100
Landlord commission fee	7,000	–	–	–	–	–	–	–	–	–
License and others	750	–	–	–	–	–	–	–	–	–
General renovation	35,625	–	–	–	–	–	–	–	–	–

(continued)

Table 4 (continued)

The Guild

Multi-year profit and loss forecast

	Sep 18 €	Oct 18 €	Nov 18 €	Dec 18 €	Jan 19 €	Feb 19 €	Mar 19 €	Apr 19 €	May 19 €	Jun 19 €
Marketing	3,031	330	330	2,291	370	330	2,291	330	330	2,291
Depreciation	374	377	374	375	376	375	374	378	373	376
	69,534	23,462	25,158	30,421	28,500	28,460	30,419	28,462	26,458	28,421
Other income (expense)										
Design commission	100	200	200	300	200	200	200	200	100	100
Workshop rent	860	860	1,500	2,000	2,200	2,400	2,600	2,800	2,100	2,100
	960	1,060	1,700	2,300	2,400	2,600	2,800	3,000	2,200	2,200
Operating profit	(41,802)	4,370	6,451	5,573	7,595	14,150	12,393	16,185	15,414	13,691
Interest expense										
Loan interest	–	145	142	138	134	130	126	122	118	115
	(41,802)	145	142	138	134	130	126	122	118	115
Net profit	(41,802)	4,225	6,309	5,435	7,461	14,020	12,267	16,063	15,296	13,576

(continued)

Table 4 (continued)

The Guild

Multi-year profit and loss forecast

	Jul 19 €	Aug 19 €	18–19 €	Sep 19–Nov 20 €	Dec 20–Feb 21 €	Mar 20–May 20 €	Jun 20–Aug 20 €	20–21 €	20–21 €
Sales									
Coffee	21,000	21,000	218,400	74,200	84,000	77,840	63,000	299,040	299,600
Food	1,560	1,560	24,570	5,655	8,775	7,410	4,680	26,520	29,640
Alcohol	25,200	25,500	273,600	78,900	81,300	83,700	86,400	330,300	360,900
	47,760	48,060	516,570	158,755	174,075	168,950	154,080	655,860	690,140
Direct costs									
Coffee	2,100	2,100	21,840	7,420	8,400	7,784	6,300	29,904	29,960
Food	468	468	7,371	1,696	2,633	2,223	1,404	7,956	8,892
Alcohol	5,040	5,100	54,720	15,780	16,260	16,740	17,280	66,060	72,180
	7,608	7,668	83,931	24,896	27,293	26,747	24,984	103,920	111,032
Gross profit	40,152	40,392	432,639	133,859	146,782	142,203	129,096	551,940	579,108
Overheads									
Salaries/Wages	21,960	21,960	266,220	72,880	80,880	77,880	71,880	303,520	320,520
Gas/Electricity/Water	395	394	4,733	14,199	14,199	14,199	14,199	56,796	56,796
Insurance	1,300	1,300	15,600	3,900	3,900	3,900	3,900	15,600	15,600
Rent	2,100	2,100	25,200	6,300	6,300	6,300	6,300	25,200	25,200
Landlord commission fee	–	–	7,000	–	–	–	–	–	–

(continued)

Table 4 (continued)

The Guild

Multi-year profit and loss forecast

	Jul 19 €	Aug 19 €	18–19 €	Sep 19–Nov 20 €	Dec 20–Feb 21 €	Mar 20–May 20 €	Jun 20–Aug 20 €	20–21 €	20–21 €
License and others	–	–	750	–	–	–	–	–	–
General renovation	–	–	35,625	–	–	–	–	–	–
Marketing	330	330	12,584	2,951	2,951	2,951	2,951	11,804	11,844
Depreciation	374	377	4,503	1,125	1,126	1,125	1,127	4,503	4,502
	26,459	26,461	372,215	101,335	109,356	106,355	100,357	417,423	434,462
Other income (expense)									
Design commission	100	100	2,000	700	1,200	1,000	600	3,500	5,000
Workshop rent	2,100	2,100	23,620	7,900	10,200	10,000	6,600	34,700	38,200
	2,200	2,200	25,620	8,600	11,400	11,000	7,200	38,200	43,200
Operating profit	15,893	16,131	86,044	41,104	48,826	46,848	35,939	172,717	187,846
Interest expense									
Loan interest	110	107	1,387	296	260	223	188	967	378
	110	107	1,387	296	260	223	188	967	378
Net profit	15,763	16,024	25,620	40,808	48,566	46,625	35,751	171,750	187,468

Table 5 Balance sheet information

All numbers in €	Sep 18	Aug 19	Aug 20	Aug 21
Fixed assets	26,130	22,001	17,498	12,996
Current assets	7,260	123,809	283,531	458,260
Creditors due within 1 year	16,282	17,491	18,161	2,381
Net current assets	−9,022	106,318	265,370	455,879
Total net assets	−16,802	109,657	281,407	468,875
Capital and reserves	−16,802	109,657	281,407	468,875

mentioned and explained in the previous section. Moreover, the first year shows increasing receipts due to higher sales. These sales cover the more or less constant costs. However, the margin between costs and sales is reduced all three month due to seasonal marketing costs.

The rising amount of money in the bank account opens the possibility the pay the loan earlier and therefore reduce payments in the following years (Table 6).

Scenario Analysis

The previous sections gave an overview of how the business of The Guild works and how it will be financed. The business itself is a unique but still flexible concept. In the following, there are considered some issues, which may occur during the operation of the business in order to prevent major shocks and react quickly to decrease losses [9].

In the case of a developing need towards a greater *range of snacks*, a first issue might occur by the fact that the Café/Bar does not provide an own kitchen. In this situation, the concept might change to different specials offered or through changing the supplier in order to satisfy the customer.

The results of the primary research shows that the need for *special drinks* is rather low. However, in case the need will appear, this challenge can be easily adopted because the Bar is fully equipped.

Nevertheless, the plan is to provide *live music* only for special events such as the opening, this could be aligned to customer needs.

The workshops are flexible in terms of executed subject. Therefore, the focus of the workshops is provided according *varying demand of customers and seasons.*

In the case a *rejection of the workshop,* there will be different opportunities of reaction.

In the first case the personnel cost would be reduced through cutting salary as well as by reducing carpenter and creative workers. This would lead to a 30% decrease of salary costs and therefore to a lower breakeven point. If the workshop is rejected, further, the concept of the workshop has the to be rethought itself. This would lead to the sale of tools and tool attachments with the value of 5095 € minus depreciation. In this case the space of the workshop would be empty and opportunities for different use.

Table 6 Cash flow forecast

The Guild

Multi-year cash flow forecast

	Sep 18 €	Oct 18 €	Nov 18 €	Dec 18 €	Jan 19 €	Feb 19 €	Mar 19 €	Apr 19 €	May 19 €
Receipts									
Coffee	11,200	11,200	14,000	16,800	16,800	21,000	21,000	22,400	21,000
Food	1,560	1,560	1,755	2,535	2,535	2,730	2,730	2,925	1,560
Alcohol	19,500	19,500	20,100	21,000	21,000	24,000	24,000	24,300	24,600
Design commission	120	240	240	360	240	240	240	240	120
Workshop rent	1,032	1,032	1,800	2,400	2,640	2,880	3,120	3,360	2,520
Personal investment	15,000	–	–	–	–	–	–	–	–
Business angel	10,000	–	–	–	–	–	–	–	–
Loan capital	50,000	–	–	–	–	–	–	–	–
	108,412	33,532	37,895	43,095	43,215	50,850	51,090	53,225	49,800
Payments									
Coffee	1,120	1,120	1,400	1,680	1,680	2,100	2,100	2,240	2,100
Food	468	468	526	761	760	820	818	878	468
Alcohol	3,900	3,900	4,020	4,200	4,200	4,800	4,800	4,860	4,920
Salaries/Wages	18,960	18,960	20,660	23,960	23,960	23,960	23,960	23,960	21,960
Gas/Electricity/Water	394	395	394	395	394	395	394	394	395
Insurance	1,300	1,300	1,300	1,300	1,300	1,300	1,300	1,300	1,300
Rent	2,100	2,100	2,100	2,100	2,100	2,100	2,100	2,100	2,100
Landlord commission fee	7,000	–	–	–	–	–	–	–	–

(continued)

Table 6 (continued)

The Guild

Multi-year cash flow forecast

	Sep 18 €	Oct 18 €	Nov 18 €	Dec 18 €	Jan 19 €	Feb 19 €	Mar 19 €	Apr 19 €	May 19 €
License and others	750	–	–	–	–	–	–	–	–
General renovation	35,625	–	–	–	–	–	–	–	–
Marketing loan payments	3,031	330	330	2,291	370	330	2,291	330	330
Tools	–	1,465	1,465	1,465	1,464	1,464	1,465	1,465	1,465
Tool attachments	3,440	–	–	–	–	–	–	–	–
Equipment cafe/bar	1,655	–	–	–	–	–	–	–	–
Electronic devices cake/bar	1,279	–	–	–	–	–	–	–	–
Bar furnishing	12,150	–	–	–	–	–	–	–	–
Hygiene settings	6,000	–	–	–	–	–	–	–	–
VAT	1,980	–	–	–	–	–	–	–	–
	–	192	–	–	1,012	–	–	1,560	–
	101,132	30,230	32,195	38,152	37,241	37,269	39,228	39,087	35,038
Net cash flow	7,260	3,302	5,700	4,943	5,974	13,581	11,862	14,138	14,762
Opening bank		7,260	10,562	16,262	21,205	27,179	40,760	52,622	66,760
Closing bank	7,260	10,562	16,262	21,205	27,179	40,760	52,622	66,760	81,522

The Guild

Multi-year cash flow forecast

	Jun 19 €	Jul 19 €	Aug 19 €	Sep 19–Nov 20 €	Dec 20–Feb 21 €	Mar 20–May 21 €	Jun 20–Aug 21 €	20–21 €	Total €
Receipts									
Coffee	21,000	21,000	21,000	74,200	84,000	77,840	63,000	299,600	817,040

(continued)

Table 6 (continued)

The Guild

Multi-year cash flow forecast

	Jun 19 €	Jul 19 €	Aug 19 €	Sep 19–Nov 20 €	Dec 20–Feb 21 €	Mar 20–May 21 €	Jun 20–Aug 21 €	20–21 €	Total €
Food	1,560	1,560	1,560	5,655	8,775	7,410	4,680	29,640	80,730
Alcohol	25,900	25,200	25,500	78,900	81,300	83,700	86,400	360,900	964,800
Design commission	120	120	120	840	1,440	1,200	720	6,000	12,600
Workshop rent	2,520	2,520	2,520	9,480	12,000	12,000	7,920	45,840	115,824
Personal Investment	–	–	–	–	–	–	–	–	15,000
Business Angel	–	–	–	–	–	–	–	–	10,000
Loan Capital	–	–	–	–	–	–	–	–	50,000
	50,100	50,400	50,700	169,075	187,755	182,150	162,720	741,980	2,065,994
Payments									
Coffee	2,100	2,100	2,100	7,420	8,400	7,784	6,300	29,960	81,704
Food	468	468	468	1,696	2,633	2,223	1,404	8,892	24,219
Alcohol	4,980	5,040	5,100	15,780	16,260	16,740	17,280	72,180	192,960
Salaries/Wages	21,960	21,960	21,960	72,880	80,880	77,880	71,880	320,520	890,260
Gas/Electricity/Water	394	395	394	14,199	14,199	14,199	14,199	56,796	118,325
Insurance	1,300	1,300	1,300	3,900	3,900	3,900	3,900	15,600	46,800
Rent	2,100	2,100	2,100	6,300	6,300	6,300	6,300	25,200	75,600

(continued)

Table 6 (continued)

The Guild

Multi-year cash flow forecast

	Jun 19 €	Jul 19 €	Aug 19 €	Sep 19–Nov 20 €	Dec 20–Feb 21 €	Mar 20–May 21 €	Jun 20–Aug 21 €	20–21 €	Total €
Landlord commission fee	–	–	–	–	–	–	–	–	7,000
License and others	–	–	–	–	–	–	–	–	750
General renovation	–	–	–	–	–	–	–	–	35,625
Marketing loan payments	2,291	330	330	2,951	2,951	2,951	2,951	11,844	36,232
Tools	1,465	1,465	1,465	4,394	4,395	4,395	4,394	175,579	51,271
Tool attachments	–	–	–	–	–	–	–	–	3,440
Equipment cafe/bar	–	–	–	–	–	–	–	–	1,655
Electronic devices cake/bar	–	–	–	–	–	–	–	–	1,279
Bar furnishing	–	–	–	–	–	–	–	–	12,150
Hygiene settings	–	–	–	–	–	–	–	–	6,000
	–	–	–	–	–	–	–	–	1,980
VAT	–	1,480	–	1,320	2,000	2,400	1,840	8,680	20,484
	37,058	36,638	35,217	130,840	141,918	138,772	130,448	567,251	1,607,734
Net cash flow	13,042	13,762	15,483	38,235	45,837	43,378	32,272	174,729	458,260
Opening bank	81,522	94,564	108,326	123,809	162,044	207,881	251,259	283,531	–
Closing bank	94,564	108,326	123,809	162,044	207,881	251,259	283,531	458,260	458,260

In this case, the workshop could be used as café and exhibition space as the rest of the Café. Furthermore, other workshops could be implemented such as cooking workshops. Another scenario would be to take advantage of the workshop for theme parties and theme evenings.

6 Conclusion and Further Research

The Guild will be a unique Café or Bar in Hamburg, Schanzenviertel, that will provide a completely new customer experience. It will distinguish itself by a complex concept consisting of three major parts: (i) Café/Bar, (ii) exhibition space, and (iii) a visible workshop. The Café itself will function as an exhibition space where you can literally "buy everything". The source of these unique designs will be young designer and "regular" people with a tendency to creativity. Furthermore, while enjoying a coffee or cold beverage the customer can watch people or The Guild own carpenter crafting the visible workshop. Customers creating in the workshop are making use of the possibility to rent a spot and use the tools under the carpenters' supervision.

The Guild will have its registered office in Hamburg, Schanzenviertel. The organisation will be registered as an equal running partnership. With the management and entrepreneur study background as well as the interest in designing, both founders bring authentic values to the company.

Nevertheless, the competition in Hamburg is fierce, The Guild provides a unique concept, which meets the need of its customers and gains therefore a competitive advantage. The target customer is aged between 20 and 35 years, has an affinity towards crafting, a need of self-fulfilment and a daily routine dominated by digital media. Furthermore, the potential customers has a proper disposable income and is likely to have his/her own flat.

With the assumptions of a slow start, the business will fast profit from its reputation, which was built through good social media appearance and the strong brand recognition. The sales will gradually grow from year one (516,755 €) to year three (690,140 €). The high financial requirements of 75,000 € will be funded by personal investment (15,000 €), Business Angel (10,000 €) and bank loan (50,000 €). The business plan gives a good understanding of the business and positive outlook to the starting phase. However, further investigations are needed in terms of choosing the right funding method including a deep examination of the state-of-the-art funding methods such as crowdfunding.

Acknowledgements The author would like to express her sincere thanks to Dustin Engels, who contributed to an early version of the business plan.

References

1. Beatty J, Samuelson S, Abril P (2018) Business law and the legal environment. Cengage, USA
2. Bret S, Bunderson S (2018) The truth about hierarchy. MIT Sloan Manag Rev 49–52
3. Burck A, Glaum M, Schnürer K (2018) Cash-Flow- Planung - Anforderungen und praktische Umsetzung im internationalen Konzern. Schmalenbachs Zeitschrift für betriebswirtschaftliche Forschung, 1–33
4. Busse V (2018) Crowdfunding—an empirical study on the entrepreneurial viewpoint. In: Fatos X, Barolli L, Gregus M (eds) Advances in intelligent networking and collaborative systems: the 10th international conference on intelligent networking and collaborative systems. Springer, Switzerland, pp 306–318
5. Gregus M, Kryvinska N (2015) Service orientation of enterprises—aspects, dimensions, technologies. Comenius University in Bratislava. ISBN: 9788022339780
6. Horvath R, Kotlebova J, Siranova M (2018) Interest rate pass-through in the euro area: Financial fragmentation, balance sheet policies and negative rates. J Financ Stab 12–21
7. Kaczor S, Kryvinska N (2013) It is all about services—fundamentals, drivers, and business models. Soc Serv Sci J Serv Sci Res Springer 5(2):125–154
8. Karlsson C, Andresson M, Cheltenham L (2018) Geographies of growth. Innovations, networks and collaborations. Reg Sci Policy Pract 65–66
9. Khosravi F, Jha-Thakur U (2018) Managing uncertainties through scenario analysis in strategic environmental assessment. J Environ Plan Manag
10. Kim C, Mauborgne R (2004) Blue ocean strategy. Havard Bus Rev
11. Klandt H (2006) Gründungsmanagement: der integrierte Unternehmensplan: Business Plan als zentrales Instrument für die Gründungsplanung. Oldenbourg Verlag, München
12. Kryvinska N (2012) Building consistent formal specification for the service enterprise agility foundation. Soc Serv Sci J Serv Sci Res Springer 4(2):235–269
13. Kryvinska N, Gregus M (2014) SOA and its business value in requirements, features, practices and methodologies. Comenius University in Bratislava. ISBN: 9788022337649
14. Lessambo F (2018) financial statements: analysis and reporting. Springer Link
15. Liu J, Liao X, Huang W, Liao X (2018) Market segmentation: a multiple criteria approach combining preferences analysis and segmentation decision. Elsevier
16. Maines L, McDaniel L (2000) Effects of comprehensive-income characteristics on nonprofessional Onvestors' judgements: the role of financial-statement presentation format. Account Rev 179–207
17. Meffert H, Burmann C, Kichgeorg M, Eisenbeiß M (2018) Marketing-Mix: Produkt- und programmpolitische Entscheidungen. Springer Link, pp 393–485
18. Mladenow A, Fuchs E, Dohmen P, Strauss C (2012) Value creation using clouds—analysis of value drivers for start-ups and small and medium sized enterprises in the textile industry. In: 2012 26th international conference on advanced information networking and applications workshops (WAINA), 26–29 March 2012, Fukuoka Institute of Technology (FIT), Fukuoka, Japan. IEEE Computer Society Digital Library, pp 1215–1220. https://doi.org/10.1109/waina. 2012.210. ISBN 978-0-7695-4652-0
19. Molnár E, Molnár R, Kryvinska N, Greguš M (2014) Web intelligence in practice. Soc Serv Sci J Serv Sci Res Springer 6(1):149–172
20. Mooi E, Sarstedt M, Mooi-Reci I (2017) The market research process. Springer Texts Bus Econ 11–25
21. Nirschl M, Steinberg L (2017) Steigende Bedeutung des influencer Marketing im Marketing-Mix von Unternehmen. Springer Link, pp 1–3
22. Omer S (2018) SWOT analysis: the tool of organizations stability (KFC) as a case study. J Process Manag New Technol 27–34
23. Reid W (2018) The meaning of company accounts. Routledge, London
24. Schawel C, Billing F (2017) Five-Forces-Modell (Strategische Unternehmensanalysekonzepte). Springer Link, pp 141–143

Printed in the United States
By Bookmasters